ARISTOTLE: *ON GENERATION AND CORRUPTION* BOOK II

Aristotle: On Generation and Corruption *Book II* is concerned with Aristotle's theory of the elements, their reciprocal transformations, and the cause of their perpetual generation and corruption. These matters are essential to Aristotle's picture of the world, making themselves felt throughout his natural science, including those portions of it that concern living things. What is more, the very inquiry Aristotle pursues in this text, with its focus on definition, generality, and causation, throws important light on his philosophy of science more generally. This volume contains eleven new essays, one for each of the chapters of this Aristotelian text, plus a general introduction and an English translation of the Greek text. It gives substantial attention to an important and neglected text and highlights its relevance to other topics of current and enduring interest.

PANOS DIMAS is Professor of Philosophy at the University of Oslo. He was the director of the Norwegian Institute at Athens and chairs the Board of the Plato Dialogue Project (PDP), for which he coedited the first two volumes, on the *Philebus* and the *Statesman*. He is currently working on a bigger project on Plato's ethics.

ANDREA FALCON is Professor Emeritus in the Department of Philosophy at Concordia University, Montreal, and currently teaches in the Department of Philosophy at the University of Milan. His books include *Aristotle and the Science of Nature: Unity without Uniformity* (Cambridge University Press, 2005) and *Aristotelianism in the First Century BCE: Xenarchus of Seleucia* (Cambridge University Press, 2012). He is the editor of the *Brill's Companion to the Reception of Aristotle in Antiquity* (2016) and has coedited (with David Lefebvre) *Aristotle's Generation of Animals: A Critical Guide* (Cambridge University Press, 2017) and (with Stasinos Stavrianeas) *Aristotle on How Animals Move: The* De incessu animalium: *Text, Translation, and Interpretative Essays* (Cambridge University Press, 2021).

SEAN KELSEY is an associate professor in the Department of Philosophy at the University of Notre Dame. He is the author of *Mind and World in Aristotle's* De Anima (Cambridge University Press, 2021).

ARISTOTLE: *ON GENERATION AND CORRUPTION* BOOK II

Introduction, Translation, and Interpretative Essays

EDITED BY

PANOS DIMAS

University of Oslo

ANDREA FALCON

Concordia University and University of Milan

SEAN KELSEY

University of Notre Dame

CAMBRIDGE
UNIVERSITY PRESS

CAMBRIDGE
UNIVERSITY PRESS

Shaftesbury Road, Cambridge CB2 8EA, United Kingdom

One Liberty Plaza, 20th Floor, New York, NY 10006, USA

477 Williamstown Road, Port Melbourne, VIC 3207, Australia

314–321, 3rd Floor, Plot 3, Splendor Forum, Jasola District Centre, New Delhi – 110025, India

103 Penang Road, #05–06/07, Visioncrest Commercial, Singapore 238467

Cambridge University Press is part of Cambridge University Press & Assessment, a department of the University of Cambridge.

We share the University's mission to contribute to society through the pursuit of education, learning and research at the highest international levels of excellence.

www.cambridge.org
Information on this title: www.cambridge.org/9781009239981

DOI: 10.1017/9781009239936

First published 2022

A catalogue record for this publication is available from the British Library.

ISBN 978-1-009-23998-1 Hardback

Contents

Figures

Tables

Notes on Contributors

ANDREAS ANAGNOSTOPOULOS is Akademischer Rat and a member of the Munich School of Ancient Philosophy (MUSAPh) at Ludwig Maximilians Universität. He has published on Aristotle's natural philosophy, metaphysics, and psychology. His work has appeared in *Oxford Studies in Ancient Philosophy*; *Phronesis*; and *Archiv für Geschichte der Philosophie*.

TIMOTHY CLARKE is Associate Professor of Philosophy at the University of California, Berkeley. He is the author of *Aristotle and the Eleatic One* (Oxford University Press, 2019). His articles include "The Argument from Relatives" (*Oxford Studies in Ancient Philosophy*, 2012) and "Aristotle and the Ancient Puzzle about Coming to Be" (*Oxford Studies in Ancient Philosophy*, 2015).

PANOS DIMAS is Professor of Philosophy at the University of Oslo, Norway. He has published on Plato, Aristotle, and Epicurus. He chairs the Steering Board of the Plato Dialogue Project (PDP) and has coedited the first two PDP volumes, on the *Philebus* and the *Statesman*, published by Oxford University Press. He is currently working on a project on Plato's ethics and a project on divisibility of magnitude in ancient thought.

ANDREA FALCON is Professor Emeritus at Concordia University, Montreal, Canada. He is currently teaching at the University of Milan, Italy. He is the author of several books on Aristotle and the Aristotelian tradition, including *Aristotle and the Science of Nature: Unity without Uniformity* (Cambridge University Press, 2005) and *Aristotelianism in the First Century BCE: Xenarchus of Seleucia* (Cambridge University Press, 2012). He is the editor of the *Brill's Companion to the Reception of Aristotle in Antiquity* (2016). He has coedited (with David Lefebvre) *Aristotle's* Generation of Animals: *A Critical Guide* (Cambridge University Press, 2017) and (with Stasinos Stavrianeas) *Aristotle on How*

Animals Move. The De incessu animalium: *Text, Translation, and Interpretative Essays* (Cambridge University Press, 2021).

CHRISTOPHER FREY is Associate Professor in the Philosophy Department at the University of South Carolina. He received his PhD in philosophy at the University of Pittsburgh. His work in the history of ancient Greek philosophy focuses primarily on Aristotle's natural philosophy; he also publishes in contemporary philosophy of mind and action. His work has appeared in *Ancient Philosophy*, *Oxford Studies in Ancient Philosophy*, *Phronesis*, *Philosophy and Phenomenological Research*, and many other venues. He is currently writing a book about Aristotle's concept of life, titled *The Principle of Life: Aristotelian Souls in an Inanimate World.*

EMILY KATZ is Assistant Professor of Philosophy at Michigan State University. Her principal research interests are ancient Greek philosophy and mathematics, particularly Aristotle's physics, epistemology, natural science, and philosophy of mathematics. Her published articles reconstruct Aristotle's philosophies of geometry and arithmetic and rehabilitate several underexamined or underappreciated texts such as *Rival Lovers* (traditionally attributed to Plato) and *Metaphysics* Books N and M.

SEAN KELSEY is Rev. John O'Brien Associate Professor of Philosophy at the University of Notre Dame. He has published on Plato and (especially) Aristotle, including, most recently, *Mind and World in Aristotle's* De anima (Cambridge University Press, 2022).

MARY KRIZAN is Assistant Professor of Philosophy at Bilkent University. She is primarily interested in Aristotle's theory of matter as it appears within his natural philosophy. Her papers have appeared in *Oxford Studies in Ancient Philosophy, Journal of the History of Philosophy, Ancient Philosophy,* and *History of Philosophy Quarterly.* She is currently working on a book on Aristotle's material elements.

SPYRIDON RANGOS is Professor of Greek Literature and Philosophy at the University of Patras, Greece. Recent papers include "*De incessu animalium* 12–13: Limb-Bending and Natural Teleology" (in A. Falcon and S. Stavrianeas, eds., *Aristotle on How Animals Move*, Cambridge University Press, 2021) and "Chance Encounters and Abrupt Endings: On the Preludes and Closures of Plato's Third Thrasyllan Tetralogy" (in

E. Kaklamanou, M. Pavlou, and A. Tsakmakis, eds., *Framing the Dialogues: How to Read Openings and Closures in Plato*, Brill, 2021).

MARGARET SCHARLE is Professor of Philosophy and Humanities at Reed College, Oregon. She works in ancient philosophy, with special interest in Aristotelian natural philosophy. Her papers have appeared in *Oxford Studies in Ancient Philosophy*; *Phronesis*; *Apeiron*; as well as several edited volumes.

STASINOS STAVRIANEAS is Assistant Professor in the Philosophy Department at the University of Patras. He has published papers on Aristotelian natural science and biology. His articles include "Spontaneous Generation in Aristotle's *Generation of Animals*" (*Rhizai*, 2009) and "Nature as Principle of Change" (in M. Leunissen, ed., *Physics: A Critical Guide*, Cambridge University Press, 2015). He is the author of a Modern Greek translation with commentary of Aristotle's *Parts of Animals* and coeditor (with Andrea Falcon) of *Aristotle on How Animals Move. The* De incessu animalium: *Text, Translation, and Interpretative Essays* (Cambridge University Press, 2021).

Preface

The volume originated in a workshop on Aristotle's *Generation and Corruption* II held at the Anargyrios and Korgialenios School of Spetses on July 3–9, 2016. This workshop was made possible by funding from the Philosophy Department at the University of Oslo, which we would like to thank for its generous support. The participants were (in alphabetical order) Andreas Anagnostopoulos, Timothy Clarke, Panos Dimas, Christopher Frey, Emily Katz, Sean Kelsey, Mary Krizan, Spyridon Rangos, Margaret Scharle, and Stasinos Stavrianeas. The group divided its work into ten two-hour sessions; each participant was assigned a chapter to present (in one case, two chapters); most presenters distributed translations of their assigned chapter or substantial portions thereof; and each presentation was followed by an extensive discussion. The quality of the presentations, and of the discussions they facilitated, was agreed to be high, so the idea was floated around of producing a volume, including a translation based on the critical edition of the Greek text established by Marwan Rashed (2005). The group was keen and agreed to proceed.

The spirit is willing, but the flesh is weak; we were fortunate, then, when in March 2018 we had the brilliant idea of asking Andrea Falcon to join us as coeditor, and more fortunate still when he agreed. The project would not have been completed without his dedication, energy, and professionalism.

We would also like to thank Andreas Anagnostopoulos and Laura Castelli, who organized a research workshop on Aristotle's *Generation and Corruption* II at the Ludwig Maximilian Universität in Munich on January 7–8, 2020, where the introduction to the volume and two other essays were presented and discussed.

Panos Dimas and Sean Kelsey

Abbreviations

Aristotle

APo	*Analytica posteriora*	*Posterior Analytics*
APr	*Analytica priora*	*Prior Analytics*
DA	*De anima*	*On the Soul*
DC	*De caelo*	*On the Heavens*
GA	*De generatione animalium*	*Generation of Animals*
GC	*De generatione et corruptione*	*Generation and Corruption*
HA	*De historia animalium*	*History of Animals*
IA	*De incessu animalium*	*On the Progression of Animals*
Juv.	*De juventute et senectute*	*On Youth and Old Age*
Long.	*De longitudine et brevitate vitae*	*On the Length and Shortness of Life*
MA	*De motu animalium*	*On the Motion of Animals*
Mem.	*De memoria*	*On Memory*
Meta.	*Metaphysica*	*Metaphysics*
Meteor.	*Meteorologica*	*Meteorology*
PA	*De partibus animalium*	*Parts of Animals*
Phys.	*Physica*	*Physics*
Poet.	*De arte poetica*	*Poetics*
Resp.	*De respiratione*	*On Respiration*
Sens.	*De sensu et sensibilibus*	*On Sense and Sense Objects*
Somn.	*De somno et vigilia*	*On Sleep and Waking*
Top.	*Topica*	*Topics*

[Aristotle]

MXG	*De Melisso Xenophane et Gorgia*	*On Melissus, Xenophanes, and Gorgias*

Philoponus

In GC	*In libros De generatione et corruptione commentaria*	*On Aristotle's* On Generation and Corruption

Plato

Tim.	*Timaeus*	*Timaeus*
Phd.	*Phaedo*	*Phaedo*
Phil.	*Philebus*	*Philebus*
Theaet.	*Theaetetus*	*Theaetetus*

Simplicius

In DC	*In Aristotelis De caelo commentaria*	*On Aristotle's* On the Heavens
In Phys.	*In Aristotelis Physicorum libros commentaria*	*On Aristotle's* Physics

Further Abbreviation

DK H. Diels and W. Kranz, *Die Fragmente der Vorsokratiker.* Weidmann Verlag, Zurich 1951^{6}

PART I

Introduction and Interpretative Essays

1

On Generation and Corruption *II*

An Introduction

Andrea Falcon

1.1 Introduction

GC consists of four pieces: an account of generation and corruption, including an attempt to distinguish these natural processes from alteration and growth (*GC* I 1–5); a discussion of contact, action and passion, and mixture (*GC* I 6–10); an elemental theory with an account of how the four sublunary elements change into one another (*GC* II 1–8); a return to the study of generation and corruption, with a special focus on the moving (that is, efficient) cause, securing it so that generation and corruption in the sublunary world be continuous and uninterrupted (*GC* II 9–11). This introduction tries to overcome the impression that *GC* is a suboptimal amalgam of these four pieces by exploring the relation between the account of generation and corruption and the elemental theory. It will emerge that the discussion of contact, action and passion, and mixture is for the sake of the subsequent elemental theory, and that this theory, along with the final discussion of the moving cause of generation and corruption, fulfill a promise made at the outset of *GC*, where Aristotle announces a search for the causes of generation and corruption.[1] Far from being a suboptimal amalgam of loosely connected investigations, *GC* is a carefully crafted work. Aristotle could have rearranged the various pieces that compose this work in a different way, but this rearrangement would have been less than optimal.

* A draft of this introduction was presented at the MUSAPh/LMU research workshop on Aristotle's *GC* held in Munich on January 7 and 8, 2020. I would like to thank Andreas Anagnostopoulos, Laura Castelli, Joe Karbowski, and Timothy Clarke for their comments on my presentation. Special thanks go to Laura Castelli (my commentator) for her helpful written feedback on an earlier version of this piece, and to Sean Kelsey for sharing his own notes on the transitional passages in *GC*. Andreas Anagnostopoulos and Sean Kelsey commented on the penultimate version of this introduction, forcing me to elaborate further my thoughts on the opening statement of *GC* and the relation between *GC* I and *GC* II.

[1] *GC* I 1, 314a1–3.

Section 1.2 offers an overview of the contents of *GC* I and *GC* II, with a focus on how the account of generation and corruption and the elemental theory are connected in the broad structure of the treatise. The upshot is that the unity of *GC* is stronger than it is often thought. Section 1.3 explores how Aristotle deals with the same themes in two other works: *DC* and *Meteorology*. This brings to the fore how close the account of generation and corruption and the elemental theory are in Aristotle's mind. Finally, Section 1.4 is a detailed summary of the contents of *GC* II, showing how the broad structure outlined in Section 1.2 helps us appreciate the overall argument within *GC* II, especially the continuous treatment of the elements in *GC* II 1–8, which is prefigured in the questions raised at the outset of *GC* I 6.

1.2 Generation, Corruption, and the Elements in *GC* I and II

GC is foundational for Aristotle's science of nature. The first lines are carefully crafted to give a precise description of the topic under investigation. The topic is generation and corruption (περὶ γενέσεως καὶ φθορᾶς), and not generation and corruption of everything but of the things that come into being and perish *naturally* (τὰ φύσει γενόμενα καὶ φθειρόμενα), where these are to be treated not individually by kinds but rather in general for all natural things alike (ὁμοίως κατὰ πάντων). We are also promised a twofold task: a search for the causes of generation and corruption (τάς αἰτίας διαιρετέον) and an account of these processes (τοὺς λόγους αὐτῶν).[2]

Contemporary readers may find the existence of generation and corruption unproblematic; as a result, they may fail to appreciate the predicament in which Aristotle finds himself at the outset of *GC*. Without a firm grasp of what generation and corruption are, including how these archetypal processes differ from other natural processes such as alteration and growth, Aristotle cannot move forward with his stated goal of providing a scientific account of them. Aristotle's predicament is not artificially created by deliberately ignoring existing accounts of these processes. What his predecessors have said on this topic makes it very difficult, if not even impossible, for him to make sense of generation and corruption. This explains why Aristotle opens his inquiry into generation and corruption

[2] Here is the opening statement in the original Greek: περὶ δὲ γενέσεως καὶ φθορᾶς τῶν φύσει γενομένων καὶ φθειρομένων, ὁμοίως κατὰ πάντων, τάς τε αἰτίας διαιρετέον καὶ τοὺς λόγους αὐτῶν (*GC* I 1, 314a1–3).

with a long, and at times convoluted, critical discussion of the views of his predecessors (*GC* I 1–2).

By the end of *GC* I 5 Aristotle is confident that he has established not only *that* generation and corruption are different from alteration and growth, but also *what* each of them is. At the outset of *GC* I 6, Aristotle negotiates the transition to the next topic on his agenda as follows:

> Since *we ought to speak first about the matter and the so-called elements* – [establishing] whether or not they exist, and whether each of them is everlasting or comes into being in some way; and if they do come into being, whether they all come into being in the same way from one another, or some one of them is primary – *we must speak beforehand about things that are currently spoken of indistinctly.* (*GC* I 6, 322b1–4)

Aristotle makes two announcements in this transitional passage. He announces that his account of generation and corruption ought to continue with a study of "the matter and the so-called elements." He adds that this study must be preceded by the treatment of a set of concepts that are relevant to any scientific account of the natural world. They are contact, action and passion, and mixture. A preliminary discussion of these concepts is needed because Aristotle's predecessors have employed them without having made the proper conceptual distinctions.[3]

Providing conceptual clarity remains a primary goal throughout the second part of *GC* I (*GC* I 6–10). Without conceptual clarity, Aristotle cannot turn to a search for the causes of generation and corruption, so he cannot fulfill the promise made in the beginning of *GC*. When we reflect on what Aristotle accomplishes in *GC* I, we see that half of the twofold task announced at the outset of *GC* is not even begun until *GC* II. We also see that Aristotle establishes a close link between the study of the elements and the account of generation and corruption and thinks of the former as a contribution to the latter – so much so that the study of the elements is presented as a natural, and indeed inevitable, development of the account of generation and corruption. This connection is confirmed by what we read at the outset of *GC* II:

> So then, with regard to mixture, contact, and acting and being acted upon, we have said how they belong to the things that change naturally. And further, with regard to simple generation and perishing, we have said how

[3] Williams 1982: 21 renders the Greek ἀδιορίστως with "in a confused way." But Aristotle's predecessors need not be confused; they only need to be unaware of the internal articulations of these concepts. Their unsystematic use of these concepts is a consequence of their not having done the required conceptual work.

> they occur, what they are of, and by what cause they occur. And similarly also with regard to alteration, we have said what being altered is and how it differs from them. *But it remains to consider the so-called elements of bodies.* (*GC* II 1, 328b26–32)

This second passage contains a precise but selective summary of the main topics discussed in *GC* I: not only a determination of what generation and corruption are, and how they differ from alteration and growth (*GC* I 1–4), but also a treatment of contact, action and passion, and mixture (*GC* I 6–10).[4] The final sentence announces the next topic on Aristotle's agenda: the so-called elements. This topic is introduced as a leftover from *GC* I – and rightly so, given that a study of "the matter and the so-called elements" was announced in *GC* I 6.

The expression "the so-called elements" is found in both transitional passages.[5] More on its significance in due course. For the time being, it is more pressing to focus on the claim (made in *GC* I 6) that the account of generation and corruption ought to continue with a study of "the matter and the so-called elements." Does Aristotle introduce two topics, matter *and* the so-called elements – or one – matter *in the sense of* the so-called elements? An answer to this question cannot be given on purely grammatical grounds. It depends on how we understand the connection between the study of the so-called elements and the study of generation and corruption. For more on this connection, we must turn to *GC* II.

As soon as we turn to *GC* II, we discover that the first and most important step in the study of the elements consists in a clarification of how we should think about matter. To this end, Aristotle singles out two main critical targets: Anaximander's *apeiron* and the Receptacle of the *Timaeus*. There are problems specific to the lines of thought leading to each of these two putative principles.[6] But both lines of thought share a common mistake: they end up positing a separable, and indeed separate, material principle, namely a self-sufficiently existing matter. Aristotle's alternative consists in adopting the explanatory strategy introduced in *Phys.* I.[7] On this strategy, matter is an existentially inseparable, but definitionally separable, potential being.

4 The discussion of how generation differs from growth (*GC* I 5) is left out of this summary. Hence, my decision to flag it as a *selective* summary of the main topics discussed in *GC* I.

5 It occurs one more time in *GC* II 1, at 329a26.

6 See the essay by Timothy Clarke in this volume.

7 The cross-reference in *GC* II 1, 329a27 is ambiguous. It can be taken to be a reference to either *DC* III or *Phys.* I. In his essay, Timothy Clarke defends the view that it is a reference to the conceptual framework introduced in *Phys.* I.

To appreciate why the study of the elements is embedded in a study of generation and corruption, we must bear in mind that "for all the naturally constituted substances, their generation and perishing are not without the perceptible bodies."[8] Aristotle's point can be restated by saying that perceptible bodies are always involved in the generation and corruption of naturally constituted substances in the sense that naturally constituted substances need to be perceptible bodies in order to be subject to the causal interactions leading to their generation and corruption. "Perceptible body" is a piece of jargon that calls for a few words of explanation. A perceptible body is not just a natural body accessible to sense perception – namely, a body that can be perceived by touch and the other senses. A perceptible body is also a natural body susceptible to entering a causal relation in which this body acts on another natural body and is acted upon by it. Such a relation, which requires two natural bodies to touch one other, is essential to our understanding of how generation and corruption are physically possible. For Aristotle, generation and corruption in all naturally constituted substances are not possible without perceptible bodies. Since the elements are the primary bodies out of which all other natural bodies are made, they are the ultimate perceptible bodies; as such, they are the ultimate source of perceptibility for the other sublunary bodies.

Emphasizing the centrality of the notion of perceptible body helps us see why Aristotle deals with contact in *GC* I 6, action and passion in *GC* I 7–9, and mixture in *GC* I 10 *before* turning to the study of the so-called elements in *GC* II. A clarification of these notions is introduced *for the subsequent discussion of the elements.* Consider the three questions that Aristotle highlights in connection with the study of the elements at the outset of *GC* I 6: (1) whether the elements exist or not; (2) whether the elements are everlasting, or rather they come into being in some way; (3) if the elements come into being, whether they all come into being from one another or rather one of them is prior to the others.[9] At the most general level, there cannot be an answer to either the second or third question on Aristotle's agenda – and so an adequate explanation of the generation and corruption of the elements – without speaking of contact, action and passion. The case of mixture is less obvious, but we should keep in mind that *GC* II 7 is concerned with elemental *mixtures*, and that these qualify as a particular case of mixture.[10]

[8] *GC* II 1, 328b32–33. See the recent discussion in Crowley 2019: 1–26. The technical meaning of perceptible bodies defended in Crowley's article is adumbrated in Buchheim 2010: 449.

[9] *GC* I 6, 322b1–4.

[10] See Andreas Anagnostopoulos in this volume.

The reading that makes *GC* I 6–10 preparatory for the study of the elements offered in the first part of *GC* II is compatible with the view that a treatment of the notions of contact, action and passion, and mixture has a theoretical significance that goes beyond the narrow boundaries of Aristotle's theory of the elements.[11] In other words, there is no contradiction in holding that the results achieved in this stretch of text play a foundational role for Aristotle's study of the natural world, and holding that a preliminary clarification of these notions is required for an optimal study of the so-called elements. At the same time, placing proper emphasis on the link between contact, action and passion, and mixture, and the study of the so-called elements helps us appreciate why Aristotle deals with these notions right before engaging in a search for the principles of perceptible bodies. Aristotle could have relocated his treatment of contact, action and passion, and mixture elsewhere. If we find his discussion of these concepts embedded in his study of generation and corruption right before the study of the so-called elements, it is because Aristotle regards this treatment as especially useful for this study.

Far from being trivial or inconsequential, this conclusion sheds light on the unity of the whole *GC*. This unity is stronger than it may initially appear. *GC* I 6–10 is added to the account of generation and corruption for the sake of the subsequent discussion of the elements. Since the latter fulfils a promise made at the outset of *GC* I, where Aristotle announces a study of the causes of generation and corruption, we see that the discussion offered in the second part of *GC* I (*GC* I 6–10) is found where it belongs.

1.3 Generation, Corruption, and the Elements Beyond *GC*

Evidence that a close link exists in Aristotle's mind between the account of generation and corruption and the study of the elements can be found outside *GC*:

> Regarding the first of the elements, we have said what its nature is and that it is not subject to generation and corruption. *It remains to speak about the other two [pairs of elements].*[12] *For those who speak about them, an investigation of the topic of generation and corruption will take place at the same time.* The reason is that either generation does not exist, or it exists only in those elements and the things made out of them. (*DC* III 1, 298b6–11)

[11] For a reading that stresses the relevance of *GC* I 6–10 beyond the boundaries of the theory of the elements, see Burnyeat 2004: 7–24 and Wildberg 2004: 219–242.

[12] That is, the light pair (fire and air) and the heavy pair (earth and water).

This passage marks the transition from celestial to sublunary physics. Aristotle refers to the celestial simple body, which is also the matter of the heavenly bodies, as "the first element." A discussion of this curious expression goes beyond the scope of this introduction.[13] More interesting for us is to observe how closely the study of the sublunary elements is tied to a study of generation and corruption. Aristotle does not envision two separate studies – namely, a study of sublunary elements followed by a study of generation and corruption. He thinks that a study of the sublunary elements is *ipso facto* a study of generation and corruption. More to the point: nothing in the above passage suggests that Aristotle is restricting his claim to the study of *elemental* generation and corruption. On the contrary, he appears to be making a stronger (and more interesting) claim: the study of the sublunary elements contributes – directly and immediately – to a study of generation and corruption *in general*. What licenses this conclusion is the claim (made at the end of the passage quoted above) that "generation does not exist, or it exists only in those elements *and the things made out of them*."

The close link that exists in Aristotle's mind between the study of the elements and the study of generation and corruption finds confirmation in how the argument unfolds in *DC* III. First, Aristotle shows that a few positions are untenable. Then, he dismisses the view of those who deny generation and corruption, as well as the alternative view that everything is subject to generation and corruption. Having ruled out these extreme positions, Aristotle concentrates on which things are subject to generation and corruption and why. It is only at that point (at the beginning of *DC* III 3) that he turns to the topic of the sublunary elements, with a focus on their number and mode of generation and corruption.

GC II builds on the results achieved in *DC* III. *GC* II 3 takes for granted that there are two pairs of elements belonging to two different regions of the sublunary world.[14] *GC* II 4 assumes that these four elements change into one another.[15] Here Aristotle refers his reader to a prior discussion on this topic. *DC* III is the only plausible candidate for this reference. This result is reached by exclusion in *DC* III. Aristotle rules out that the sublunary elements are eternal and that they are generated from something else, either corporeal or incorporeal. There remains only that they are

[13] For a full discussion of the expressions "first element," "first body," and "first substance," see Falcon 2005: 113–121.

[14] *GC* II 3, 330a30–33.

[15] *GC* II 4, 331a7–10.

generated from one another. The manner in which they are generated from one another is the focus of *GC* II 4.

The conceptual affinities between *DC* III and *GC* II do not extend to include *DC* IV. If we bracket the question of whether the solitary δέ in the opening sentence of *GC* refers to the solitary μέν in the last sentence of *DC* IV, and we concentrate on the conceptual resources employed in these works, we see that *DC* IV and *GC* I–II are conceptually discontinuous to the extent that the discussion of the heavy and light offered in *DC* IV does not contribute to the study offered in *GC* I–II. Although the two projects are not totally unrelated, they make use of different conceptual tools. Consider the following passage from *GC* II 2, where Aristotle tells us why the light and the heavy are not among the contraries that are relevant to his project:

> Here are the pairs of contrarieties that relate to touch: hot-cold, dry-moist, heavy-light, hard-soft, viscous-brittle, rough-smooth, coarse-fine. *Of these contrarieties, heavy and light are neither active nor passive. The reason is that things are not called heavy or light because they act on something else or they are acted upon by something else*; by contrast, the elements must act on or be acted upon by one another, since they mix and change into one another. (*GC* II 2, 329b17–23)

Aristotle could not be clearer as to the reason why *DC* IV and *GC* II are distinct projects: the heavy and the light are neither agents nor patients, so they do not contribute directly to our understanding of how elemental generation and corruption take place.

Reflecting on the relationship between *GC* II and *DC* III helps us realize that the discussion of the elements advanced in *GC* II must be placed in a larger context. It is time to elaborate further on this larger context by returning to the expression "*the so-called elements.*" This phrase occurs three times in *GC* II 1. By contrast, the qualification "so-called" is dropped at the beginning of *DC* III 1. There, Aristotle is happy to speak of the elements of bodies. In fact, he is also eager to extend the use of the term "element" to the celestial simple body, which is called "the first element." More to the point: in *DC* III 3, Aristotle provides us with his own definition of element: an element is the primary constituent into which a body is divided.[16] There is no evidence that this definition is dropped in *GC* II. On the contrary, it is *prima facie* plausible that this definition is tacitly at work in *GC* II – especially if we take the view that *GC* II builds

[16] *DC* III 3, 302a14–19.

on the results achieved in *DC* III. But this only makes it more pressing for us to try to understand what meaning should be assigned to the qualification "*so-called*" when it is added to "*elements.*"

Timothy Crowley has recently argued that Aristotle does not mean to cast doubt on the elemental status of earth, water, air, and fire when he speaks of them as the so-called elements.[17] And yet, their status was at the time disputed. In the *Timaeus*, Plato criticizes the view that earth, water, air, and fire are the elements of everything. On his account, earth, water, air, and fire should not even be compared to syllables – that is, to the first level of complexity one obtains when one combines the letters of the alphabet (στοιχεῖα).[18] While Plato acknowledges the use of the term "element" in connection with earth, water, air, and fire, he also rejects it. For him, the true elements are the elementary triangles employed to build the four regular solids associated with earth, water, air, and fire. By using the expression "the so-called elements" Aristotle establishes contact with the very use of the term that Plato rejects. While Plato rejects this use, Aristotle endorses it.

It is a further question whether Plato and Aristotle are reacting to a common rather than technical use of the term "element." Crowley thinks that both philosophers are reacting to a common use.[19] While this scenario is possible, it is also difficult to prove due to the absence of textual evidence. What matters most for the sake of this introduction is that the larger context in which we are expected to place the account of the elements offered in *GC* II includes Plato's criticism of the use of the term "element." In *DC* III, Aristotle dismisses any geometrical reconstruction of the elements and their mutual transformation of the sort attempted in the *Timaeus.* Aristotle does not offer a positive account of how the elements change into one another in *DC* III. He does so in *GC* II. From *GC* II we learn that an analysis of elemental change requires a study of how the basic contrarieties, namely hot-cold and dry-wet, are involved in the explanation of elemental intertransformations. This does not entail that these contraries are the true elements, and that earth, water, air, and fire are only *prima facie* elements. For Aristotle, earth, water, air, and fire remain the elements once the analysis of elemental change is in place. In fact, the search for the principles announced at the beginning of *GC* II 2 is an indirect vindication

17 Crowley 2008: 223–242.

18 *Tim.* 48 B–C. Crowley 2005: 367–394 (especially 378–384) argues against the traditional reading of the passage. By his lights, there is no metaphorical use of "elements" in this passage. I remain skeptical.

19 Crowley 2005: 367–394 combined with Crowley 2008: 223–242.

of the view that earth, water, air, and fire are the elements of all sublunary bodies to the extent that their analysis in terms of primary qualities is not a threat to their status as the primary bodies.

There is another text that is immediately relevant to a reflection on the larger context in which we should situate the investigation conducted in *GC* II. This is the map of natural investigations sketched out at the beginning of the *Meteorology*. This map confirms the results reached so far; it also prompts a further reflection on how Aristotle envisions the relation between the study of generation and corruption and the study of the elements. Let us recall the first part of this famous map:

> We have already spoken about the first causes of nature and all natural motion; about the stars ordered in the motion of the heavens, *and about the bodily elements, [establishing] how many they are and what their nature is, and how they change into one another, and about generation and corruption in general.* There remains to speak about meteorology. (*Meteor.* I 1, 338a20–26)

The words highlighted in italics can be read in two ways. At first sight, Aristotle seems to outline three distinct investigations: while the first is concerned with the number and nature of the bodily elements, the second deals with how they change into one another, and the third is about generation and corruption at the most general level. Upon reflection, however, it is far from obvious that the first two investigations are distinct in Aristotle's mind. On the contrary, one can take the view that the number and nature of the elements ought to be approached via a study of how they change into one another. On this view, Aristotle is envisioning two rather than three investigations: while the first is concerned with the number and nature of the elements, including a study of how they change into one another, the second is about generation and corruption in general. Either way, the study of generation and corruption is here regarded as a logical part of a larger unit that includes the study of the elements. This result confirms the close connection that Aristotle establishes between the study of the sublunary elements and the general study of generation and corruption in *GC* I and *GC* II.

The programmatic passage from the beginning of the *Meteorology* helps us see that there are two ways to look at the study of the so-called elements announced in *GC* I 6 (and repeated in *GC* II 1). We can think of the study of the elements as embedded in the study of generation and corruption and as a logical part of that study. On this reading, we engage in a study of the so-called elements because we want to get clear about generation and corruption. We may be able to establish that generation and corruption are natural processes different from, and not reducible to alteration and growth, without engaging in a study of the elements. We may even be able to say what generation and corruption are without engaging in such a study. But the study of the elements is instrumental to fulfilling the promise made right at the outset of *GC* where we are promised a determination of the nature *and causes* of generation and corruption.

But there is another way to look at the relation between the study of the elements and the study of generation and corruption. The passage from the *Meteorology* is evidence that the study of the elements begins in earnest before the study of generation and corruption. Moreover, this passage suggests that the study of generation and corruption is embedded in a study of the elements. When we take this larger context into account, we are left with the strong impression that Aristotle engages in a study of generation and corruption to gain clarity about the elements rather than *vice versa*. If we return to the opening lines of *DC* III 1, and we read this transitional passage without having in mind what Aristotle says in *GC*, we see that Aristotle remains neutral with respect to these two readings. He only states that the study of the elements is *ipso facto* a study of generation and corruption, without further stating that we engage in the study of the elements to gain clarity on the topic of generation and corruption. To someone objecting that this second reading is contingent on our pretending not to know what Aristotle says in *GC*, we should reply that Aristotle expects us to approach his writings on natural philosophy in a definite order. Surely, he expects us to read *DC* III before *GC* I and *GC* II.

In conclusion, there are two ways to look at Aristotle's study of the elements vis-à-vis his study of generation and corruption. They are not equivalent since they motivate the theoretical need for these studies in different ways. We do not have to choose between them, but we must appreciate that they afford alternative insights into the project attempted in the *GC*. While the first places an emphasis on the foundational role of generation and corruption as archetypal natural processes, the second stresses the role of the elements in the composition of all natural bodies.[20]

[20] In a footnote to his contribution to the Symposium Aristotelicum on *GC* I, Jacques Brunschwig briefly remarks that it is not at all obvious that the study of generation and corruption in general should follow the study of elemental generation and corruption, and that the reverse order has some justification (Brunschwig 2004: 30n15). This section tries to unpack what he may have meant in his short (and rather enigmatic) footnote.

1.4 *GC* II in Outline

The first two sections of this introduction have been devoted to exploring the context (or rather contexts) in which *GC* II ought to be placed. It is now time to concentrate on how the overall arguments unfold within *GC* II.

As we turn to this argument, we should recall that at the beginning of *GC* I 6 Aristotle has promised a study of matter *and* the so-called elements. In *GC* II 1, Aristotle does not conflate matter with the elements. He sides with those who speak of a matter of the elements, but contrary to the line of thought that ends up positing the existence of a separable, and indeed separate, material principle, Aristotle argues that matter is inseparable. By his lights, matter never exists on its own independently of the perceptible bodies. Rather, it combines with the different perceptible qualities to constitute the perceptible bodies. Matter so understood is potentially a perceptible body.[21] When Aristotle says that something is potentially a perceptible body, he means to say that something is potentially a body that is susceptible to action and passion.

The link between perceptibility and susceptibility to action and passion comes to the fore in *GC* II 2. Aristotle reduces all tangible properties to four tangible contraries, organized in two pairings: hot-cold and dry-wet. The reduction performed in *GC* II 2 does not eliminate the richness and complexity of the perceptible world. Aristotle does not mean to suggest that there is nothing more to the other qualities than the primary ones. He only means to give us a rational way to see how all the qualities can be traced back to four basic tangible properties, which are either active or passive. The emphasis on tangibility must be understood in light of the assumption that action and passion require contact. *GC* II 3 shows that each of the four elements is characterized by two of these four tangible contraries: fire is hot and dry, air is wet and hot, water is cold and wet, and earth is dry and cold.

By the end of *GC* II 3, Aristotle has an answer to the first of the three questions outlined in the transitional passage at the outset of *GC* I 6. They are: (1) whether the elements exist or not; (2) whether the elements are everlasting or come into being in some way; and (3) if the elements come into being, whether they all come into being from one another or one of

[21] *GC* II 1, 329a33. The position of δυνάμει in the syntagm τὸ δυνάμει σῶμα αἰσθητόν suggests that it qualifies "perceptible body/σῶμα αἰσθητόν."

them is prior.[22] At the outset of *GC* II 4, Aristotle turns to the second question. He shows that the elements are not eternal; rather, they come into being from one another, since each of them consists of a combination of an active and a passive property. The upshot of *GC* II 4 is that there are no barriers in the nature of the elements to their mutual transformation. Put differently, all elements are of a nature to change into one another.[23]

What comes next is an attempt to answer the third and final question on Aristotle's agenda. To this end, in the first part of *GC* II 5, Aristotle assumes that there exists a single body that is the matter of generation and corruption in perceptible bodies. First, he shows that this body cannot be one of the four elements. Then, he argues that it cannot be a body above and beyond the four elements. Such a body would have to be thicker than fire and air and thinner than water and earth. Aristotle's analysis of this putative intermediate body shows that either this body will have none of the four basic contraries (hot and cold, and dry and wet) or it will have all of them. Since there cannot be a body that has no perceptible properties, we can safely rule out the first possibility. We can also rule out the second possibility since the explanation of generation and corruption would become impossible. In the end, we are left with perceptible properties distributed over four elemental bodies. In other words, we are back to our original four elemental bodies.

At this point, there are still two theoretical scenarios to consider. In the first, the four elemental bodies change into one another but one of them is the source of the other three. In the second part of *GC* II 5, Aristotle shows that neither of the two extreme bodies (fire and earth) nor one of the two intermediate bodies (air and water) can be the ultimate source of the other elemental bodies. Having established that the four elements must enjoy the same ontological status, Aristotle turns to the second scenario. In this alternative scenario, the four primary bodies behave in the way Empedocles thinks they do: they mix without being able to change into one another.

Aristotle is centrally concerned with this scenario in *GC* II 6. His main objection is that the elements would not be commensurable with one another. We could not even say what Empedocles claims about them – namely, that "they are equal." Dealing with this Empedoclean scenario turns into an extensive criticism of Empedocles' views on growth, nature, motion, and the soul. The relevance of this extensive criticism to the overall argument is not immediately obvious. What is clear is that, by

[22] *GC* I 6, 322b1–4. [23] *GC* II 4, 332a1.

the end of *GC* II 6, Aristotle has fulfilled all the promises made in the transitional passage at the outset of *GC* I 6. By the end of *GC* II 6, Aristotle has demonstrated (1) that the elements exist; (2) that they come into being from one another; and (3) that none of them is more fundamental than the others. In retrospect, we can safely say that Aristotle has answered all the questions raised at the beginning of *GC* I 6, and that these questions shape and control the way in which his overall argument unfolds in *GC* II 1–6.

Aristotle turns to the generation of homoeomerous bodies (lit. like-parted bodies) such as flesh, bone, and marrow in *GC* II 7. Aristotle takes these bodies to be elemental mixtures and at the same time to form a class of bodies ontologically distinct from the four elements. At the very heart of his whole discussion in *GC* II 7 is an *aporia* that can be introduced with the help of the following question: How in the world can something that is over and above the elements come into being as a result of their combination?[24] Aristotle shows that his elemental theory can account for the generation of homoeomerous bodies in terms of mixture. By contrast, the view that the four elements mix with one another without being able to change into one another cannot achieve this result. If we read the whole chapter as an attempt to adjudicate between two competing accounts of the elements considering their ability to resolve the above *aporia*, we see that the account of the generation of the homoeomerous bodies is a vindication of the (Aristotelian) account advocating the mutual transformation of the four elements. We also see that the alternative account is not an abstract theory of the elements; rather, it is the Empedoclean account criticized in *GC* II 6. At this stage of his argument, Aristotle takes the Empedoclean theory to be the only plausible alternative to his own. Finally, we see that Aristotle has not really turned away from the topic of the elements. Rather, he has expanded on it by looking at them as the ingredients of what he regards as *mixtures* of the elements.

Aristotle is still concerned with the elements in *GC* II 8. He shows that each of them is present in every *mixed* body. Here Aristotle develops further the discussion of elemental mixtures started in *GC* II 7. He goes beyond anything done up to this point because the mixed bodies he is concerned with are not just the elemental mixtures discussed in *GC* II 7; rather, they are all the mixed bodies inhabiting the region around the center of the universe. Insofar as this region is the place where the natural processes of generation and corruption take place, *GC* II 8 prepares the grounds for the final section of the *GC* II (*GC* II 9–11). At the same time,

[24] *GC* II 7, 334a21–23.

Aristotle does not really explore the cosmological implications of the thesis defended in the chapter – namely, that *all* the mixed bodies that inhabit the region around the center are composed of *all* the simple bodies. Since the four elements remain Aristotle's main concern in *GC* II 8, the chapter is best understood as completing the search for the *material* principles of perceptible bodies.

GC II 1–8 turns out to be a single textual unit in which Aristotle offers a comprehensive investigation of the material principles of perceptible bodies – their existence, number, and structure, how they change into one another, and how they give rise to and are present in the mixed bodies. If the above reconstruction of Aristotle's argument is on the right track, this argument is tighter than is usually appreciated. As we go beyond the results reached in *GC* II 1–8, it is helpful to recall, once more, the transitional passage that announces a study of the so-called elements at the outset of *GC* I 6:

> Since we ought to speak *first* (πρῶτον) about the matter and the so-called elements – [establishing] whether or not they exist, whether each of them is everlasting or comes into being in some way; and if they do come into being, whether they all come into being in the same way from one another or some one of them is primary – we must speak *beforehand* (πρότερον) about things that are currently spoken of indistinctly. (*GC* I 6, 322b1–4)

The sequence outlined in this passage is the following: an extended discussion of contact, action and passion, and mixture (*GC* I 6–10), followed by a study of matter and the so-called elements (*GC* II 1–8), which is the *first instalment* of a larger project. It is time to look in some detail at what is left on Aristotle's agenda beyond the study of the elements.

If we consider *GC* II 1–8 to be in service of the second half of the task announced at the outset of *GC* – a search for the *causes* of generation and corruption – we can take the stretch of text that begins at *GC* II 9 to be a continuation of the same task. Up to this point, Aristotle has concerned himself with the *material* principles of generation and corruption; he is now ready to move beyond those principles, with a special focus on the *moving* (aka *efficient*) principle of generation and corruption. What justifies this new focus is the observation that though everyone before him has dreamt of this principle, no one has been able to give an adequate account of it (*GC* II 9, 335b7–8).

The stretch of text that begins at *GC* II 9 is best understood as being part of the attempt to fulfill the promise made right at the beginning of the

GC, where Aristotle has announced his plan *to distinguish* the different causes of generation and corruption in a way that applies to *all* the things that naturally come to be and perish *alike*.[25] Aristotle's choice of words suggests that he is not envisioning a mere enumeration of the causes of generation and corruption. Rather, he wants to distinguish and explain them in a way that encompasses everything that naturally comes to be and perishes alike. In *GC* II 9, Aristotle shows that although his predecessors singled out either the material or the formal/final cause, they only dreamed of the moving cause. His articulation of the four causes is meant to carve out the conceptual space for this third kind of cause, which becomes the focus of his attention in *GC* II 10. The most important result secured in this chapter is the discovery that the *proximate* moving cause of the uninterrupted chain of generation and corruption is the circular motion of the sun along the ecliptic.[26] In Aristotle's view, this is a very important result since it explains *how* generation and corruption in the sublunary world can be everlasting.

GC II 11 begins with the observation that there is an unbroken chain of generation and corruption, alteration, and in general change, in the sublunary world. The idea seems to be that when we concentrate on any unbroken sequence of change taking place in the sublunary world, we *see* that this sequence is not interrupted by intervals of rest.[27] This claim is far from being obvious, let alone uncontroversial. For us to be able to see an unbroken chain of motion, our observation of the world around us ought to be illuminated by what the results have established so far. Recall that Aristotle has given us the *material* cause contributing to the explanation of why generation and perishing are continuous in *GC* I 3, where he has argued that the generation of something is always the perishing of something else.[28] He has then completed his explanation in *GC* II 10, where he has given us the *moving* cause of the eternity of generation and perishing, which is the motion of sun along the ecliptic.[29] As a result, the observation that opens the chapter, and prompts the question that concerns Aristotle in *GC* II 11, is informed at the very least by the results secured in *GC* I 3

[25] Let us recall, again, the Greek text: περὶ δὲ γενέσεως καὶ φθορᾶς τῶν φύσει γενομένων καὶ φθειρομένων, ὁμοίως κατὰ πάντων, τάς τε αἰτίας διαιρετέον καὶ τοὺς λόγους αὐτῶν (314a1–3). The focus is now on the underscored words, where Aristotle is promising, in addition to an account of generation and corruption (τοὺς λόγους αὐτῶν), an articulation of their *causes*.

[26] *GC* II 10, 336a32.

[27] *GC* II 11, 337a35: ὁρῶμεν. For the use of (οὐχ) ὁρῶμεν in the sense of making (failing to make) an observation, see *GC* I 3, 323a23; *GC* I 9, 327a16.

[28] *GC* I 3, 318a9–10 combined with 318a23–27.

[29] *GC* II 10, 336a15–23 (repeated at 336b25–26).

and *GC* II 10. More to the point: it is only at this juncture of the argument – that is, after *GC* II 10 – that the question prompted by the initial observation becomes really pressing. The question is whether something will come to be *of necessity* in this unbroken chain of generation and perishing. Aristotle's answer to this question is affirmative: provided that they are cyclical, and on the ground that in this case their cycle is without interruption, the coming to be of a few things in the sublunary world is of necessity. For instance, there will be of necessity a summer after the spring, and there will be of necessity a sunny and dry season after a cloudy and rainy one.

1.5 A Coda on Prime Matter

Some of the contributors to this volume share the view that Aristotle is committed to the existence of prime matter and its necessary role in the explanation of elemental change. However, they do not place the same emphasis on prime matter, let alone understand prime matter in the same way. For this reason, it is appropriate to end this introduction with a brief overview of what they say (and do not say) on the topic of prime matter. This overview will offer a first orientation to the reader who has an interest in this especially thorny issue.

In his essay on *GC* II 1, Timothy Clarke argues that the chapter supports the view that Aristotle is committed to prime matter. By Clarke's lights, the final section of the chapter (329a24–b6) is evidence that Aristotle posits an ultimate material substrate that (i) underlies the process of elemental intertransformation and (ii) partly constitutes each of the four sublunary elements. Clarke argues that the traditional (that is, prime matter) interpretation is the most natural understanding of what Aristotle says in *GC* II 1. He allows that other, revisionary interpretations of the passage are possible, but argues that these revisionary interpretations are less plausible than the traditional interpretation because they struggle to accommodate certain important textual details.

Emily Katz mentions prime matter only once in her essay. By her lights, nothing in *GC* II 3 rules out prime matter; but neither does anything in the chapter commit Aristotle to the existence of prime matter. One of the things that Emily Katz does argue for in her essay is that the primary differentiae are not the *elements* of the elementary bodies. But she takes that to be entirely compatible with the view that the elementary bodies are somehow *constituted by* the differentiae and possibly also prime matter. That is so because the elements that constitute a thing must belong to the same higher kind as the thing itself. Since neither prime matter nor the differentiae are bodies, they cannot count as the elements of the elementary bodies. But that does not mean that the elementary bodies cannot have them as their constituents. *GC* II 3 does not rule out that the elementary bodies have the differentiae and prime matter as their constituents; it only rules out that the differentiae (and indeed prime matter) are the elements of the elementary bodies.

GC II 4 has grounded many scholars' arguments against Aristotle's commitment to prime matter.[30] These scholars have read the chapter to show that any individual change from one element into another requires a contrary to underlie the change, thereby obviating the need for prime matter to underlie such changes. By contrast, traditionalists have defended Aristotle's commitment to prime matter by reading the text to show that some individual changes do not require an underlying contrary, thereby reclaiming the role for prime matter. Margaret Scharle's reading of the chapter undermines the assumption that stands behind both sides of the argument. On her interpretation, the chapter focuses not on comparing individual changes from one element into another, but rather on comparing complete cycles of changes by which all four elements come to be, either singly or in pairs. While her reading of *GC* II 4 remains neutral on the issue of prime matter, Scharle undermines the anti-traditionalist attempts to attack Aristotle's commitment to it based on *GC* II 4. Her interpretation of the chapter is consistent with her view that the best evidence for Aristotle's commitment to prime matter can be found not in his account of elemental transformation, but rather in consideration embedded in *Phys.* II 1's conception of what it is for an element to have a nature.[31]

In her essay on *GC* II 5, Mary Krizan argues that (1) a primary contrary is a common matter in each instance of elemental transformation and (2) there is no body beyond the four simple bodies. Contrary to what may initially appear, these results do not by themselves imply a rejection of prime matter; they only imply that *if* prime matter exists, it cannot itself be a body (or a kind of extension). More to the point: *GC* II 5 is neutral between two positions: (a) the primary contraries are the only matter for the simple bodies and (b) the primary contraries are matter for the simple bodies in one way, and traditional prime matter is matter in another way. In short, according to Krizan, *GC* II 5 does not offer a decisive case for or

[30] Most notably, Gill 1989: 77 and Lewis 1996: 23. [31] Scharle 2009: 326–345.

against prime matter; rather, it leaves room for primary contraries to function as constituents of simple bodies.

In his essay on *GC* II 7, Andreas Anagnostopoulos presupposes that there is a single material principle underlying the elements. He takes Aristotle's description of matter as common and as enabling elemental transformation, as well as the account of mutual affectability from *GC* I 7, which crucially informs *GC* II 7's explanation of elemental mixture, to be most naturally understood in terms of such a principle. In addition, he holds that the chapter speaks against certain specific revisionary conceptions of elemental matter. The fact that Aristotle entertains, as distinct initial distinct options for what comes to be from fire and earth which share a common matter: (i) one or another of fire and earth; (ii) a composition; or (iii) the matter, suggests that matter is distinct from each element in a particular case. Anagnostopoulos also rejects the idea that any elemental contrary remains robustly present in elemental mixtures, and this casts doubt on the idea that in every change involving the elements, some contrary is available to play the role of matter by persisting through the change.

2

On Generation and Corruption *II 1*

Timothy Clarke

2.1 Introduction

The first eight chapters of *GC* II are an investigation of "the so-called elements," the primary bodies that make up all the more complex bodies of the sublunary world. For Aristotle these elements are the traditional quartet of earth, air, fire, and water. (The heavens are composed of a fifth, *sui generis* body, discussed in *DC* I–II.) In the course of this investigation, Aristotle defends the view that there are four, and just four, primary sublunary bodies. He gives an account of their intertransformations and argues that none of them is prior to any of the others. He goes on to explain how it is that these primary bodies give rise to other, compound bodies.

Chapter 1 introduces this project. It is an important chapter for understanding how the treatment of the elements fits into the overall program of the *GC*, and so an important chapter for understanding the unity of the treatise. It is also a crucial text for understanding Aristotle's theory of matter, and especially for the question of whether he accepts such a thing as "prime" matter.

The chapter can be divided into three sections. The first section (328b26–329a5) gives a brief summary of *GC* I and explains why we need to investigate the so-called elements.[1] In the second section (329a5–24), Aristotle broaches the question of whether there is a kind of matter "beyond" these elements. He criticizes two earlier theories that (he thinks) give an affirmative answer to this question. The first is Anaximander's theory

* For their comments on earlier versions of this essay, I would like to thank Andreas Anagnostopoulos, Laura Castelli, Andrea Falcon, Joseph Karbowski, Sean Kelsey, Emily Perry, and audiences in Spetses and Munich.

[1] For the phrase "the so-called elements," see Andrea Falcon's introductory essay. With Falcon, I think that the point of the phrase is to signal that these bodies' status as elements is disputed (see Plato, *Tim.* 48 B6–C2). I do not think that Aristotle himself wants to deny that they are really elements: see *GC* II 1, 329a5–8 (to be discussed below). Cf. also Crowley 2008: 240–242 and Buchheim 2010: 448–449.

of the *apeiron*; the second is Plato's theory of the Receptacle in the *Timaeus*. In the final section (329a24–b6), Aristotle sets out his own position. According to his theory, the primary bodies themselves have matter, but this matter is incapable of separate existence. The interpretation of this final section is contested, but I shall argue that it strongly supports the view that Aristotle accepted the existence of prime matter – an ultimate material substratum that partially constitutes each of the primary sublunary bodies and which underlies the process of elemental intertransformation.

2.2 The Need for an Investigation of the Elements

Back in *GC* I 6, Aristotle had spoken of the need to discuss "the matter and the so-called elements." There he had said that before we come to this topic we first have to look at the topics of action and affection and of mixture. These notions play a central role in our predecessors' explanations of generation and perishing, but have not yet received a proper elucidation.[2] By the time we get to *GC* II, this preliminary stage of the inquiry has been completed.[3] Aristotle starts the book by summarizing (most of) *GC* I and then makes the promised transition to the study of the elements:

> So then, with regard to mixture, contact, and acting and being acted upon, we have said how they belong to the things that change naturally. And further, with regard to simple generation and perishing, we have said how they occur, what they are of, and by what cause they occur. And similarly also with regard to alteration, we have said what being altered is and how it differs from them [*sc.* generation and perishing]. But it remains to consider the so-called elements of bodies. The reason is that, for all the naturally constituted substances, their generation and perishing are not without the perceptible bodies.[4] And of these some people say that the underlying matter is one, positing, for example, air or fire or something intermediate between these, which is both a body and separable, while other people say that it is more than one in number, some speaking of fire and earth, while others speak of these and of air as a third, and others also speak of water as a fourth of these, as does Empedocles. From these things, when they are combined and separated or altered, they say that the generation and perishing of things result. (*GC* II 1, 328b26–329a5)

[2] See *GC* I 6, 322b1–26.

[3] Action and affection were discussed in *GC* I 7–9, mixture in *GC* I 10. The analysis of contact in *GC* I 6 was itself a preliminary to these discussions (see 322b21–25).

[4] Rashed puts a paragraph break at this point in his edition, which suggests that he takes Aristotle's explanation of why "it remains to consider the so-called elements" to end here. By contrast, I take the explanation to comprise the whole μὲν γάρ [. . .] δέ complex at 328b32–329a5.

Aristotle is here giving us something that was missing in *GC* I 6. Back then he said that we need to discuss the so-called elements, but he did not say why. We now get an explanation of why this topic belongs on the agenda of the *GC*.

So what is the explanation? It starts with the opaque claim that "for all the naturally constituted substances, their generation and perishing are not without the perceptible bodies" (328b32–33). The "naturally constituted substances" are presumably *complex* natural substances: animals, plants, and their (compound) bodily parts.[5] Aristotle is telling us that the generation and perishing of such substances necessarily involve perceptible bodies. On the face of it, this is a puzzling claim. For surely, one might think, complex natural substances *just are* perceptible bodies, and so *of course* their generation and perishing do not occur "without the perceptible bodies."[6] So Aristotle seems to be saying something trivial, and which should go without saying.

One response to this puzzle would be to reconsider what is meant by "perceptible bodies." The phrase is naturally taken as referring to perceptible bodies in general, but we might try adopting a restrictive reading on which it refers *only* to the so-called elements. On this restrictive reading, the sentence is not trivial: it tells us that the generation and perishing of complex natural substances always involve the so-called elements – and this is a substantive claim. However, such a reading is not very plausible.[7] First, Aristotle has clear and unambiguous ways of referring to the elements, and there is no obvious reason why he should now refer to them in this unprecedented and misleading way. Second, he goes on (at 328b33–329a3) to present Empedocles' theory of the four elements as a theory of the underlying matter *of* the perceptible bodies.[8] This suggests that "the perceptible bodies" cannot refer exclusively to the elements. Empedocles' theory of the four elements is a theory of the underlying matter of perceptible bodies in general, and not a theory of the underlying matter of the elements themselves.

The solution to the puzzle, I think, is to see that Aristotle would deny that *all* complex natural substances "just are" perceptible bodies. Consider the following sentence from the beginning of the *DC*:

[5] Elsewhere Aristotle includes the eternal heavenly bodies among the "naturally constituted substances" (see *PA* I 5, 644b22–24), but here he must be thinking only of those complex natural substances that come into and go out of existence.

[6] See Williams 1982: 152.

[7] Cf. Joachim 1922: 191; Crowley 2019: 7–9.

[8] The antecedent of τούτων ("of these") at 328b33 is clearly "the perceptible bodies."

> For, of naturally constituted things, some *are* bodies and magnitudes, some *have* body and magnitude, and some are *principles* of the things that have them. (*DC* I 1, 268a4–6)

He does not give examples, but it is plausible that the first group includes the (compound) parts of animals and plants – flesh and bone, bark and sap, and so on. These things are bodies and magnitudes. The second group presumably includes animals and plants themselves. I have a body, but it is not the case that I simply am a body. Rather, I am a composite of body and soul. (The same goes for the tree in my garden.) The third group is likely to consist of souls; these are the principles of the animate natural substances that have bodies.[9] Similarly in *DC* III 1, Aristotle makes it clear that he does not want to identify all natural substances with their bodies: "[A]ll natural substances either are bodies or come to be with bodies and magnitudes" (298b3–4).

This is why it is not trivial to say that "for all the naturally constituted substances, their generation and perishing are not without the perceptible bodies." I take Aristotle's point to be that whenever a complex natural substance comes to be or passes away, this process at the very least *involves* a perceptible body coming to be or passing away. Specifically, it involves the perceptible body *of* that natural substance coming to be or passing away. This is true even if the substance and its body are strictly speaking nonidentical.[10]

How does this help explain why we need to investigate the so-called elements? We can now move to the second part of the explanation (328b33–329a5). Different philosophers have taken different views about the matter of the perceptible bodies. Some posit a single body (air or fire, say) as the ultimate matter for all other bodies – as does Anaximenes, for example. Others, like Empedocles, posit a plurality of basic elements. But despite this disagreement, there is widespread agreement that generation and perishing, or what we *call* generation and perishing, come about when these primary bodies undergo some sort of change (such as alteration, or combination and separation). So, if we want to give an account of the generation and perishing of complex natural substances, we will need an account of the generation and perishing of perceptible bodies; and if we want an account of the generation and perishing of perceptible bodies, we will need a theory of the primary bodies that are the underlying subjects of

[9] This analysis of *DC* 268a4–6 follows that of Falcon 2016: 428–429.

[10] For a suggestion along similar lines, cf. Joachim 1922: 192–193. For a different solution, see Crowley 2019: 20–21.

the relevant changes. This is why, in the context of the *GC*, we need a theory of the elements.

Notice that this rationale for studying the elements is valid even if (as Aristotle thinks) they themselves come into and go out of existence, and even if they are not absolutely basic. Even if the primary bodies are in some sense derivative entities, it is still the case that identifying these bodies will form an important part of our overall account of generation and perishing in the natural world.

2.3 Criticisms of Anaximander and the *Timaeus*

The second section of the chapter turns to the question of whether there is a more basic kind of matter beyond the elements:

> So let us agree that is correct to say that the primary ones are principles and elements – the things from which, by their changing either by combination and separation or by some other kind of change, generation and perishing result. However, those who posit a single matter beyond the aforementioned things, and this as bodily and separable, are in error. For without perceptible contrariety it is impossible for this body to exist.[11] For it is necessary for this *apeiron*, which some say is the principle, to be light or heavy or cold or hot. And the account in the *Timaeus* has no determinacy. For he has not said clearly, of the "all-receiver," whether it is separate from the elements. And nor even does he make any use of it, despite having said that it is something underlying the so-called elements, prior to them, as gold is related to golden products. (And further, this too is not well said, when said in this way. Instead, the things of which there is alteration are like this; but with regard to things of which there is generation and perishing, it is impossible to refer to them as that from which they have come. And yet he says that truest by far is to say that each of them is "gold.") But he carries out the analysis of the elements, which are solids, as far as plane figures. But it is impossible for the "nurse" and the primary matter to be the plane figures. (*GC* II 1, 329a5–24)

Aristotle begins by emphasizing a point of common ground between his theory and the natural philosophical tradition. He endorses the general schema described in the previous section. Generation and perishing take place when "the primary ones" – I take this to mean the primary *bodies* – undergo some sort of change (whatever this might be).[12] While there are different views about the number and nature of these bodies, they are

[11] Reading αἰσθητῆς at 329a11; Rashed prints αἰσθητόν. I discuss these two variants below.

[12] I follow Solmsen 1958: 249n25 and Rashed 2005: 52 in taking τὰ πρῶτα at 329a5 to refer to the primary bodies.

rightly regarded as principles and elements of the things whose generation and perishing result from their change (329a5–8).

So the primary bodies are the principles and elements of the more complex bodies. Aristotle's next question is whether these primary bodies are absolutely basic, or whether they too have matter. In the third section of the chapter, he puts forward his own theory of the matter of the elements. In the present section he criticizes two earlier theories.

His first target is a view on which the familiar elements are generated from a distinct kind of matter that is "bodily and separable." He takes Anaximander's theory of the *apeiron* (the "infinite" or "boundless") to be a representative of this sort of view.[13] On Aristotle's interpretation, the *apeiron* is the material principle of the familiar elements, the primordial substance from which they and everything else were first generated.[14] The *apeiron* is "separable" from – in the sense of being capable of existing independently of – the things that were generated out of it. It existed before they did, and it continues to exist in its original form outside the heavens.[15]

The argument against Anaximander is very brief, taking up just two sentences.[16] Its interpretation is complicated by the fact that the manuscripts are divided at 329a11.[17] The reading of family **a** is the neuter accusative αἰσθητόν (EMW);[18] the reading of family **b** is the feminine genitive αἰσθητῆς (HJ^{1}V). These variants give us the following two versions of the argument:

(A) For without contrariety it is impossible for this body to be perceptible [αἰσθητόν]. For it is necessary for this *apeiron*, which some say is the principle, to be light or heavy or cold or hot.

(B) For without perceptible [αἰσθητῆς] contrariety it is impossible for this body to exist. For it is necessary for this *apeiron*, which some say is the principle, to be light or heavy or cold or hot.

13 Aristotle does not mention Anaximander by name, but it is widely accepted that he is the target here.

14 See *Phys.* III 4, 203b10–15; III 5, 204b22–24. Cf. also Simplicius, *In Phys.* 24.13–18.

15 For the *apeiron* as surrounding the heavens, see, e.g., *Phys.* III 4, 203b11; *DC* III 5, 303b11–13; *GC* II 5, 332a24–25.

16 Some commentators (e.g., Williams 1982: 164; Wedin 2013: 19n4) think that Aristotle reprises this argument later on, at *GC* II 5, 332a20–26. However, these arguments seem to me to be distinct and to be making different points. The purpose of the later argument is to show that there cannot be any additional element besides earth, air, fire, and water. The purpose of the present argument, as I read it, is to show that the elements (however many they may be) cannot have emerged from a body that is free from perceptible contrariety.

17 See Rashed's apparatus *ad locum*.

18 This was also the reading of Ḥunayn's Greek exemplar.

We can make sense of either text. Nonetheless, I have a preference for version (B), for two reasons.

First of all, on version (A), the alleged problem for Anaximander is that the *apeiron* cannot be *perceptible*. However, in the absence of further argument, it is not obvious why this shows that his theory is mistaken. Indeed, one might think that the imperceptibility of the *apeiron* is something that Anaximander could happily embrace; it would allow him to explain why we never actually *see* the *apeiron*.[19] On version (B), the alleged problem is that the *apeiron* cannot *exist*. This is unquestionably a problem for Anaximander: no further argument is necessary.

A second reason for preferring version (B) is that it makes better sense of the second sentence ("For it is necessary. . ."). On version (B), the second sentence straightforwardly explains the first. If it is necessary for the *apeiron* to have some perceptible quality (such as light or heavy or cold or hot), then it straightforwardly follows that without such qualities it cannot exist. On version (A), it is unclear exactly how the second sentence is meant to explain the first. I think the best option would be to interpret Aristotle as meaning that, *if it is to be perceptible*, it is necessary for the *apeiron* to be light or heavy, cold or hot, and so on. (Hence, if the *apeiron* lacks such qualities, it cannot be perceptible.) The difficulty is that this is not what he actually says. So version (A) requires a somewhat strained interpretation of the second sentence.

Let us suppose, then, that version (B) is the right one. In that case, Aristotle's starting point is that Anaximander wanted the *apeiron* to be free of perceptible contrariety. And the objection is simply that – *pace* Anaximander – it must have *some* perceptible quality; therefore, the *apeiron*, as conceived by Anaximander, cannot exist.[20]

One question is why Aristotle claims that it is necessary for the *apeiron* to have some perceptible quality. I think there are two possibilities here. The simplest explanation is that he is assuming that all bodies, insofar as they are bodies, have perceptible qualities.[21] Another possibility is that he

[19] This means that, if we go with version (A), the argument needs to be completed. For example, we could supply the further premise that there cannot be an imperceptible body. If it is impossible for the *apeiron* to be perceptible, and if there cannot be an imperceptible body, then it follows that the *apeiron* cannot exist. For a different way of completing the argument, see King 1956: 380n28.

[20] It is sometimes suggested that the argument involves an additional step: given that the *apeiron* must have some such perceptible quality, it must be identical to one of the four familiar elements (or to some compound of them). Therefore, it cannot exist separately from them. See Philoponus, *In GC* 208.15–18; Williams 1982: 153; cf. *DC* III 6, 305a22–25. I do not think that we need this additional step; the argument makes good sense without it.

[21] See, e.g., *DA* III 12, 434b12, where he says that every body is tangible.

is thinking of the *apeiron*'s role as a principle. The *apeiron* is supposed to be the source of the familiar elements: they are generated from it. Now, if some portion of the *apeiron* takes on the characteristics of fire (say) after previously lacking them, then it presumably becomes hotter than it was before. But in that case the original portion of the *apeiron* was colder than the fire that is generated from it. And so we can conclude that the *apeiron* is characterized by a quality on the hot–cold spectrum.

Another question is why Aristotle thinks that Anaximander wanted to say that the *apeiron* is free from perceptible contrariety. Here, a passage from the *Physics* is helpful:

> But nor is it possible for an infinite body to be one and simple, whether it is, as some say,[22] the body beyond the elements, from which they generate these, or whether it is taken without qualification. For there are some who make this [*sc.* this body "beyond the elements"] the infinite, rather than air or water, so that the others may not be destroyed by the one of them that is infinite. For they [*sc.* the elements] have contrariety relative to one another; for example, air is cold, water is wet, and fire is hot. If one of them were infinite, then the others would already have been destroyed. But they say that, in actual fact, [the infinite] is something else from which these [have come to be]. (*Phys.* III 5, 204b22–29)

The passage explains Anaximander's motivation for positing the *apeiron* as his material principle, rather than air or water (the principles favored by his fellow Milesians). The thought is that he wanted his primordial stuff to be infinite in extent.[23] This meant that he could not identify it with any of the familiar elements. These elements are opposed to one another, which means that they are liable to destroy one another. (For example, water destroys fire and vice versa, depending on their respective quantities.) So if the infinite principle were any one of the familiar elements, it would already have overwhelmed and destroyed all of the others. Hence Anaximander's decision to make his principle a different kind of body, one lacking the perceptible qualities that characterize the familiar elements.

Following the criticism of Anaximander, Aristotle turns to the theory of the Receptacle in Plato's *Timaeus*. Timaeus introduces the Receptacle in preparation for his account of the nature and genesis of the four primary bodies (see *Tim.* 48 B3–49 A6). Aristotle takes it that the Receptacle is

[22] As in *GC* II 1, Anaximander is not here mentioned by name, but he is clearly the principal referent. Cf. Simplicius, *In Phys.* 479.32–480.4.

[23] Cf. *Phys.* III 4, 203b18–20.

intended to serve as the matter of these bodies.[24] On this view, the Receptacle is a material substrate which persists through their inter-transformations – water turning into air, air into fire, and so on.

There is an important difference between Aristotle's criticisms of Anaximander and of the *Timaeus*. His objection to Anaximander was that the theory is mistaken; no such body can exist. His main criticism of the *Timaeus*, by contrast, is not that the theory is mistaken, but rather that it is indeterminate on crucial points.

As I read the passage, Aristotle offers two reasons for this judgement of indeterminacy. The first is that the precise relationship between the Receptacle and the primary bodies is left unclear. Specifically, Aristotle claims that it is unclear whether the Receptacle is *separate* from the primary bodies. The second reason is that Timaeus does not in fact make use of the Receptacle in his account of the primary bodies, despite having originally introduced it for this purpose.

Start with the complaint that is unclear whether the Receptacle is separate. The complaint, I take it, is that it is unclear whether the Receptacle, or some region of it, can exist independently of the four primary bodies that it "receives."[25] Now, Aristotle does not say why he finds this unclear. And one's initial thought might be that, *contra* Aristotle, there is good evidence that the Receptacle *is* supposed to be separate from the elements. After all, the Receptacle is ungenerated and indestructible (*Tim.* 52 A8–B1), whereas the primary bodies are constructed by the Demiurge. Timaeus tells us that, prior to the Demiurge's intervention, the universe contained nothing worthy of the names "fire," "water," and so on (69 B6–8). This suggests that the Receptacle preceded the elements, which seems straightforwardly to establish its separateness from them. The Receptacle is apparently capable of existing in their absence, just like Anaximander's *apeiron*.[26]

[24] Whether he is right about this is controversial. For a helpful discussion, defending a negative answer, see Beere 2016: 201–226.

[25] More precisely, I think the complaint must be that it is unclear whether some region of the Receptacle can exist independently of the primary bodies *or any of their fragments*. Timaeus evidently thinks that the primary bodies can be splintered into parts or fragments (μέρη, τμήματα) – the elementary triangles. Given that these detached fragments exist within the Receptacle, it follows that there are regions of the Receptacle that are not occupied by any of the four primary bodies.

[26] This argument assumes a "literalist" reading of Timaeus' creation story – a reading on which we are to take seriously the description of the cosmos as having a temporal beginning. Aristotle himself favors such a reading: see *DC* I 10, 280a28–32.

In Aristotle's defense, however, it is doubtful that this settles the question of the Receptacle's separateness. For while Timaeus does say (at 69 B6–8) that there was nothing worthy of the names of the four primary bodies in the precosmic Receptacle, he also speaks as though these bodies were already present back then, albeit in an inferior, vestigial form. At 53 A2–B7, he says that even before the Demiurge entered the scene, inferior versions of the four kinds (the "traces") had come to occupy their own different regions.[27] So the question is whether any region of the Receptacle ever exists – or is able to exist – without being occupied by *some version* of one of the primary bodies.[28] And here it is tempting to agree with Aristotle – it is hard to see that there is anything in the text of the *Timaeus* that would allow us to decide this question one way or the other.[29]

Aristotle's second complaint is that Timaeus does not "make use" of the Receptacle. The objection is that while (as Aristotle thinks) Timaeus initially brings in the Receptacle to function as the substrate that underlies elemental intertransformations on the model of some gold's relation to a succession of different artefacts formed out of it (50 A4–B6), he then goes on to analyze the primary bodies in terms of plane figures, the elementary triangles (53 B7–57 C8). The four bodies are solid geometrical structures built up from these triangles. When one of the bodies changes into another, underlying this process are the triangles, which dissociate and recombine in different configurations.[30] Now, the Receptacle cannot be identified with these triangles.[31] So Aristotle's objection is that it is really the triangles that do the

[27] Cf. *DC* III 2, 300b17–18, where Aristotle refers to the elements as having existed in the disorderly, precosmic universe.

[28] Or by some fragment thereof. For the additional qualification, see footnote 25 above.

[29] For a different explanation of why Aristotle thinks that it is unclear whether the Receptacle is separate from the elements, see Williams 1982: 153. On Williams's view, the complaint is that the Receptacle is assigned different and incompatible functions: the function of a *container* and the function of a *stuff.* "If it is a container, as the word 'omnirecipient' indicates, it must be 'separated' from the things it contains. But if it is the stuff out of which the elementary particles are fashioned it is not so separable." I am not convinced that Aristotle would see these different functions as bearing on the question of the Receptacle's separateness. The fact that the Receptacle "receives" the elements does not imply that it can exist without being occupied by them. Conversely, if it is the stuff out of which they are made, there is no reason to think that it cannot exist in their absence.

[30] Aristotle has already discussed this theory of elemental intertransformation in *DC* III 7–8.

[31] I take this to be the point of the final sentence of the passage: "it is impossible for the 'nurse' and the primary matter to be the plane figures" (329a23–24). Aristotle is sometimes taken to be claiming here that it is *in fact* impossible for plane figures to serve as the ultimate matter for bodies (e.g. by Philoponus, *In GC* 210.7–16). This is certainly something that he believes (see especially *DC* III 1, 298b33–300a19). Given the context, however, it is more likely that he means that, *as a matter of Platonic exegesis*, it is impossible for the Receptacle (the "nurse of all becoming" as it is described at *Tim.* 49 A5–7 and 52 D4–5) to be identified with the triangles. Among other reasons, this is because the triangles are generated, whereas the Receptacle is ungenerated.

work of matter in Timaeus' theory of the elements. The Receptacle falls out of the picture: it fails to do the work that it was originally introduced to do.

In addition to these two complaints, Aristotle adds a further, parenthetical criticism of Timaeus' gold analogy (329a17–21). Timaeus asks us to imagine a person continually forming and reforming a series of different shapes from gold.[32] Suppose one were to point to one of the shapes and ask what it is. Timaeus tells us that "by far the safest answer with respect to truth" would be to answer "gold" rather than "a triangle" (50 A4–B5). In the same way, if one were to point to some currently fiery region of the Receptacle, and ask the same question, the safest answer would be to say that it is "the Receptacle" rather than "fire." In Aristotle's view, this is a mistake. Timaeus' "safe" answer suggests the materialist thesis that what a golden artefact "really" is is just *gold* – gold that currently happens to be in such-and-such a configuration. Aristotle thinks that we should instead say that the artefact is not "gold" but "golden."[33] This indicates that the artefact is composed of gold but avoids the mistaken materialist implication about what it "really" is.

This parenthesis does not contribute to Aristotle's main objection to the *Timaeus*, namely that its theory of matter is indeterminate. But it is still an important point for him to make in the present context. As we will see, he agrees with the *Timaeus* (or with the *Timaeus* as he reads it) that there is a kind of matter that underlies elemental intertransformation. But if he were to accept the materialist implication of Timaeus' recommended way of speaking – that this portion of fire, say, is really just its underlying matter – then he could not allow that the primary bodies genuinely come to be and pass away. Elemental interchange will merely be the alteration of the matter, and not generation and perishing proper. Since Aristotle wants to maintain that the primary bodies genuinely come into and go out of existence, it is important for him to distance himself from the *Timaeus* on this point.

2.4 Aristotle's Theory: Matter, Contrariety, and the Primary Bodies

The final section of the chapter (329a24–b6) presents Aristotle's positive view. He agrees with Anaximander and the *Timaeus* that there is a kind of matter of the elements, but he insists that this matter is "inseparable."

[32] Aristotle thinks that we are meant to imagine someone making a succession of differently shaped golden artefacts. For a different reading of the analogy, see McCabe 1994: 182 and Harte 2002: 255–258.

[33] Cf. *Phys.* VII 3, 245b9–12; *Meta.* VII 7, 1033a5–23; and IX 7, 1049a18–b2, with discussion in Beere 2006: 304–316.

This passage has played a central role in the modern debate over prime matter in Aristotle.[34] As I shall use the term,[35] "prime matter" (if there is such a thing) is the ultimate material substrate of everything in the sublunary world. It is a common constituent of the primary bodies, and the underlying subject of their intertransformations. Considered by itself, it is imperceptible and incorporeal, but it combines with different perceptible qualities so as to constitute the different primary bodies. It never exists on its own, independently of these bodies and the compounds formed out of them.

Does Aristotle affirm the existence of this sort of matter in the final section of *GC* II 1? The traditional view is that he does,[36] but the issue is controversial. Two main types of revisionary interpretation have been offered in recent times, both of which deny that Aristotle thinks that there is any persisting substrate in cases of elemental intertransformation.

On the first revisionary interpretation, when he talks about the "matter" of the elements, he is simply referring to the elements themselves.[37] The primary bodies are the matter *for each other*, in the sense that they are the (non-persisting) subjects of their intertransformations. So, for example, when air changes into water, the matter for this change is air, and when water changes into air, the matter for this change is water. There is no more basic kind of matter that persists through these changes.

According to the second revisionary interpretation, the "matter" of the elements is in fact a persisting *property* which plays a matter-like role.[38] To understand this interpretation, we need to anticipate some later developments in *GC* II. Aristotle argues that each of the four primary bodies is characterized by a pair of perceptible contraries: fire by hot and dry, air by hot and wet, water by cold and wet, and earth by cold and dry (*GC* II 3, 330b3–5). He proceeds to explain elemental transformations in terms of the exchange of these contraries. For example, when (hot, wet) air turns into (cold, wet) water, the hot is replaced by the cold, while the wet persists (*GC* II 4, 331a29–32). The second revisionary interpretation says that,

[34] See, e.g., King 1956: 381–383; Solmsen 1958: 248–250; Lacey 1965: 461–462; Charlton 1970: 132–136; Robinson 1974: 176–183; Williams 1982: 214; Charlton 1983: 201–202, Furth 1988: 226–227; Gill 1989: 243–247; Broadie 2004: 140–142; Rashed 2005: xcv–c; Buchheim 2010: 136–137; Krizan 2013: 221–222; Ebrey 2020: 88–94.

[35] Cf. Lewis 2008: 124–128.

[36] See Philoponus, *In GC* 210.17–212.2; Averroes, *In GC* 88.12–89.15 Eichner.

[37] For this interpretation, see King 1956: 381–383; Charlton 1970: 135–136; Charlton 1983: 201; Broadie 2004: 141–142; Rashed 2005: xcix–c.

[38] For this interpretation, see Furth 1988: 226–227 and Gill 1989: 243–247.

because it persists, the wet effectively plays the role of "matter" in this process. There is no underlying subject that remains through the change.

In what follows I want to argue that neither of these revisionary interpretations can do full justice to the details of 329a24–b6, and that tradition was right to see the passage as evidence of Aristotle's commitment to prime matter. The text is complicated, and so to make my case I will need to go through it sentence by sentence.

> (i) But we say that there is a certain matter of the perceptible bodies, yet this is not separable, but is always with contrariety, from which the so-called elements come to be. (*GC* II 1, 329a24–26)

This first sentence asserts the existence of a matter, or a kind of matter, of the perceptible bodies,[39] and says – at least as I read the relative clause – that it is from this matter that the so-called elements come to be.[40] This kind of matter is "not separable, but is always with contrariety." That is, it is incapable of existing on its own, free of the perceptible qualities that characterize the elements and their compounds. Aristotle does not spell out why it must be regarded as inseparable, but this is presumably meant to be a corollary of the preceding argument against Anaximander's *apeiron*.

> (ii) And about them we have made more precise determinations elsewhere. (*GC* II 1, 329a27)

The meaning of this sentence is not immediately obvious. It is initially tempting to take him to be referring back to "the so-called elements" – the nearest plural noun phrase.[41] But this makes it hard to understand the cross-reference. If the reference is to another discussion of the elements, then it must presumably be to *DC* III. But it is not easy to see why he should describe that earlier discussion as "more precise" than *GC* II 1–8. The inquiry into the sublunary elements in the *DC* was preliminary and incomplete. It is only in *GC* II that we get determinate answers to the key questions of what elements there are and how they change into one another.

[39] I take it that it is matter "of the perceptible bodies" in the sense that it is the ultimate matter of perceptible bodies in general – or at any rate of the sublunary perceptible bodies.

[40] The clause is ambiguous. I take the antecedent of "which" to be "a certain matter," but it might instead be "contrariety." (For the latter construal see King 1956: 381; Broadie 2004: 140n53.) I do not think that much turns on this; either way it is clear that Aristotle thinks that the matter in question is the matter of the elements (see 329a27–32).

[41] See Williams 1982: 155–156; Broadie 2004: 140.

If the antecedent of "them" is not "the so-called elements," then what are the other options? A common suggestion is that "them" picks up "matter" (a24) and "contrariety" (a26), and that the cross-reference is to *Phys.* I.[42] However, this suggestion faces a problem of its own. The matter referred to at 329a24 is the matter involved in elemental interchange. It seems unlikely that Aristotle would be pointing us back to *Phys.* I for a more precise discussion of *that* kind of matter. His concern in *Phys.* I was with matter in general.

My view is that Aristotle *is* pointing us back to *Phys.* I. But I do not think that he is saying that we should look to *Phys.* I for a more precise discussion of matter and contrariety. Rather, I think that the most likely antecedent of "them" is "the perceptible bodies" at 329a25. His claim is that, if we want a more precise account of perceptible bodies in general (an account of their principles), then we should look to *Phys.* I.

> (iii) However, since it is also in this way that the primary bodies are from the matter, it is necessary for us to make determinations about these too,[43] regarding as a principle and as primary the matter which is inseparable but which underlies the contraries; for neither is the hot the matter for the cold, nor is this the matter for the hot, but the underlying thing is the matter for both. (*GC* II 1, 329a27–32)

In *Phys.* I, Aristotle argued that, in every case of generation, an underlying subject – the matter – goes from having a privation to having the opposed form. The generated entity is always a composite of the matter and the form.[44] For example, a statue comes to be when the bronze goes from being unshaped to being shaped, and the statue is composed of the bronze and the shape.

I take him now to be claiming that the primary bodies "are from the matter" in the same way as perceptible bodies in general. This is most naturally read as a claim about the primary bodies' composition: just like other perceptible bodies, the primary bodies are themselves composed of matter.[45] He proceeds to give an (extremely condensed) argument for this

[42] See, e.g., Philoponus, *In GC* 210.24–26; Joachim, 1922: 198–199; Charlton 1970: 131; Gill 1989: 244.

[43] I think that "these" (τούτων) are the primary bodies. Philoponus takes "these" to mean "these contrarieties" (*In GC* 211.7–8), but this is difficult given that the plural ἐναντιώσεις does not occur in the preceding lines.

[44] See especially *Phys.* I 7, 190b10–23.

[45] We should distinguish the claim that the primary bodies "are from" matter (329a28–29) from the claim that they "come to be from" matter (329a26). The latter is a claim about their generation; the former is most plausibly a claim about their composition or constitution (once they have been generated). On the difference between these two locutions ("X *is* from Y"; "X *comes to be* from Y") see Lorenz 2019: 264–266, discussing *Phys.* I 7, 190b17–29.

view. The basic idea is that we must regard the primary bodies as being composed of matter, because this matter is what "underlies the contraries."

To make sense of this argument, we should begin by reminding ourselves of some of the key results of the preliminary inquiry into the sublunary elements in *DC* III. There he argued that the elements are generated, not eternal.[46] He went on to argue that they come to be from each other,[47] and that this happens by their intertransformation.[48]

Consider, then, a case of elemental intertransformation, as when air turns into water, or water into air. Aristotle's general account of change in *Phys.* I is that change takes place between contraries or properties that are intermediate between contraries.[49] So our case of elemental interchange should be understood in the same way – for example, as a change from hot to cold, or from cold to hot.[50] However, if this is so, then we must also posit, in addition to these contrary properties, a third thing: the matter which underlies the contraries. As Aristotle puts it, "neither is the hot the matter for the cold, nor is this [*sc.* the cold] the matter for the hot, but the underlying thing is the matter for both." For such a change to take place, there needs to be something that starts off hot and becomes cold, or that starts off cold and becomes hot. The contrary properties cannot play this role themselves.[51]

This is why we can conclude that each of the primary bodies is composed of ("is from") matter. The generation of a primary body is a process whereby some matter loses one property and gains another. So we can infer that the generated body is partly composed of the matter that is the underlying subject of this process.

If this account of Aristotle's reasoning is correct, then the present lines are incompatible with the first revisionary interpretation of the matter of the elements – the view on which the elements are the matter for each other. The air that changes into water does not go from being hot to being

46 See *DC* III 6, 304b23–305a14.

47 See *DC* III 6, 305a14–32.

48 See *DC* III 7, 305a33–b28. Cf. also the argument against Empedocles at *GC* I 1, 314b15–26, which is recalled at 329a35–b2.

49 See *Phys.* I 5, 188a30–b26.

50 Note that Aristotle has not yet established that heat and cold are two of the four basic tangible contraries involved in elemental interchange. (This will be established in the following chapters.) So his present suggestion – that certain cases of elemental interchange involve a change between these two properties – should for now be seen as provisional.

51 The same point is made in general terms in *Meta.* XII 1–2, 1069b3–9: "If change is from opposites or intermediates, and not from all opposites (for voice is not white) but from the contrary, then there must be something which subsists, which changes into the contrariety. For the contraries do not change. Further, something remains [through the change], whereas the contrary does not remain. So there is some third thing besides the contraries: matter." Cf. *Phys.* I 6, 189a21–b1; *Phys.* I 7, 190b29–191a5.

cold; nor is the resulting water composed of the air. The matter must be something else, something which persists through the change, and which is a component of the end result.

These lines also count against the second revisionary interpretation – the view on which the "matter" in elemental interchange is the persisting contrary property. Aristotle is explicit in *Phys.* I that the matter that underlies the process of generation is not a contrary property: "[T]he underlying thing is different [from the contraries]; for this is not a contrary" (*Phys.* I 7, 190b33–35). For him to abandon this idea in the case of the elements would be a major shift. But instead he appeals to the continuity between his theory of the elements and the general theory of *Phys.* I, which suggests that his position has not changed.

The present lines therefore support the view that Aristotle believed in prime matter. They indicate that he thinks that the primary bodies are composed of matter. This matter is the subject of – and persists through – the process whereby one primary body perishes and another comes to be. And since the primary bodies can all change into one another, we can infer that they share a single, common matter.

> (iv) So that, first of all, the potential perceptible body is a principle; second, the contrarieties [are principles] – I mean, for example, heat and cold; and only thirdly [are] fire, water and such things [principles], for these change into one another, and it is not as Empedocles and others say (for there would be no alteration), while the contrarieties do not change.[52] (*GC* II 1, 329a32–b3)

Aristotle now draws a further consequence of the preceding argument. It shows that we must distinguish three levels of principles:

(1) the matter, now described as "the potential perceptible body";
(2) the contrarieties that are the termini of elemental interchange (heat and cold, and so on); and
(3) the primary bodies (fire, water, and so on).

The matter and the contrarieties are the principles of the primary bodies; the primary bodies are the principles of complex bodies.

Once again, these lines are hard to accommodate on the revisionary interpretations of the matter of the elements. The distinction between (1) and (3) is a problem for the view that this matter is nothing other than the primary bodies themselves.[53] In response to this problem, it has been

[52] On the grammar of this sentence, see Solmsen 1958: 248n24.

[53] Cf. Buchheim 2010: 136.

suggested that the primary bodies actually play two distinct roles within the schema. Considered as the matter *for one another*, they belong at level (1); considered as the matter *for complex bodies*, they belong at level (3).[54] However, if this is Aristotle's position, then it is puzzling that he only speaks of "fire, water and such things" when he comes to level (3). The implication of his wording is that these entities have not already appeared at levels (1) and (2).

The schema also presents a difficulty for the second revisionary interpretation. It is clear why Aristotle might want to distinguish the persisting contrary from the non-persisting contraries that are the termini of elemental interchange. But it is less clear why he would describe only the former as "the potential perceptible body." Take the example of (hot, wet) air changing into (cold, wet) water. One of the initial contraries (the wet) persists, the other (the hot) is replaced. But why pick out the wet, and not the hot, as "the potential perceptible body"? We might say that the wet is a "potential perceptible body" in the sense that it is capable of combining with the cold to constitute water.[55] But in that case the hot would seem to be a "potential perceptible body" too, given that it is capable of combining with the dry to constitute fire.

We can make better sense of the expression if Aristotle is referring to prime matter. Taken on its own, prime matter is merely potentially a perceptible body. It only qualifies as an actual perceptible body by virtue of its being characterized by certain perceptible qualities.

> (v) But nonetheless, even so, it should be said which and how many are principles of body; for the other philosophers posit them and make use of them, and say nothing as to why these or this many. (*GC* II 1, 329b3–6)

Aristotle has just argued (at 329a35–b3) that we must distinguish the primary bodies from the contrarieties.[56] And of course it is the former that are our primary objects of concern. Nonetheless, we now need to turn our attention to the contrarieties and ask "which and how many" of them are the principles of the primary bodies. This is the project that Aristotle will undertake in the next chapter. The reason why we need to look into this topic is that it will give us a way of determining the number and nature of the primary bodies themselves.

[54] See Broadie 2004: 141–142; cf. Charlton 1983: 201.

[55] See Gill 1989: 246–247.

[56] The argument is very simple: the contrarieties do not change, whereas the primary bodies do, so they cannot be the same thing.

3

On Generation and Corruption *II 2*

Stasinos Stavrianeas

3.1 Introduction

Aristotle concludes *GC* II 1 by enumerating three types of principles governing the constitution of perceptible bodies: (a) the matter that underlies the contrarieties; (b) the contrarieties such as heat and cold; and finally (c) the elements such as fire, water, and the like.[1] This chapter focuses on contrarieties, with the aim of determining which of them constitute principles for the composition of perceptible bodies. This task is announced in the closing lines of *GC* II 1: "the other philosophers posit them and make use of them, and say nothing as to why (I) these or (II) this many."[2] This statement concerns both (b) contrarieties and (c) elements. So there are at least four questions that need to be addressed: (bI) which contrarieties are principles of perceptible bodies; (bII) why they are as many as they are; (cI) which elements should be posited as principles; and (cII) why they are as many as they are. *GC* II 2 is concerned with contrarieties and deals with questions (bI) and (bII), specifying which contrarieties must be posited as principles, namely hot and cold, moist and dry, and explaining why they cannot be fewer than four. *GC* II 3–4 answer questions (cI) and (cII).

GC II 2 divides into four sections dealing with the following topics: (a) location of the contrarieties constituting the principles of perceptible bodies in the kind "tangible body" (329b6–17); (b) first reduction of tangible contrarieties eligible to serve as principles of perceptible bodies (depending on whether or not they can affect and be affected by one another) (329b17–31); (c) second reduction of pairs of contraries eligible

* I would like to thank the organizers of the Spetses Conference on *GC* II as well as the participants for the creative discussion and friendly atmosphere during the workshop. I would also like to thank Panos Dimas, Andrea Falcon, Sean Kelsey, and Spyros Rangos for reading and commenting on previous drafts of this paper. I apologize for any remaining obscurities or inaccuracies.

1 *GC* II 1, 329a30–35. 2 *GC* II 1, 329b4–6.

to serve as principles of perceptible bodies (depending on whether or not they are produced by some other contraries) (329b31–330b24); and (d) explanation of why the principles are four in number (330b24–29). I divide my discussion accordingly.

3.2 Tangible Contrarieties as Principles of Perceptible Bodies (329b6–16)

This first section contains an argument for the claim that the contrarieties we are looking for are *tangible*, as well as a counterargument to a possible objection that other kinds of perceptible contrarieties (for instance, the contrarieties relating to vision) are prior to the tangible ones. I will give a brief review of these arguments and then address a few puzzling points.

The first argument turns on defining perceptible bodies as being essentially tangible. The result is that the contrarieties we are looking for are contrarieties of bodies *qua* tangible bodies – that is, they relate to the sense of touch. Aristotle motivates this result by claiming that it is tangible contrarieties that produce the forms and principles of bodies – namely, that differentiate them into kinds (329b8–11). That is why, he says, the contrarieties that relate to other senses (e.g., to sight) do not produce any element (329b11–13). He concedes that vision is prior to touch, and that (therefore) the subject (ὑποκείμενον) of vision is prior to the subject of touch. Nonetheless, he insists, the quality (πάθος) that makes bodies visible does not belong to them *qua* tangible. The result is that, even if the quality that makes bodies visible is naturally prior to the quality that makes them tangible, still, the contrarieties he is seeking here are to be found among tangible contrarieties (329b14–16).

I would like to discuss first the identification of perceptible body with tangible body and then the claim that tangible contrarieties produce not only the elements of body but also its forms or principles (329b9–13). The crucial role that tangible qualities play in the definition of bodies resurfaces in the *DA*: "the differences of body *qua* body are tangible; and I mean the differences that define the elements" (*DA* II 11, 423b27–28). But what reason might support this claim? Philoponus suggests that tangible qualities are closest to the matter of the elements, so they are the most common among simple bodies (*In GC* 213.20–24). Some interpreters further support this suggestion by arguing that no body exists without tangible qualities, while other perceptible qualities do not belong universally to bodies.[3] This argument, however, is not explanatory. There must be a reason why tangible contrarieties are more basic and so more common. A second group of interpreters point to the universality of the sense of touch, which is present in all animals, by contrast with the other senses, which are not.[4] To me, this line of interpretation seems to get the order of explanation the wrong way around. The sense of touch is more widespread among animals because tangible properties are the most common perceptible properties rather than the reverse. That said, however, this second line of interpretation is instructive because it suggests that Aristotle's reference to the sense of touch may, after all, be misleading. For Aristotle's reference to touch may be merely a heuristic device for detecting what contrarieties to look for, instead of offering a reason why tangible properties are fundamental. In fact, as we will see in due course, the sense of touch does not play any role in defining the contrarieties Aristotle identifies as principles of body. Instead, Aristotle defines those contrarieties in terms of the active and passive capacities they bestow on the bodies constituted out of them.

Given this approach to tangible contrarieties, it is puzzling that commentators have not turned to the discussion of contact advanced in *GC* I 6. There Aristotle considers the view that the generation of the elements is due to aggregation and segregation, or action and passion, a thesis universally accepted by previous thinkers (322b6–9).[5] Such processes of aggregation and segregation require an active power (say, the capacity to heat) and a corresponding passive power (the capacity to be heated), while they also require a single underlying subject that undergoes the change (322b19–20). Aristotle distinguishes two types of process, action and passion (i.e., alteration) and mixture, and holds that contact or touch (ἁφή) is a necessary condition for both:[6]

> It is necessary for those things which are involved in mixing to be capable of contact with one another (ἀλλήλων ἁπτικά), and the same holds for anything which properly speaking acts on, or is affected by, another. (*GC* I 6, 322b27–30, Williams' translation)

[3] Joachim 1922: 202 and Rashed 2005: 155. I note, in passing, that this claim is arguably refuted by the fact that, for Aristotle, superlunary bodies are visible but not tangible.

[4] Williams 1982: 157.

[5] For the rendering of ποιεῖν/πάσχειν as "affecting/acting" and "being affected/ being acted upon" as well as the different uses of affection/passion (πάθος), see Wildberg 2004: 221–223.

[6] The difference is explained further on in *GC* I. Here suffice it to say that in alteration one factor is active and the other passive (*GC* I 7, 324a18–19), while in mixing every ingredient is both active and passive (*GC* I 10, 328b20–22). See also the essay by Anagnostopoulos in this volume.

Notice that the things involved here must be in mutual contact. An analogous condition is introduced in *GC* II 2 concerning the mutual active/passive roles of the elements in mixing and changing into one another (*GC* II 2, 329b21–23). Now, of course, acting and being acted upon are different from touching and being touched. But given what follows in *GC* I 6 concerning the various senses of "contact," it seems that mutual contact is for Aristotle something more robust than what *we* mean by this expression. It is a kind of contact specifically involved in mixing and affecting, different from the contact that obtains, for instance, between mover and moved, or between any two things that have just their extremities together (*GC* I 6, 323a4). Aristotle distinguishes between the way movers (the genus under which agents fall) must be in contact with the things that they move from the way affecting agents (the species of movers that interests us here) must be in contact with the things they act upon by pointing out that the action of these agents results in some affection on the part of the things they act upon. This characteristic leads to the following claim:

> [T]here is a sense in which the things which cause motion will touch the things they move [*sc.* in the sense labeled "general" below] and a sense in which they will not [*sc.* in the sense labeled "mutual contact" below]. But the definition of "being in contact" in general is applicable to things which have a position and are capable the one of moving and the other of being moved, whereas that of "mutual contact" is applicable to things capable of moving and being moved in which there is acting and being affected. (*GC* I 6, 323a21–25, Williams' translation)

Thus, Aristotle tries to define a sense of "touch" that will apply to a particular aspect of the interaction between bodies: the one that results in an affection in the patient, though not just any type of affection, such as a change of color. For since the affection will be a result of mutual touching, it must be first and foremost a tangible affection (whatever other types of affection come to be realized in the patient, either coextensive with the change in the tangible affection or not). Hence, contrarieties with respect to touch are primary with respect to mixing and alteration. Their mutual touching, which is defined as a mutual action/passion, is the type of contact whereby bodies mix with, and alter and are altered by, one another. And as we will see below, tangible contrarieties will be defined in terms of reciprocal action between the touching bodies.

One final comment concerning the initial argument in *GC* II 2 for the restriction of perceptible qualities to tangible ones is in order. Aristotle says

(i) that not all contrarieties produce (ποιοῦσιν) forms and principles of body, but rather only tangible contrarieties do and (ii) that tangible contrarieties produce the elements (329b9, 13). These two claims appear alongside one another within the same argument and so it seems that the latter explicates the former. Some commentators, however, take the reference to forms as alluding to the contrarieties informing prime matter, thus *constituting* the "forms" of the primary perceptible bodies.[7] But then, the use of the verb for *action* (ποιεῖν) sounds exaggerated. It makes more sense to assume that the contrarieties, which are inseparable from the material principle, produce, by their own action, the elements that come to be from them (cf. *GC* II 1, 329a25–26). These elements are the different forms or kinds of perceptible body and are, therefore, also principles of perceptible body (*GC* II 1, 329b29–35). This idea is echoed elsewhere in the Aristotelian corpus (e.g., in the discussion of three levels of composition in the constitution of natural bodies offered in *PA* II). The first level consists in the composition of the elements from the basic contrary powers which are the matter of composite bodies (*PA* II 1, 646a16–20).[8] Here the contrarieties (along with the matter from which they are inseparable) are said to be the matter of (i.e., *to produce*)[9] the higher-level composites, including the elements. We can make sense of the production of the forms of the kind "material body" by the contraries: the contraries *in* matter produce the elements and other uniform bodies by producing the tangible affections characteristic of those bodies.[10] As we shall see below (Section 3.4), this is what Aristotle achieves in *GC* II 2. He shows how the other qualitative differences of bodies can be *produced by* the four basic contrarieties.

Let us turn to the objection Aristotle answers in the second part of our passage (329b14–16). The objection is motivated by the idea that visible qualities are prior by nature to tangible ones;[11] hence, the subject of visible

[7] Joachim 1922: 202.

[8] One could object here that it is not the elements that are produced from the basic contrary powers but rather the inorganic uniform, i.e., homoeomerous materials in general. Still, a case can be made that the latter encompasses the impure forms of the elements as well.

[9] The term used in *PA* II 1 is σύνθεσις (composition). In the ensuing lines, when Aristotle comments on the priority relations among the levels of composition, he glosses σύνθεσις with γένεσις and uses the verb ποιεῖν to express the idea that every generated thing produces its generation out of something and into something. Thus, we may safely assume that the three levels of composition are also cases of production.

[10] The tangible affections thus produced are defining or essential properties of the uniform bodies *qua* tangible bodies.

[11] The priority of sight has to do with their importance in conveying a richer variety of information concerning the perceptible bodies. Sight reveals a greater variety of differences between perceptible

contrarieties should also be prior to the subject of tangible contrarieties. Aristotle does not deny this conclusion; he only points out that visible contrarieties do not belong to the tangible body as such but rather in virtue of something else – namely, in virtue of the fact that a certain quality (presumably, the quality that makes bodies visible)[12] belongs to what is a tangible or perceptible body as such. This observation concerning visible contrarieties must be generalized over the perceptible contrarieties that relate to all other senses (see the reference to contrarieties that relate to taste at 329b12). None of them relates to the perceptible body *qua* tangible. Thus *perceptible* body seems to be narrowly construed. I want to suggest that Aristotle's goal in thus dealing with this objection is to bring forward the tacit distinction between a narrower and a wider sense of perceptible body.[13] In the narrower sense, a perceptible body is of a kind produced by specifically tangible contrarieties (compare the notion of contact or touch in the strict sense as defined in *GC* I 6). This narrower sense must be distinguished from a wider sense according to which anything that can be perceived through any sense modality is a perceptible body. Of course, there is at least another way to understand the objection. One can read it as a cautionary note: visible contrarieties are indeed prior by nature and more valuable compared to tangible ones, and this claim is not undermined by the identification of perceptible body with tangible body. I do not see a way of deciding between these two options.

3.3 Aristotle's First Reduction (329b16–32)

In this section, Aristotle lists a variety of tangible contrarieties that are candidates for being the principles of tangible bodies (i.e., perceptible body in the narrow sense of "perceptible"); then, he introduces a necessary condition for contrarieties that fit the bill (the pair of contraries must be such that the one is able to act upon and be acted upon by the other); finally, he defines the two pairs of contraries (hot and cold, dry and moist)

bodies (*Meta.* I 1, 98a24–27), and it is also clearer than touch (*NE* X 5, 1175b36–1176a1). Moreover, while touch is necessary, perhaps conditionally necessary, for living an animal life, sight as well as the other sense modalities belong to the animal in order to secure its wellbeing or living well (*DA* III 13, 435b19–22, *Sens.* 1, 436b8–16). So perhaps Aristotle thinks they are prior in the order of nature in the sense in which the more perfect, valuable, and desirable is prior to what is less so (Joachim 1922: 203). See also Code 2008: 236–237.

12 For the subject of sight, see *DA* II 7, 418a26, namely (a) colors and (b) a nameless quality that causes things to be seen in the dark, with Joachim (1922: 203).

13 Cf. Verdenius-Waszink 1966: 37–38.

which by the end of the chapter will be his favorite pairs of tangible contrarieties from which all other tangible contrarieties follow or derive.

The section starts with the question that will occupy Aristotle for the rest of the chapter – namely, which tangible contrarieties are the primary principles of tangible bodies. Aristotle's list includes seven contrarieties: (1) hot-cold; (2) dry-moist; (3) heavy-light; (4) hard-soft; (5) viscous-brittle; (6) rough-smooth; and (7) coarse-fine.[14] Pair (3) is ruled out on the grounds that presumably rule out pair (6) as well (though (6) is not ruled out explicitly).[15] In the next section, pairs (4), (5), and (7) are reduced to (2). Since (1) is irreducible to (2) (330a25–29), the two of them are the successful candidates for the title of primary principles of tangible bodies.[16]

Aristotle introduces a second necessary condition for contrarieties that define perceptible bodies (in addition to the condition that they must relate to touch): they must be capable of acting on one another (*GC* II 2, 329b23–24).[17] While it is true that, strictly speaking, the active-passive relation is attributed to the elements themselves, not the contrarieties, Aristotle is presumably thinking that it is the contrarieties that bestow such capacities on the elements. This condition is not new. In *GC* I 6, we are told that alteration and mixing require the presence of such contrarieties. But this condition is now invoked to exclude the heavy and the light, since they are neither active nor passive (329b21). There are two questions concerning this exclusion. First, it is not clear whether Aristotle means that the heavy and the light are not active and passive in relation to one another or, more generally, in relation to other things.[18] Second, there is a slight tension with the claim made in *GC* I 6 323a9–10, where we are told that the bodies that are in contact are heavy or light, and that such bodies are in fact active or passive.

Let us address the perceived tension first. As Joachim points outs, although heavy and light bodies may be passive or active, they are so not

[14] To be sure, this list includes fewer contrarieties than the eighteen mentioned, and briefly discussed, in *Meteor.* IV 8 (although not all of the latter relate to touch). This fact may be explained if we assume that *GC* II 2 does not aim at an exhaustive treatment of tangible contrarieties but presents us with a merely indicative list of them.

[15] This is Philoponus' view (*In GC* 214.31 ff.) concerning the rough and the smooth, adopted by Joachim (1922: 204). In *PA* II 2 the rough and the smooth, as well as the heavy and the light, are excluded from the contrarieties controlling the generation and perishing of living substances (648b2–4).

[16] As we will see in the present section and the next, although tangible contrarieties are reduced to the dry and the moist, the activity of the hot and the cold is necessary for their production.

[17] Since only contrarieties that relate to touch can affect one another in the way this was defined in *GC* I 6, the second necessary condition seems to be a specification of the first condition.

[18] Williams 1982: 58.

because they are heavy or light.[19] Rather, they are heavy or light because of their capacity to move toward their natural places; in other words, they have a merely passive capacity to be acted upon (*Phys.* VIII 4, 255b35–256a3). And yet this may not be directly relevant here. For Aristotle is excluding the possibility that heavy and light act as causes "by producing something or being affected by something else" (329b21–22). Such changes fall under the species of action/passion and, unlike changes of place, they result in a new affection in the entity affected. Heavy and light by contrast are powers responsible only for changes of place. Hence, they do not qualify as contrarieties of the sort that Aristotle is looking for in our chapter.

These considerations resolve the worry with respect to the suspected inconsistency with *GC* I 6; they also help answer the question concerning the inability of the heavy and the light to act or to be acted upon. The heavy and the light are capacities to be acted upon; as such it is questionable whether they can act on other bodies. And even if the heavy and the light could be said to act upon one another, their activity relates to motion and not to changes that produce an affection, that is, cases of action/passion in the strict sense. Still, we would like to be clear as to whether Aristotle means to point out merely that heavy and light cannot act on one another, or (more strongly) that they cannot act on anything (i.e., on any body of any quality whatsoever). An answer to this question is important for understanding in what way the contrarieties that are capable of acting and being acted upon – hot and cold, dry and moist – are capable of doing so. What Aristotle says in our section suggests that the hot, the cold, the dry, and the moist can act and be acted upon, not because one contrary acts on the other (i.e., the hot on the cold, and the dry on the moist), but rather because one pair of contraries is capable of acting (the hot and the cold) while the other is capable of being acted upon by the first (the dry and the moist).[20] Although this is the reading our passage explicitly suggests, and it is also supported by what we read in *Meteor.* IV 1 concerning the two basic contrarieties,[21] Aristotle cannot rule out that, at least in

[19] Joachim 1922: 204.

[20] See *GC* II 2, 329b24–26. Aristotle seems to be speaking not of the qualities themselves but rather of the qualified *stuffs* which are said to be active/passive (ποιητικόν/παθητικόν). But as Joachim comments: "he [*sc.* Aristotle] is speaking of [...] qualified stuffs [...] but he is attending to the qualities and trying to determine these in abstraction from the stuff which they qualify" (Joachim 1922: 201). Following this observation, I will be referring to the hot/the cold etc. as qualities that "produce" elementary bodies.

[21] Cf. Philoponus, *In GC* 216.13–16. See *Meteor.* IV 1, 378b10–18.

some sense, opposites can act on one another as well.[22] For, both in the *GC* and in *Meteor.* IV, we find claims that lead to the idea that the hot and the cold are mutually affecting or act and are acted upon by one another.[23] So let us look at the evidence, respectively, for the claim that action and passion operate across pairs of contraries, and the claim that action and passion operate between the contraries that constitute such pairs.

The claim that the contrary pairs hot-cold and moist-dry are related as active to passive capacities is followed by the definitions of these contrarieties (329b26–32). The hot is that which associates bodies belonging to the same kind, while the cold is that which associates bodies that may or may not belong to the same kind. The dry is defined as that which is determined by a limit of its own, while the moist as that which cannot be determined by a limit of its own but still can be easily delimited by another body (e.g., it is because wine is moist that it takes the shape of the wineskin). In other words, the members of the first pair of contraries are defined by an active capacity, while the members of the second pair are defined by a passive capacity, namely, by the relative difficulty or ease with which they can receive boundaries. This picture becomes more detailed throughout *Meteor.* IV, but it suffices to have a look at the introductory remarks of that book to get a sense of the explanatory strategy adopted by Aristotle. There, the hot and the cold are named as the two active powers in the constitution of simple bodies (and the mixtures thereof), while the dry and the moist are named as the two corresponding passive powers.[24] Aristotle confirms this by appealing to empirical evidence:

> It is always the hot and the cold that are observed to determine, combine and change things both of the same kind [hot and cold] and of different kinds [cold], as well as moistening, drying, hardening and softening: things dry and moist, on the other hand are the subjects of determination and the other changes just enumerated, both in isolation and in combination with each other. (*Meteor.* IV 1, 378b15–20, Lee's translation modified)[25]

These contraries are operating in the generation, alteration, and decay of natural substances thus:

[22] On this see Joachim 1922: 204–207.

[23] For *GC*, see: *GC* I, 7 323b29–31; I 10, 328a18–19; II 4, 331a29–30. For the *Meteorology*, see *Meteor.* IV 5, 382b4–6.

[24] *Meteor.* IV 1, 378b10–14; 378b26.

[25] Aristotle offers a second argument stemming from the definitions of these four terms (*Meteor.* IV 1, 378b20–26). These definitions are condensed versions of those offered in *GC* II 2.

> Simple natural generation is a change effected by these properties [sc. the active ones, hot and cold], when present in the right proportions, in the matter underlying the particular natural thing, this matter being the passive properties of which we have spoken. The hot and cold produce change by mastering (κρατοῦντα) the matter. (*Meteor.* IV 1, 378b32–379a2, Lee's translation)

Although this is a general description, it gives an accurate picture of the structural roles of active and passive capacities in Aristotelian natural changes. The hot and the cold act as efficient causes, vaguely corresponding to the operation of the Empedoclean powers of Love and Strife, and determining the material available: material constituted by the other pair of contraries, the dry and the moist, and which takes the form of the mixtures and compounds we encounter in animate and inanimate bodies.

With these results in place, we can now turn to the evidence for mutual action and passion on the part of contraries within a single contrariety. We have a direct endorsement of this position in *GC* I 7:

> Agent and patient are contraries and coming to be is to the contrary. So it is necessary that the patient change into the agent, because this is the way in which coming to be will be to the contrary. (*GC* I 7, 324a10–13, Williams' translation)

Aristotle illustrates the above principle by taking as an example the activity of the hot on the cold and *vice versa*. And since Aristotle generalizes over all contraries, the same must hold for the pair moist/dry. Indeed there is some evidence that they can interact as active and passive capacities in Aristotle's discussion of elemental transformation:

> Out of fire there will be air when one quality changes (for if the former was hot and dry, the latter hot and wet, the result if dry should be overcome by wet will be air). (*GC* II 4, 331a27–30)

In this example the role of the active factor is attributed to the wet/moist, which acts on the dry. Evidently, the two contraries can both serve as active or passive power in any given change. This suffices to establish that there is a double aspect in qualitative changes (and in the generation of the simple bodies): (a) the two contrarieties, the hot and the cold on the one hand, and the dry and the moist on the other, represent the two opposing powers in such changes, namely the active and passive sides correspondingly; (b) the two contraries inside each pair can alternate the role of active or passive power of such changes. But now we would like to know how these pairs of contrary powers interact and thus understand what operating factors control each instance of natural change. Aristotle does not address this question head-on in the *GC*. But in the discussion of action and passion (*GC* I 7) and mixture (*GC* I 10), he gives some indications concerning his views on the operation of these factors.

In *GC* I 7, Aristotle discusses his predecessors' views, which he divides into those who hold that like is affected by like (atomists) and those who hold that it is "unlike and different things which act on or are affected by each other" (323b6–7). Each side is saddled with a number of impossibilities or puzzles. Aristotle's preferred solution uses elements of both:

> It is necessary that in one sense agent and patient should be the same and in another sense different and unlike one another [. . .] patient and agent are the same and like in genus but unlike in species, and it is contraries that are like this. (*GC* I 7, 324a2–5 Williams' translation)

Aristotle then specifies that fire heats and cold things cool, and generally what is active makes the patient like itself (324a9–10), without giving an explicit example for the role of the patient.[26] And rightly so, because first a distinction is in order:

> Sometimes it is the substratum which we speak of as being affected, as when we say that the human being gets well or gets hot, or any other things of this kind; sometimes though, we say that what is cold gets hot, what is ill gets well. (*GC* I 7, 324a15–17, William's translation)

For instance, when the heat in the air affects water and turns its cold moisture into hot moisture, thus producing air from water, we could say either that the heat acts on the coldness of water as its contrary, or that it acts on the moisture as the substratum of the change. This distinction can be applied analogously to the efficient cause: we could say either that the active power is the heat in the air or air itself. The above passage illustrates the double aspect of active and passive powers in natural qualitative changes, which reflects a standard Aristotelian doctrine distinguishing in any change the substratum of that change from the contrary which belongs to it before the change and is the target of that change.[27] So although strictly speaking the hot and the cold are both active powers, while the dry and the moist are passive powers, in a different sense the hot acts on the cold and *vice versa*.[28] This is confirmed by a passage from *GC* I 10:

[26] He only adds that it is necessary that the patient changes into the agent because this is the way in which coming to be will be to the contrary (324a11–12).

[27] For a similar distinction, see *Phys.* I 7, 190b29–35, with the difference, however, that the *Physics* focuses only on the passive side of the change.

[28] That contraries destroy one another is repeated in *Phys.* I 9, 192a21. The point that they act on one another is introduced in *Phys.* I 5, 188a30–34. Contraries interact not so as to produce something

> There are, as we say, some beings which are capable of acting and some which are capable of being affected by these others. Now (i) for some things [*sc.* those that are capable of acting] there is conversion, those, namely, which share the same matter, being able both to act upon one another and to be affected by one another. (ii) Some things, on the other hand, though they act, are incapable of being affected, those, namely, which do not have the same matter. So of these things there is no mixing. (328a17–22, Williams' translation)

The passage introduces a division (already presented in *GC* I 7) between (i) efficient causes that, though active, are also themselves affected (e.g., the food an organism consumes) and (ii) efficient causes that are active but are not themselves affected (e.g., the doctor who administers the drug). The crucial point is that for entities under (i) there is conversion: they can mutually affect one another. And this observation follows the division of beings into those that are capable of acting and those that are capable of being affected. So here again we see that there is a double aspect to action and passion. Hot and cold are capacities to act on the corresponding passive properties, yet at the same time they mutually affect one another.

Let us turn, finally, to the definition of the four elementary powers. The definitions of the hot and the cold turn on their capacity to combine bodies (in the former case, bodies of the same kind, while in the latter, bodies that may or may not belong to the same kind, cf. *Meteor.* IV 1, 378b20–23). The activity of the hot is twofold: it combines materials of the same kind and by the same token it also dissociates materials of different kinds. So heating is a selective process that can purify the materials on which it operates. However, it seems that the hot cannot bring about this purifying result all on its own. According to *Meteor.* IV 1:

> Simple and natural generation are brought about by these powers [*sc.* the hot and the cold] when they are present in the right proportion (λόγον), in the matter underlying by nature each change. (*Meteor.* IV 1, 378b32–379a1, Lee's translation)

In other words, the heating activity is already a function of a proportion between the hot and the cold.[29] Too much heat may dry completely the

out of one another; rather, by acting on a third thing they produce a new affection in it (*Phys.* I 6, 189a24–26). What was said above concerning the hot and the cold must also hold for the passive pair of contraries, the dry and the moist. Although they constitute the passive pair of contraries, they act on one another either when the dry displaces the moist (as when air becomes fire or water becomes earth) or the moist displaces the dry (in the reverse cases).

[29] This is true of the well-proportioned, goal-directed heating aiming at the realization of a thing's nature (e.g., a living being) or of some latent form (e.g., in the production of an artifact, *Meteor.* IV

available material while an inferior type of heat may lead to decay.[30] The cold seems to be here a regulator of the efficient power of the hot, which can temper it so that the right amount of heat is employed.

The definitions of the dry and the moist turn on their capacity to be delimited, to be molded into a particular shape (cf. *Meteor.* IV 1, 378b24–25). These two passive capacities are related to one another in a way that may in fact license the claim that the two contraries are mutually affecting one another, alternating between them the role of acting on and being affected. This reading is supported by an Aristotelian comment on putrefaction:

> Things that putrefy [*sc.* the result of the corruption of the natural body] begin by being moist and end by being dry. For the moist and the dry were the matter, and the operation of the active qualities caused the dry to be determined by the moist. (*Meteor.* IV 1, 379a9–11)

Here the process of corruption is said to follow the steps of generation in the opposite order: while in the case of generation moisture affects dryness and thereby shapes the body that is being generated, in the case of corruption the moisture recedes and thereby leaves the material completely dry. Once more, we see that the two contraries, moist and dry, in addition to being what is acted on in a natural change, can also play the role of what does the acting in natural change, by prevailing on one another. Depending on which one dominates, the plasticity of the natural body is determined.

3.4 Aristotle's Second Reduction (329b32–330a24)

This section includes a detailed reduction of a few pairs of tangible qualities to the two basic ones: hot/cold and dry/moist. If, as I have just been arguing, none of the two basic pairs can accomplish its role as active or passive in natural change without the other, then the reduction of the rest of tangible contrarieties is due to the joint operation of hot and cold, moist and dry.[31]

2, 379b25–35; *GA* II 1, 734b31–735a2). Heating that is not goal-directed, or is not well-proportioned, produces an affection tending toward the destruction of the affected entity (*Meteor.* IV 1, 379a16–26, IV 3–4).

30 *Meteor.* IV 1, 379a1–3.

31 As seems to be Philoponus' understanding (*In GC* 219.24 ff.) and as suggested by Zabarella and endorsed by Joachim 1922: 208; see also Kalfas 2011: 159.

The section divides into two subsections: (a) a reduction of fine and coarse, viscous and brittle, and hard and soft, to the dry and the moist (329b32–330a12); (b) a reduction of the various senses of "dry" and "moist" to the primary senses of these terms (330a12–24).

Aristotle's argument for the reduction of the fine and the coarse to the moist and the dry uses as a middle term the capacity of a body to take the shape of what it touches (its being ἀναπληστικόν). Such a capacity belongs to the fine, namely to what has small particles, which tends to fill up any vessel that may contain it. Philoponus' illustration is a portion of flour poured inside a vessel full of nuts.[32] The flour fills up every empty space inside the vessel. The premises of the syllogism must be: (1) whatever is able to fill things is moist;[33] (2) whatever is fine is able to fill things; hence, (3) whatever is fine is moist. Aristotle does not produce an explanation of the reduction of the opposite property, the coarse to the dry, either because he considers this reduction obvious or because he assumes the principle that contrary properties belong to the contraries.[34]

The poles of the next pair of contraries, viscous and brittle, relate to the moist and the dry respectively, but for slightly different reasons in each case. The viscous belongs to the moist because it is an affection of it, while the brittle belongs to the dry because it is a property that belongs to the dry in its extreme form, that is, when no moisture is left in it. To be sure, this may be an affection too; but, more interestingly, it is an excess of dryness, which is not the case for the viscous in relation to the moist. In addition to this disanalogy, which seems unique to this pair of contrarieties, the pair viscous/brittle seems to be absent from the *Meteorology*, and it is not clear that we can identify it with one of the other contrarieties listed there. In *Meteor.* IV 6, 382b31 ff., and again in *Meteor.* IV 8, 385a22–33, Aristotle distinguishes two forms of solidification depending on whether (a) the cold crushes out the hot while the moist evaporates with it (as in bodies where water predominates, e.g., ice, lead, bronze) or (b) the heat dries out the moist in them (as in bodies where earth predominates e.g., clay, soda, salt). Again, the brittle must belong to bodies under (b), so to bodies made primarily of earth.[35] In the *Meteorology*, however, Aristotle does not

[32] Philoponus, *In GC* 220.6–8.

[33] *Pace* Williams 1982: 159, who argues that Aristotle's text justifies only the claim that "whatever is wet is able to fill things."

[34] Philoponus, *In GC* 220.13.

[35] A passage from *PA* II 9 about the cartilaginous nature of the spines of selachian fish confirms this suggestion: selachian fish need a spine that is not brittle; moreover, all the earthy material available is consumed in their skin so that none is left for a different kind of spine, which presumably would

introduce any special term for this particular property of earthy compounds, classifying it instead under the general heading of solidification. So one wonders whether the study of the passive properties of material bodies recorded in the *Meteorology* has led Aristotle to revise his views and terminology regarding the contrariety between solidification and liquefaction as more basic than the viscous-brittle. Indeed, the term for brittle (κραῦρον) is not used in *Meteorology*. By contrast, the term for viscous (γλίσχρον) is listed in the catalogue of passive properties, its contrary now being the ψαθυρόν (fragile):

> A thing is viscous when it is ductile as well as being liquid and soft. And this characteristic belongs to all bodies with interlocking parts, whose composition is that of chains; for they admit of considerable extension and contraction. (*Meteor.* IV 8, 387a11–15; Lee's translation)

This passage may suggest that the pair viscosity-brittleness in *GC* II corresponds to the pair viscosity-fragility in *Meteor.* IV, which does not have as central a role there as it does in *GC* II. However, this suggestion seems dubious to me. I will argue that the contrariety viscous-brittle is replaced by, or becomes one version of, the general contrariety that opposes things that solidify (πηκτά) to things that are impervious to solidification (ἄπηκτα).

There are two key facts that support this proposal. First, brittle must be identified as a property belonging to bodies in which earth predominates, since brittleness is the capacity of a body to lose its moisture due to excessive heat. This is a property of what is capable of solidification (πηκτόν). Second, the contrary property, viscosity, is the capacity of a body to retain its moisture despite the activity of heat. Oil is the only example of viscous substance in our passage from the *GC*. It does not dry up because of its viscosity and because the air within it prevents evaporation (384a2–3); it does not solidify because (again) it contains a great deal of air within it (385b4–5). Two kinds of bodies are incapable of solidification (ἄπηκτα): (a) those that contain no watery moisture and are not watery, and in which heat and earth predominate rather than water, such as honey and must (they are in a kind of ferment) and (b) those in which, though they do contain water, air predominates, such as oil and quicksilver (385b1–5). An indication that oil contains air is that it is not boiled off by fire since its viscosity prevents evaporation (384a1–2). It is this second class

then possess the property of being brittle (*PA* II 9, 655a22–28). Brittleness then appears to be a dispositional property of earthy compounds.

of things that seems to correspond to viscous substances. In *Sens.* 4, oil is said to extend beyond water because it is viscous, while water is fragile (ψαθυρόν); that is also why it is more difficult to hold water in the hand than oil (441a26–28). It seems, then, that viscous substances are distinguished by two different properties. On the one hand, viscous bodies contain a great deal of air; as a result, heat cannot cause the moisture they contain to evaporate due to their viscous constitution. This resistance to solidification makes viscous substances useful for the constitution of certain animal parts (marrow and sinew).[36] Viscosity is a property that makes a body impervious to solidification; as a result, viscosity is contrary to what is capable of solidification. On the other hand, viscous substances are ductile and moist since they have interlocking parts; this property makes them useful in the constitution of living bodies, for example, for holding together the animal bodies (*GA* II 3, 737b1; cf. *GC* II 8, 335a1–2), or for bestowing some oiliness to their parts so that they are not easily dried up (*Long.* 6, 467a8). This resistance to drying makes viscous substances ductile and opposes them to fragile things (ψαθυρά). It seems then that in *Meteor.* IV and in the biological writings the primary contrariety is things that solidify and things that are impervious to solidification (πηκτόν-ἄπηκτον), with viscosity as a kind of imperviousness to solidification with multiple roles to play in the constitution of homoeomerous bodies.

The final contrariety reduced to the moist and the dry is the pair soft-hard. What opposes the two properties is whether something yields into itself (with the qualification that it yields without changing its position). This qualification serves to distinguish the soft from the moist, since the moist also yields, but by changing its position.[37] Does this license the further claim that the soft is "characteristic of" (i.e., reducible to) the moist (*GC* II 2, 330a8)? One way to support this claim is to hold that yielding is a quality of the moist in general, while soft bodies possess the more specific characteristic of yielding without being displaced. This makes the quality of softness a determination of the moist. However, this cannot be right. The moist is a characteristic of bodies that, insofar as they are moist, yield by changing their position; in this respect, the moist is opposed to the soft,

[36] From *PA* II 6, 652a18–19 we learn that the marrow inside the spine is viscous so that it can be stretched.

[37] The same definition and qualification are repeated in *Meteor.* IV 4, 382a11–14, but the qualification is applied to water. This difference is understandable. In *GC* II, Aristotle is interested in reducing the soft to the moist (and not to watery stuffs). In *Meteor.* IV, he is content to use the moist body *par excellence* (water) as an example.

in which case the non-displacement of the soft cannot be a further determination of a property characteristic of the moist. So some other quality must be responsible for it. The best candidate is hardness as produced by the dry which takes part in the constitution of the body. In *Meteor.* IV 4, Aristotle reaffirms that all bodies are compounds out of the two primary passive properties (381b24–25), the moist which is easily determined, and the dry which is determined with difficulty (381b29–30).[38] Since the moist and the dry belong primarily to water and earth, all natural bodies must contain both water and earth, in which case hardness and softness belong of necessity to all bodies (382a4–5). From this we can infer that a body which is moist will yield (be determined by another body that touches it), although, to the extent that it contains some minimum dryness, it will yield without being displaced (because at the same time even this minimal dryness bestows a limit on it, a point beyond which it does not recede). The result is that soft bodies possess some limit of their own (as opposed to moist bodies such as water); to that extent, soft bodies are the result of solidification, at least to some degree.[39] But there are also bodies that contain no dryness at all, like water, and Aristotle labels them ὑγρά (moist/liquids), namely, bodies that are not soft. Hence, he divides natural bodies into three groups: soft, hard, and moist/liquid (ὑγρά).[40] Yet the soft is also "characteristic of the moist" inasmuch as it is the moist that influences the extent to which the soft has its own limit: while an absolutely dry body is completely determined (that is, solidified) by its own limit (it is in this sense that we have to understand the claim that the hard is solidified, 330a11), a soft body's own determination is as weak as the quality and quantity of moisture in it allows.

Having discussed the reduction of the three main passive tangible properties to moist and dry, Aristotle distinguishes a few secondary senses of the moist and the dry in order to show again the dependence of the corresponding tangible properties to what is absolutely moist or dry, as defined at 329b30–32. He distinguishes three pairs of secondary senses of the dry and the moist. The moist side of the contrarieties consists in: (i) the

[38] There is some tension here since the claim appears to imply that what is said to be completely dry (330a6–7) must contain some minimal moisture. One possible remedy is to appeal to the distinction (made in *GC* II 3, 330b21–29) between "mixed" and "simple" bodies, where the former are everyday perceptible versions of the elements, while the latter are the elements themselves. See Emily Katz in this volume. The "mixed" versions will always contain a trace of the opposite quality (so that a "completely dry" body can contain traces of moisture), even if this does not make a difference for its possessing derivable properties such as brittleness (and this may be the reason why it is said to be completely dry).

[39] *Meteor.* IV. 5, 382a22–27.

[40] *Meteor.* IV 8, 388a28–29.

damp (διερόν), that is, what has foreign moisture on its surface; (ii) the sodden (βεβρεγμένον), namely, what has foreign moisture penetrating on the inside (but moisture may also be found on its surface); (iii) the moist as that which possesses its own proper moisture. Things are not as clear concerning the contraries of these three properties, that is, the secondary senses of the dry (neither their nomenclature nor their character is clear). Aristotle says that the dry is opposed to the damp (330b16), and again that it is deprived of the moisture characteristic of the sodden; but he does not distinguish between these two cases. Perhaps this is understandable. On the one hand, if the opposite of the damp is what has lost foreign moisture from its surface, either it still possesses foreign moisture in its core (in which case it is not dry but sodden) or it does not (in which case it is altogether dry). On the other hand, if the opposite of the sodden is what has lost the foreign moisture from its core, then either it still possesses foreign moisture on its surface (in which case it is not dry but damp) or it does not (in which case it is again altogether dry). So it seems that the two senses of the dry opposites to the damp and the sodden come down to the same thing described in different terms (lacking foreign moisture from its core and lacking foreign moisture from its surface).[41] As for the third derivative sense of the dry, there is a term for it: the solidified (πεπηγός).[42] Solidification consists in a body losing its own moisture. According to the theory exposed in *Meteorology* this loss can take any one of two forms: either (a) the hot dries the moisture of a body in which water predominates, for example, dried mud or (b) the cold expels the heat of a body in which earth predominates, for example, cold freezes water (*Meteor.* IV 8, 385a23–28). This completes the list of derivative senses of moist and dry (see Table 3.1 on page 57).

The three secondary pairs of contraries of the moist and the dry are derived from what is said to be moist or dry unqualifiedly; and this could be interpreted as simply meaning that the unqualifiedly moist (or dry) is part of their definition.[43] So let us try to substitute the terms "moist" or "dry" with the corresponding definitions in the descriptions of the

[41] This may appear to violate the rule that there is one contrary *per* contrary. However, there are two descriptions focusing on different aspects of the dry body.

[42] This contrariety is used in *MA* 8, 702a9–10. It is also worth mentioning the definition of solidification (of water into ice) in *APo.* II 12, 95a17: utter lack of heat. This confirms that the reduction of tangible qualities in the second half of our chapter is not to the moist and the dry, but rather to the operation of both pairs of basic contrarieties. The same point can be made in more detail if one turns to *Meteor.* IV 6, which is devoted to liquefaction and solidification.

[43] Joachim 1922: 212.

Table 3.1 *The various senses of moist and dry*

Primary		Secondary	
Moist (υγρόν): what is easily limited	**Moist** (υγρόν): what possesses its own moisture	**Sodden** (βεβρεγμένον): what has foreign moisture penetrating to its core	**Damp** (διερόν): what has foreign moisture on its surface
Dry (ξηρόν): what cannot be easily limited	**Dry/Solidified** (ξηρόν/πεπηγός): what is deprived of its own moisture	**Dry** (ξηρόν): what is deprived of foreign moisture	

secondary senses of the terms by taking as an example the description of the sodden. The resulting formula must be the following:

> sodden $=_{\text{def}}$ what has in its core a foreign capacity to be easily delimited though it cannot be determined by a limit of its own.

If this is roughly what Aristotle intends here, his reduction of perceptible qualities in *GC* II 2, even if incomplete, manages to reduce the rich variety of affections characterizing homoeomerous bodies to four basic contraries. More to the point: it also reduces the complexity and diversity of tangible properties in natural bodies to a few properties that have to do with the shaping of matter: aggregation and segregation on the one hand, and capacity of being delimited or not on the other. This project is handled in a much more detailed manner in *Meteorology* IV. However, the directions for how to reduce natural perceptible affections to fewer, simpler, and thus more basic properties relating to how material bodies are shaped is outlined in *GC* II 2.

3.5 Concluding Remarks

The chapter closes with an enumeration of the four basic tangible contraries constituting the principles of elemental bodies (and composites or mixtures out of them). This answers the question broached toward the end of *GC* II 1. While all other tangible contrarieties are reduced to two basic pairs (the hot and the cold, the dry and the moist), these basic pairs cannot be reduced to one another.

GC II 3 begins with the statement that the elements are four in number.[44] If Aristotle's analysis concerning contrarieties is right, these can be combined in four possible ways only, and thus can constitute only four different combinations (hot/moist, hot/dry, cold/moist, cold/dry), or equivalently types of simple bodies. But it is not clear that this is exactly Aristotle's argument. What is important, however, is that which Aristotle achieves in *GC* II 2: the reduction of several qualitative features of simple bodies to a few basic capacities, namely the capacity to associate bodies (and, in the case of heat, which is a two-headed capacity, also to dissociate bodies), and the capacity to be resistant or susceptible to external determination.

[44] Most commentators take στοιχεῖα in lines 330a30 and 330a33–34 to refer to the four sublunary elements. But the case for taking the word as referring to the primary contraries is defended in Crowley 2013: 168–169. See also Emily Katz in this volume.

4

On Generation and Corruption *II 3*

Emily Katz

4.1 Introduction

At the start of *GC* II 1, Aristotle tells us that "it remains to consider the so-called elements of bodies" (328b31–32).[1] In *GC* II 3, he offers an initial justification of three theses about these elements: (i) that there are four primary bodies; (ii) that these are earth, air, fire, and water; and (iii) that each is associated with two primary differentiae (though strictly with just one). Aristotle's establishment and defense of (i)–(iii) draws on two theses he has established in the first two chapters of *GC* II: that qualities like hot and cold are somehow prior to the primary sensible bodies (*GC* II 1, 329a34–36), and that the four tangible qualities hot, cold, dry, and moist are the primary differentiae of sensible bodies (*GC* II 2).

My main aim is to determine where, how, and how well Aristotle supports (i)–(iii) in *GC* II 3. This will require addressing several puzzles, among which: whether 330a30–34 indicates that the primary bodies are not really elements for Aristotle (Section 4.2); whether the "apparent simple bodies" are the primary bodies or something else (Section 4.3); and how to understand the "fiery" and "airy" things that are not fire and air (respectively) but somehow like them (330b21–30) (Section 4.5).

There is so much to puzzle over in this short chapter in part due to its highly condensed nature. I argue that despite his laconic presentation, Aristotle offers an impressive variety of justifications for the chapter's main contentions: he appeals to combinatorial mathematics, to what is "in accordance with argument," to empirical observation, and to his

* I am grateful for the questions and insights of the participants in the *GC* II reading group, for the work of the editors of this volume, and for Sean Kelsey's extensive written comments. This paper has been much improved by them.

[1] Translations of *GC* are from this volume's collective translation. Translations of other Aristotelian texts are from Barnes 1984. See Crowley 2008 for a persuasive case that καλούμενα ("so-called" or "what are called") does not suggest that Aristotle thinks these are not really elements.

predecessors. I also find that the primary bodies are indeed genuine elements of sensible bodies for Aristotle; that the "apparent simple bodies" are the everyday counterparts of the elemental bodies; and that the 330b21–30 contrast between fiery and airy things on the one hand and fire and air on the other is a contrast between the elemental bodies and their everyday counterparts.

4.2 What Are the "Elements"?

The first puzzle appears immediately: It has seemed to most commentators that by "elements" (στοιχεῖα) at *GC* II 3, 330a30 and 330a33–34, Aristotle means not earth, air, fire, and water but rather the primary differentiae hot, cold, dry, moist.[2] This is the most natural reading, given the way *GC* II 2 ends and *GC* II 3 begins (330a29–30):

> End of *GC* II 2: "As a result, it is necessary that there be these four [*sc.* the primary differentiae or elementary qualities]."
>
> Start of *GC* II 3: "Since the elements are four [. . .]."

Many conclude that by calling the primary differentiae "elements," Aristotle is indicating that since the differentiae are more basic than the elementary bodies, the elementary bodies are not genuine elements.[3] Some argue that the differentiae are more basic in the sense that they are physical components of the elementary bodies;[4] others make the differentiae components of a conceptual analysis of the elementary bodies.[5]

Timothy Crowley recognizes "elements" at 330b30 and 330b33–34 most naturally refers to the differentiae.[6] But he worries that such an identification causes serious theoretical difficulties: since Aristotle insists that the "elements of sensible bodies are bodies" (*Phys.* IV 1, 209a17–18), and since differentiae are not bodies, how can differentiae be elements of sensible bodies?[7] Hence Crowley argues that the passage must be read such that the "elements" here as elsewhere are just elemental earth, air, fire, and water.[8] He proposes that "at 330a30, having just concluded that there are four differentiae (*GC* II 2, 330a29), Aristotle is reminding us that there are four στοιχεῖα. Thus, the

[2] See, e.g., Joachim 1922: 213; Tricot 1951: 104; Williams 1982: 160; Frede 2004: 303; Rashed 2005: 56n2.

[3] E.g., Tricot 1951: 104 n.2; Rashed 2005: 152.

[4] E.g., Kahn 1960: 120; see also 124.

[5] E.g., Joachim argues that the primary bodies "presuppose (logically, though not temporally) more primitive 'constitutive moments'," namely prime matter and the primary differentiae (Joachim 1922: 137).

[6] Crowley 2013: 167.

[7] Crowley 2013: 163–166.

[8] Crowley 2013: 167–172.

number of the differentiae matches the number of the στοιχεῖα."[9] Crowley recognizes that the second occurrence of "elements," at 330a33–34, is more problematic for anyone wishing to deny that Aristotle is calling the primary differentiae "elements." Aristotle writes:

> [I]t is clear that the pairings of the elements will be four: hot and dry and hot and moist, and again cold and moist and cold and dry. (*GC* II 3, 330a33–330b1)

Crowley argues that the 330a34 phrase "pairings of the elements" (αἱ τῶν στοιχείων συζεύξεις) can be read so that it means not that the differentiae are themselves elements being paired, but rather that their pairings "are said of, or belong to, the στοιχεῖα," that is, to elemental earth, air, fire, and water.[10] He paraphrases 330a33–34 as follows: "it is clear that the pairings of (the differentiae that are said of) the στοιχεῖα will be four." Crowley admits that if this is what Aristotle means, then the phrase αἱ τῶν στοιχείων συζεύξεις is somewhat misleading. But he finds confirmation for his reading in the fact that in numerous places, "hot and cold, wet and dry are conceived of as differentiae, rather than as στοιχεῖα."[11]

So we face an interpretative dilemma. If we opt for the more natural reading of 330a30–b1, according to which the elements are the differentiae, we seem to saddle Aristotle with an inconsistency: even though he maintains elsewhere that the elements of sensible bodies must be bodies, and even though the differentiae are not bodies, here he insists that these differentiae are really the elements of sensible bodies. But if we attempt to avoid the theoretical inconsistency in the ingenious way Crowley suggests, we must bite the bullet on a reading of 330a30–b1 that even Crowley concedes is strained.

I see a third way: the primary bodies can be the elements of sensible bodies and the elements of 330a30 and 330a33–34 can be the differentiae. I start with a brief analysis of Aristotle's definition of element (στοιχεῖον). According to his initial definition in *Meta.* V 3:

> We call an element that which is the primary component immanent in a thing, indivisible in kind into other kinds. (*Meta.* V 3, 1014a26–27)[12]

That is, the conditions for *E*'s being an element are:

(1) *E* is a component of some compound, *C*
(2) *E* is not divisible into further entities different in kind from *E*.

9 Crowley 2013: 168. 10 Crowley 2013: 168–169. 11 Crowley 2013: 169–170.

12 I have deleted Ross' "and" before "indivisible" because it is not in the Greek and obscures the fact that "indivisible in kind into other kinds" explains "primary."

I take (2) to be what makes a component "primary" in the relevant sense. Aristotle goes on to say that if *E* is divisible, it is only divisible into parts that are the same in kind as *E* (τὰ μόρια ὁμοειδῆ), "as a part of water is water (while a part of the syllable is not a syllable)" (*Meta.* V 3, 1014a30–31).

It should be obvious why (1) is a condition for *E*'s being an element of *C*. But what is the reason for (2)? It seems to come from Aristotle's insistence elsewhere that *C* must not be the same as *E*. For example, at *Meta.* VII 17, 1041b12–14, he writes: "the syllable is not its elements, *ba* is not the same as *b* and *a*, nor is flesh fire and earth."[13] And at *Meta.* XII, 4 1070b15–16, he insists that "the product must be different from the elements." The difference between the syllable or flesh (the products) on the one hand, and letters or fire and earth (the elements) on the other, is that the product is not just the elements, but "also something else" (ἀλλὰ καὶ ἕτερόν τι; *Meta.* VII 17, 1041b16–19). The syllable and its letters are different kinds of things: the syllables *ba* and *ab* are *organized* (compounds of) letters. By contrast, a mass of water can only be divided into parts that are the same in kind as the mass itself; hence the constituent parts of a mass of water are not really elements of that mass of water. That is, (a smaller quantity of) water is not an element of (a larger quantity of) water.

When we consider (2) together with its motivation (that *C* must not be the same as *E*), we can see that the requirement goes in both directions, so to speak. That is:

(2) (a) *E* is indivisible into entities different in kind from *E*, and (b) *E* is somehow different from *C*.

It is easy to see how this picture has led many to conclude that the primary differentiae are the true elements of sensible bodies. While, for example, earth and fire are constituents of flesh, they are themselves somehow constituted by the primary differentiae (e.g., earth by cold and dry). Since the primary bodies admit of further analysis into primary differentiae, and since the latter are different in kind from the former, the primary bodies fail to meet condition (2). Hence, they are not genuine elements. By contrast, the primary differentiae fulfill both conditions: (1) they are constituents of the elementary bodies; (2a) they are not divisible into further entities different in kind; and (2b) they are not the same as the elementary bodies; hence they are the true elements.

[13] See also *Meta.* XII 4, 1070b5–6.

However, Aristotle's clarificatory examples, which he gives immediately after the *Meta.* V 3 definition, show that there is an implicit third condition for being an element. He writes:

> [F]or example, the elements of speech are the parts of which speech consists and into which it is ultimately divided, while they are no longer divided into other forms of speech different in kind from them. If they are divided, their parts are of the same kind. (*Meta.* V 3, 1014a27–30)

That is, letters are the proper elements of speech because they are (1) the components of speech that cannot be further divided into (3) *forms of speech* (2a) different in kind from letters. In other words, letters are the basic *speech*-constituents of speech. So we see here a third condition for *E*'s being an element:

(3) *E* belongs to the same higher kind or genus as *C*.[14]

We are now in a position to be more precise about the formulation of (2). Condition (3) requires that both the elements and that of which they are the elements belong to the same *higher* kind, for example, speech (τὸ κοινὸν ἡ φωνή; see, e.g., *Meta.* III 3, 998a23–25);[15] yet condition (2b) requires that they be somehow different. The natural way to understand (2b) in light of (3) is that the letters must be different in *sub-kind* (of speech) from that of which they are the elements (syllables, words, and so on). And the natural way to understand (2a) in light of (3) is that the letters must not be divisible into speech-entities different in sub-kind from *E*. Thus:

(2a) *E* is indivisible into entities of the same kind but different sub-kind as *E*, and (2b) *E* belongs to a different sub-kind than does *C*.

When he shifts from the elements of speech to the elements of bodies, Aristotle is explicit that the cases are similar (ὁμοίως; *Meta.* V 3, 1014a31). The elements of bodies are the components of bodies that cannot themselves be further divided into other *bodies* differing in sub-kind.[16] And as

[14] In his 2005 paper, Crowley recognizes what I have labeled conditions (1) and (2), but not (3) (Crowley 2005: 370–373); in his 2013 paper, he recognizes that the elements "of any class of things should be the same in kind as the things of which they are principles" (Crowley 2013: 163), i.e., condition (3). Crubellier recognizes conditions (1), (2b), and (3) (Crubellier 2000: 143). However, he does not formulate (2a) – that is, his analysis does not limit *E*'s indivisibility to indivisibility into further entities *different in kind* from *E*. Hence, he describes elements as "absolutely simple items" and takes this to preclude the primary bodies' being genuine elements (Crubellier 2000: 142–143).

[15] See also *Meta.* XII 4, 1070a33–b4.

[16] The elements of bodies are "the things into which bodies are ultimately divided, while they are no longer divided into other bodies differing in kind" (τὰ τῶν σωμάτων στοιχεῖα [. . .] εἰς ἃ διαιρεῖται

we have seen, that the elements of bodies must themselves be bodies is also clear from *Phys.* IV 1 209a17–18.[17]

Thus, even if the elementary qualities are somehow more basic than the elementary bodies, so that elementary bodies are somehow analyzable into them, the elementary qualities are not other *bodies* differing in sub-kind from elementary bodies. Hence, they do not fulfill condition (3) and do not undermine the elementary bodies' status as the elements of bodies.

And since being *absolutely* simple is not a condition for being an element, earth, air, fire, and water can be the elements of bodies even if they are somehow composite – so long as they are not composed of other bodies differing in kind from them.[18] This is why flesh, for example, cannot be an element: flesh can be divided into component bodies (earth and fire) different in kind from it.

The last step is to defend the natural reading of 330a30–b1, according to which the "elements" at 330a30 and 330a33–34 are the primary differentiae. Crowley objects to this reading on the grounds that Aristotle would not "suddenly switch from calling hot, cold, dry and wet 'contraries' or 'differentiae' (διαφοραί) as he has been doing up to this point, to calling them στοιχεῖα" here at 330a30 and 330a33–34.[19] I agree with Crowley that such an "unannounced switch in terminology" would be troubling.

However, we need not presuppose a switch in terminology if we recognize that στοιχεῖον is for Aristotle a relative term.[20] An element is always an element relative to something (*E* is always *of* some *C*) – and

τὰ σώματα ἔσχατα, ἐκεῖνα δὲ μηκέτ' εἰς ἄλλα εἴδει διαφέροντα·) (*Meta.* V 3, 1014a32–34). I have modified Ross' translation: while he reads the neuter plural ἄλλα as the substantive "other things," it is better read as modifying an implied τὰ σώματα (appearing earlier in the same line), so that we can understand it as ἄλλα σώματα and translate it "other bodies." This reading makes 1014a34 parallel to 1014a29, where we have the feminine plural ἄλλας φωνάς ("other speech") rather than ἄλλα (which in that context would have to mean "other things").

17 See also *DC* II 3, 302a15–18.

18 Although Kahn decides based on *Meta.* V 3 that the elementary bodies cannot be elements (Kahn 1960: 120, 124), Aristotle does not there (or anywhere) maintain that elements must be *absolutely simple*. They must only be simple or primary, i.e., indivisible, in the special way that involves meeting conditions (2) and (3).

19 Crowley 2013: 167.

20 Elements and that of which they are elements fit the following *Meta.* V 15 description of a kind of correlatives: "in general that which contains something else many times to that which is contained many times in something else" (1020b27–28). Note that this does not imply that the entities that are elements relative to what they compose belong only to the category of relatives. Entities belonging to other categories of being are also relatives in one respect or another. For example, all beings are relatives insofar as they are knowable (*Cat.* 7, *Meta.* V 15); yet the known entities themselves (e.g., horses, colors) belong to other categories of being and are only relative *qua* knowable.

different kinds of things will have different kinds of elements. Although Aristotle allows an analogical sense in which all things have the same elements (*Meta.* XII 4, 1070a31–33, 1070b10–19), strictly "the elements are different in different things" (ἄλλων δ' ἄλλα, 1070b17; 1070b26–27). Letters are not elements *tout court*; they are the elements of speech; similarly, earth, air, etc., are not elements *tout court*; they are the elements of perceptible bodies. Although Aristotle sometimes specifies that he is speaking of the elements of speech (e.g., *Meta.* V 3 1014a27–28) or of bodies (e.g., *Meteor.* I 1, 338a22–23), he often uses the term "element" alone when he means specifically the elements of bodies, counting on the context to make it clear what it is in relation to which earth, air, fire, and water are elements.

So when Aristotle calls the differentiae elements at *GC* II 3, 330a30 and 330a33–34, the question we need to ask is: Given the immediate context, *relative to what* is Aristotle calling the differentiae elements? And the answer is pretty clearly that they are elements relative to the pairings he is discussing at *GC* II 3, 330a30–330b1.[21] Hot, cold, moist, and dry are the elements of the pairings hot–dry, hot–moist, cold–dry, cold–moist: they (1) are components of these pairings; (2a) are indivisible into further entities different in kind from themselves; (2b) belong to a different sub-kind than the pairings; and (3) belong to the same higher kind as their pairings.[22]

Note that the claim that, for example, hot and dry are elements of the pairing hot–dry is distinct from the claim that hot and dry are elements of fire. This is because fire is a body, while the pairing hot–dry is, in the context of 330a30–b1, just one of the possible ways in which two members of a set of four terms may be selected so as to form a subset of two.

[21] It is possible that he means that they are the elements of the tangible qualities. I hesitate because while the *GC* II 2 discussion indicates that hot, cold, moist, and dry are (2a) indivisible into further tangible qualities, (2b) belong to a different sub-kind of tangible qualities than, e.g., brittle and viscous, and (3) are, like brittle and viscous, tangible qualities, it is unclear whether they are in the relevant sense (1) *constituents* of these other tangible qualities. He says that the other tangible qualities are "derived from them" (ἐκ τούτων, 329b33) and "of" them (τοῦ ὑγροῦ and τοῦ ξηροῦ, 330a4), and that they are all "reduced to" them (ἀνάγονται εἰς, 330a25).

[22] That is, (3) singletons and pairs are both subsets of the four terms, but (2b) they are different subsets of the four terms. I should say that it may not be necessary to insist that Aristotle is using "element" at 330a30 and 330a33–34 in the strictest sense (the sense that involves conditions (1)–(3)) – that is, he may only be using the term in the more general sense "first component immanent in each thing" (*Meta.* V 14, 1014b14–15) (and see also his concession at *GC* II 1, 329a5–8). To address Crowley's concern, it is enough to recognize that "element" – even in a looser sense – is always a relative term.

(That is, it is what we have come to call a "combination" in the technical mathematical sense.)

Thus, we need not worry with Crowley that if we take the elements of 330a30 and 330a33–34 to be the differentiae, then we attribute to Aristotle an unannounced terminological switch from calling hot, cold, moist, and dry differentiae in *GC* II 2 to calling them elements at the start of *GC* II 3. Rather, now in *GC* II 3 he is simply putting hot, cold, moist, and dry into relation with things that they constitute as elements: two-member subsets of the four-member set of primary differentiae, that is, their pairings. The primary differentiae are properly called elements *of their pairings*.

In short, when Aristotle calls the primary differentiae elements at the start of *GC* II 3, he is not implying that even though the primary bodies are sometimes called elements, the *real* elements of sensible bodies are the primary differentiae. Since condition (3) means that only bodies can be elements of bodies, the primary differentiae – which are not bodies – are not in competition with the primary bodies for the status of elements of bodies.

4.3 What Are the "Apparent Simple Bodies"?

With this solution in hand, we can move on to the rest of the passage. Aristotle uses a mathematical and a metaphysical principle in conjunction to establish that the primary differentiae form the four pairs hot–dry, hot–moist, cold–moist, and cold–dry. The mathematical principle is that there are six pair combinations of a set of four terms; the metaphysical principle is that contraries do not pair. Since there are four pair combinations of a set of four terms when each term is non-combinable with one other in the set, the primary differentiae form four pairs. In addition to ruling out the pairs hot–cold and moist–dry, the principle that contraries cannot be combined also rules out any grouping of differentiae greater than two.[23] So this principle explains both why there is no element that is either hot–cold or moist–dry and also why *only* two primary differentiae are associated with each element.

Aristotle then establishes thesis (iii) by assigning two elementary qualities each to "the apparent simple bodies" (τὰ ἁπλᾶ φαινόμενα σώματα)

[23] That is, it rules out the four trios hot–cold–dry, hot–cold–wet, wet–dry–hot and wet–dry–cold and the quartet hot–cold–wet–dry. Thanks to Sean Kelsey for this suggestion.

fire, air, water, and earth (330b1–7). There are two puzzles in this next section of the text.

First, what are the apparent simple bodies with which Aristotle is associating pairings of primary differentiae? They are either:

(a) elemental earth, air, fire, and water, or
(b) what we might call "everyday" earth, air, fire, and water – e.g., garden soil, atmospheric air, hearth fire, and lake water.

Second, what is the rationale for Aristotle's assignment of certain pairs of primary differentiae to certain bodies? He claims that the pairings correspond to the bodies "reasonably" (εὐλόγως) and "in accordance with argument" (κατὰ λόγον).[24] κατὰ λόγον justifications are the subject of some debate in the literature: some argue that they should be understood as dialectical arguments, while others argue that since many such arguments appeal to empirically established general principles, κατὰ λόγον justifications are not always dialectical.[25] But regardless of which view we take, there is a problem: there is neither a clear appeal to a general principle in 330b1–7, nor does the discussion of predecessors that follows (330b7–21) justify the 330b3–5 thesis associating specific differentia–pairs with specific bodies. How then should we understand Aristotle's appeals in this passage to what is reasonable and in accordance with argument?

We can begin with the first puzzle. I will argue (Section 4.5) that later in the chapter, when Aristotle speaks of mixed "fire and air and each of the things mentioned" (330b21–22), he is referring to the apparent simple bodies here at 330b2. Since I will also argue that "mixed" fire and air are not the elementary bodies but rather their everyday counterparts, the apparent simple bodies of 330b2 will be the everyday bodies. That the apparent simple bodies are really the everyday bodies is also supported by the fact that the air Aristotle mentions at 330b4 is "a sort of vapor" or "like vapor" (οἷον ἀτμίς). Vapor is not one of the elemental bodies.

But perhaps my claim that the bodies mentioned at 330b2 are the everyday bodies is perplexing. In *GC* II 3 Aristotle is supposed to be giving an account of the elements of sensible bodies, that is, of *elemental* earth, air, fire, and water. One might then reasonably ask why he begins by associating the primary differentiae with *everyday* earth, air, fire, and water.

[24] Bolton shows that when Aristotle appeals to what is reasonable (in contrast to what is apparent to perception), he often uses εὔλογον and κατὰ λόγον interchangeably (Bolton 2009: 59–60).

[25] For representative views, see Bolton 2009: 61; Karbowski 2014: 25–38; Falcon-Leunissen 2015: 217, 222–223, 240; and Karbowski 2016: 135–137, 154.

The answer addresses both puzzles. Although Aristotle states that the pairings of the primary differentiae correspond to earth, air, fire, and water "in accordance with argument," and that their distribution among these bodies is "reasonable," the passage in fact appeals to *empirical* evidence in support of the 330b3–5 thesis associating specific pairings with specific bodies: Aristotle associates the primary differentiae with the earth, air, fire, and water of everyday experience.

While Aristotle's treatment is very condensed, he makes the empirical nature of his reasoning clear when he states that "air is hot and moist (for air is a sort of vapor)" (330b4). Since vapor is an observable phenomenon, this indicates that Aristotle is drawing on empirical evidence to support his assignment of the differentiae to everyday earth, air, fire, and water.[26] He is explicit about this procedure in *GC* II 4, where, for example, he argues that his view about how elemental fire comes to be from air and earth is "confirmed also by perception" (331b24–26). A virtue of this reading, then, is that it accords with his procedure in the very next chapter, which he begins by stating that "it is evident according to perception" (καὶ κατὰ τὴν αἴσθησιν φαίνεται) that the elements change into one another (331a8–9). It also fits with his explicit procedure when, in *GC* II 5, he returns to the *GC* II 3 question about the number and nature of the primary bodies and argues that: "If then, there is nothing perceptible prior to these [i.e., earth, air, fire, and water], these will be everything" (332a26–27).

In short, Aristotle's rationale – the reason he associates particular differentia pairs with particular elements – is grounded in what we can observe about the "apparent simple bodies,"[27] that is, the elemental bodies' everyday counterparts. Aristotle starts (in characteristic fashion) from what is most readily and widely accessible to the senses and proceeds to what is

[26] Joachim also makes note of this: "It is evident to perception that 'air' is hot and moist, if 'air' is understood in Aristotle's sense as 'a sort of ἀτμίς'" (Joachim 1922: 213).

[27] While τὰ ἁπλᾶ φαινόμενα σώματα is often translated as "the apparently simple bodies," i.e., the bodies that appear to be but are not in fact simple (Rashed 2005: 55; Williams 1982: 40, 160; Tricot 1951: 105 (with n.1); Joachim 1922: 213), "apparent simple bodies" better reflects the fact that the Greek leaves it open whether Aristotle means that the bodies in question are the *appearing* (i.e., observable) bodies (φαινόμενα σώματα), or the merely *apparently* simple (ἁπλᾶ φαινόμενα) bodies. If I am right that later in this chapter (330b21–22) Aristotle maintains that these "apparent simple bodies" are in fact "mixed," then the phrase is likely meant to emphasize the fact that the bodies in question are only apparently simple. However, since the bodies in question are the everyday fire, air, etc., that appear to us – that we can directly observe – the fact that these bodies can be *apparently* simple is grounded in the fact that they *appear* to us at all. So Aristotle's choice of the word φαινόμενα is apt: it indicates that these are not really the simplest of the bodies while also indicating that his reasoning in the lines that follow is empirical in nature (330b3–5).

less so. It is apparent to our senses which two qualities are associated with each of the versions of the elements directly accessible to our senses, that is, everyday earth, air, fire, and water. If (as he maintains) the elements are like (τοιαῦτα, 330b23) their everyday counterparts, then we can learn about the qualities of the former via our observation of the qualities of the latter.

How should we understand Aristotle's claim that the distribution of the primary differentiae among the primary bodies is "reasonable" (330b6)? I suggest that he just means that since the primary differentiae are associated with the primary bodies in pairs, and since contraries cannot pair (the general principle cited in 330a31), the associations he establishes on empirical grounds in 330b3–5 are reasonable.[28] Finally, how should we understand Aristotle's claim at 330b7 that "the number of them" is "in accordance with argument"? "The number of them" (τὸ πλῆθος αὐτῶν) is ambiguous: it could refer either to the number of differentiae or to the number of primary bodies. I argue (Section 4.4) for the latter, so that this appeal to what is "in accordance with argument" is meant to support Aristotle's thesis that there are four primary bodies (thesis (i)).[29]

4.4 The Survey of Predecessors

Aristotle's next move (330b7–21) is to give a survey of his predecessors. Since he begins with a γάρ clause, we should expect the survey to somehow support what he has just established. Some argue that it is meant to support what I have called thesis (iii): Joachim claims that the survey shows that "in all previous theories the number of the 'simple bodies' depended upon the number of elementary qualities recognized."[30]

But this reading is not well supported by the text. Aristotle begins by arguing that those who have one primary body (e.g., air, for Anaximenes) make the other kind of principle – what he calls the producers

[28] Falcon and Leunissen argue that Aristotle often combines an appeal to what is reasonable (or according to argument) with an appeal to what can be observed (Falcon-Leunissen 2015: 224–225). I take this passage to be an example of this procedure.

[29] In *Meteor.* II 4, Aristotle claims that vapor is naturally *cold* and moist because it comes from water (360a22–24; see also *GC* II 8, 367a34). But when Aristotle says that vapor is cold, he is always contrasting it with another exhalation, smoke, which is hot and dry. This suggests that vapor is cold relative to smoke. The upshot: in *GC* II 3, vapor and air are likened to one another (οἷον), which indicates that vapor is being contrasted with water. Since vapor is hot relative to water, if air is a kind of vapor (or like the vapor produced from heated water) then air should be hot rather than cold. Aristotle's appeal here to his meteorological findings may be needed in the case of air because its qualities are less empirically obvious than are those of fire and water.

[30] Joachim 1922: 213–14. See also Tricot 1951: 105n3.

(δημιουργοῦντα) – two: the rare and dense, which come down to the hot and the cold (330b9–13).[31] On this view, which Aristotle mentions elsewhere,[32] the one primary body functions as underlying matter while the primary qualities (hot and cold) are principles that produce the other elementary bodies from out of it. Thus, Aristotle is not showing that the number of primary bodies (one) depends on the number of primary qualities (two). If he is showing anything about the relation between the number of primary bodies and the number of elementary qualities in his predecessors' views, it is rather that the latter depends on the former. That is, those who begin with only one primary body as their material principle will need more than one producer principle if they are to generate the other elementary bodies from this first body.[33]

Nor does the rest of the survey (330b13–21) make any mention of primary qualities. Instead, it mentions only the number of primary bodies or material principles (one, two, three, or four) in each of the predecessors' theories, and shows that there is a sense in which all (except the first thinker) make their primary bodies or material principles two.[34]

I propose that the survey justifies the chapter's other two theses, namely (i) that there are four primary bodies and (ii) that these are earth, air, fire, and water. The way the survey is introduced indicates that Aristotle is justifying (i). The immediate reason he gives for his claim that "the number of them is in accordance with argument" (330b7) is: "For everyone who makes the simple bodies elements makes either one or two or three or four" (330b7–9). As for what the survey in fact does, its clearest upshots are (a) that Aristotle's predecessors all posit one to four primary bodies; (b) that for each thinker the candidate primary bodies include some or all of Aristotle's four (earth, air, fire, and water), but no others; and (c) that most in effect make their primary bodies (material principles) two (while the first thinker makes his producer principles two).

Now given that one of the survey's upshots is that most of these thinkers have ended up with *two* primary bodies (c), one might reasonably ask how the survey can function as support for Aristotle's thesis that there are *four*

[31] Simplicius (preserving a fragment of Theophrastus) attributes this view to Anaximenes (*In Phys.* 24.26–25.1).

[32] See *Phys.* I 4, 187a12–16 and *Meta.* I 5, 985b10–13.

[33] When he returns to this matter in *GC* II 5, he reasons in both directions (see 332a34–b5).

[34] The first thinker (330b9–13) has two *producer* principles (hot and cold or rare and dense) but only one material principle (air, if this is Anaximenes' view).

primary bodies (thesis (i)). But perhaps (c) does not so much support this thesis as show that previous thinkers have been rather confused, or at least unsystematic, about the actual number of primary bodies to which each is committed. Upshot (c) may then be much like one of the upshots of Aristotle's *Meta.* I 3–9 survey of other thinkers' views about causes: his predecessors have also sought what Aristotle is presently seeking, but they have sought after it "vaguely"; and "though in a sense [the things sought] have all been described before, in a sense they have not been described at all" (*Meta.* I 10, 993a14–15). If this is right, then upshots (a) and (b) must be what make Aristotle's thesis about the number of primary bodies "in accordance with argument." (a) supports thesis (i) by showing that no previous thinker posits any *more* than four primary bodies. And (b) supports theses (i) and (ii) as follows: while for example, Anaximenes (presumably the thinker referred to at 330b9–13) makes air prior to the other bodies, and for example, Parmenides makes fire and earth prior to air and water (insofar as the latter are mixtures of the former), they posit no candidates for primary bodies *other than* earth, air, fire, and water. As in his *Meta.* I 3–9 survey of his predecessors' views about causes, then, the overall purpose of the survey is either to find another candidate or to "be more convinced of the correctness of those which we now maintain" (983b5–6).

4.5 Fire and Fiery Things, Air and Airy Things

GC II 3, 330b21–30 presents us with several new puzzles. In this passage, Aristotle sets up two contrasts: one between "fire and air and each of the things mentioned," which are mixed (μικτόν), and "the simple things" (τὰ ἁπλᾶ) (330b21–23); and another between fire itself and something that "resembles fire" (πυρὶ ὅμοιον), which is "fiery" (πυροειδές) (as well as between air, earth, and water and the things resembling air, earth, and water) (330b23–25). What is not yet clear is:

(1) Where are "the things mentioned" (i.e., mixed fire and air) mentioned?
(2) What are the mixed fire and air of 330b21–22 and what are the "simple things" with which they are contrasted (τὰ ἁπλᾶ, 330b23)?
(3) What are the things resembling fire or air, that is, the fiery or airy things?

We can begin with (1). Most commentators take "the things mentioned" to refer to the fire, air, water, and earth of the 330b1–7 discussion. Crowley proposes an alternative according to which "fire and air and each of the things mentioned" (330b21–22) are not "the things mentioned earlier in this chapter," but rather "the things mentioned by my predecessors."[35] But such a reading causes a problem for the lines that follow, as we can see when we attempt to paraphrase 330b23–25 with "fire" and "air" meaning "*what my predecessors called 'fire' and 'air'.*" Aristotle goes on to say that the simple body that "resembles *fire*" (τῷ πυρὶ ὅμοιον) is not *fire* (οὐ πῦρ) but fiery (πυροειδές) (and likewise for air and the airy, 330b23–25). If "fire" means "*what my predecessors called 'fire',*" then we must paraphrase as follows: the simple body *resembling what my predecessors called "fire"* is not *what my predecessors called "fire"* but fiery. Since this makes little sense, I think the straightforward reading must be correct: "the things mentioned" are just the fire, air, and so on Aristotle mentioned at 330b1–7.[36]

(2) Given my answer to (1), if the mixed fire and air of 330b21–22 are everyday fire and air, then the "apparent simple bodies" of 330b1–7 are everyday fire, air. etc. Since I have already indicated (Section 4.3) that I endorse the consequent, it is time to justify this by establishing the antecedent.

Let us consider the alternative, namely that the mixed fire and air of 330b21–22 are elemental fire and air and the "simple things" are the primary differentiae or qualities (hot, cold, dry, moist). As we have seen (Section 4.2), there is a sense in which the elemental bodies are not simple, since each is associated with, or perhaps even somehow constituted by, two primary differentiae. However, Aristotle is here contrasting simple not with composite (συγκείμενον or σύνθετος) but with mixed (μικτόν), and the elemental bodies are not *mixtures* of primary qualities. As he makes clear in his *GC* I 10 account of mixture, what is mixed must be capable of existing separately (327b15–19, 327b27–29), and hot, cold, moist, and

[35] Crowley 2013: 173–174. Crowley offers this alternative to accommodate his identification of the apparent simple bodies of 330b2 with the simple things of 330b23. He reasons that since the 330b2–4 apparent or observable fire is just hot and dry, it cannot be identified with the mixed fire of 330b24–26, which is a boiling or excess of heat. Hence observable fire must instead be identified with simple fire (2013: 177). However, in the earlier passage (about the apparent simple bodies) Aristotle need not be stating that the fire in question just is (to be identified with) the hot and dry; he could easily be stating that the fire in question is (has the properties) hot and dry. I have argued (Section 4.2) that Aristotle is in that passage, in characteristic fashion, using what we can observe about e.g., the perceptible fire of our everyday experience (observable fire) as evidence for his view about the nature of elemental fire. Perceptual observation tells us that everyday fire has the properties dry and hot, and Aristotle takes this as strong evidence that the two qualities that should be associated with fire are the hot and the dry.

[36] Thanks to Sean Kelsey for noticing this problem. To be clear, I am only denying that mixed "fire and air and each of the things mentioned" (330b21–22) means "just what my predecessors called fire, air, etc. *and not* the things mentioned earlier in this chapter." So I can still maintain that the things mentioned earlier in the chapter at 330b1–7 (which I take to be the everyday bodies) are indeed what Aristotle's predecessors called fire, air, and so on.

dry lack this capacity. Further, the cold and moist in water (for example) persist actually in the water, whereas if they had been mixed so as to produce water, then they would persist only in potentiality (*GC* I 10, 327b29–31). But water is *actually* both cold and moist. Finally, there is no suggestion that the primary differentiae cold and moist somehow moderate one another[37] in elementary water, as they would if they were mixed (*GC* II 7, 334b10–11). Hence water cannot be a mixture of cold and moist.[38] For these reasons, the "mixed" fire and air of 330b21–22 cannot be the elemental bodies and so must rather be the "simple things." It is then natural to think that everyday fire, air, and so on are the mixed things with which the elemental bodies are being contrasted.

But perhaps the case is not yet clinched. Dorothea Frede maintains that the mixed air and fire of 330b21–22 are elemental air and fire and that the elemental bodies are indeed "mixtures" of the primary qualities hot, cold, moist, and dry.[39] She argues that in *GC* Aristotle sometimes "uses the term mixture in a misleading way"; she cites this particular passage as an example of this atypical use of "mixture."[40] On her account, Aristotle speaks of "mixed" here "because of the lack of a better term to designate the relation between the elementary qualities."[41]

[37] Or, as Dorothea Frede puts it, "mixture must consist in a mutual 'toning down' of the opposed qualities" (Frede 2004: 306). But while there are intermediate states between hot and cold, there is no intermediate between cold and moist (since these are not contrary to one another). Hence, there is little reason to think that mixing moist with cold alone will "tone down" the moist, or *vice versa*. And when the elementary bodies transform into one another (e.g., air into water) this involves no toning down of their qualities, but rather an outright exchange of qualities.

[38] But see Krizan 2018 for the argument that at least in *GC* II 7, Aristotle "employs an extended sense of mixing" that applies to the primary differentiae and that explains the formation of the homoeomers (192). Given space constraints, I will not weigh in on *GC* II 7; but regardless of whether in *GC* II 7 Aristotle writes of the mixing of primary differentiae, the identification of mixed air and fire with elemental air and fire in *GC* II 3 causes problems in the lines that follow (as I explain below).

[39] Frede 2004: 303–304.

[40] Frede 2004: 304 and 304n37. Frede indicates that *GC* II 1, 329a34 and II 2, 329b11 are also places where Aristotle "speaks of 'mixing' of elemental properties" (303n35), but I cannot there locate μίξις, μικτός, μίγνυμι, or any cognates or synonyms. I find Aristotle referring to mixing in (1) a reference to *GC* I 10 at the start of *GC* II 1 (328b26) and (2) a statement that "the elements" must mix and change into one another at *GC* II 2, 329b23–24. Since the primary differentiae do not change into one another (*GC* II 1, 329b1–2), they are not the "elements" being said to mix in this passage (2). So if μικτόν at *GC* II 3, 330b22 has the atypical sense Frede suggests, this is the first time in *GC* II that Aristotle uses the language of mixing in this way.

[41] Frede 2004: 304. This reading may be motivated by the supposition that since the primary bodies are really "composites of different qualities," the primary qualities rather than the primary bodies must be the proper elements of bodies (303n36). I have argued (Section 4.2) that even if the primary bodies are somehow constituted by the primary qualities, nevertheless the primary bodies – and *not* the primary qualities – are the proper elements of sensible bodies.

But if, at 330b21–22, Aristotle were trying to designate the relation between the primary differentiae, he could easily have used the less misleading "composite" (συγκείμενον or σύνθετος). The context calls for nothing more specific than this (as it might for example if the mutual transformation of the elements were under discussion); and he freely uses "set together" or "associate" (συνίστημι) at 331a3 when describing the relation between the primary differentiae in a given elemental body. Hence, it is unlikely that he uses "mixed" here simply for lack of a better term. More troubling, though, is how Frede's reading would play out in the lines that follow. Consider 330b23–25: if the simple things are the primary differentiae, then the thing resembling fire would have to be *either* the hot *or* the dry (at least at this point in the chapter); and likewise the thing resembling air would have to be either the hot or the moist. Even taking into account Aristotle's later claim that each elemental body is strictly *of* just one primary quality (ἑνὸς ἕκαστόν ἐστι, 331a3–6), "what resembles air" or the "airy" is a very peculiar way to designate the moist. Further, on this reading the fire that Aristotle goes on to describe as an "excess of heat" (330b25–26) will have to be elemental fire. But as I will argue shortly (when I address (3)), because the fire of the 330b25–26 analogy is paired with ice, it cannot be elemental fire and must rather be everyday fire.

For these reasons, I think the answer to (2) is clearly that the mixed bodies of 330b21–22 are everyday fire, air, etc., and that the 330b23 "simple things" with which they are contrasted are elemental fire, air, etc. On this reading, we have no need to suppose that Aristotle is using "mixed" in a novel and misleading way at 330b22. The everyday bodies are indeed mixed: there is some earth from the wood in the hearth's fire, water in the air rising from the lake, and so on.[42]

This brings us to (3): what are the "fiery" and "airy" things of 330b23–25? Are they the mixed everyday bodies, or are they the simple elemental bodies? It is natural[43] to think that the elemental body is what is *really* fire, and the everyday body only somehow fiery or fire-like (πυροειδές). But this is incorrect.

[42] And if the "mixed bodies" of *GC* II 8, 334b30–335a9 include not just e.g., flesh but also hearth fire, lake water, etc., then the everyday bodies are mixtures of all four elemental bodies. See Andrea Falcon's essay on *GC* II 8 in this volume.

[43] Natural to just about anyone reading *GC*, that is, since they will also have spent a good deal of time reading Plato. Verdenius and Waszink explain how this might happen (Verdenius-Waszink 1966: 59).

First, the structure of the sentence indicates clearly that the simple thing, that is, the elemental body, is not fire, but the fiery thing: Aristotle's statement at 330b23 that (a) the simple things are (b) such as mixed fire and air (c) but not the same is parallel to his statement at 330b23–24 that (a) the thing resembling fire (b) is fiery (c) but not fire.[44] Second, the 330b25–26 analogy shows that the fire in question cannot be an elemental body. Aristotle writes that "fire is excess of heat, just as (ὥσπερ) ice is [excess] of coldness." In the analogy fire: heat :: ice : coldness, fire, and ice correspond, so that the fire in question should be the same kind of thing as ice. Since ice is not an elemental body, neither is this fire that is the "excess of heat." Both are everyday bodies. That the fire and ice of the analogy are the same kind of thing is also clear from the fact that at 330b28–29 Aristotle takes himself to be warranted in inferring something about this kind of fire from something that holds of ice: *if* ice is a congealing of the moist and cold, *then* fire will be a boiling of the hot and dry.

And as Crowley also notes, that everyday fire is the fire of 330b24 and elemental fire the fiery thing of 330b24 is made even clearer in the *Meteorology* (I 3, 340b22–23), where Aristotle states that the elemental body we call "fire" is not *really* fire.[45] He sometimes refers to everyday fire as "flame" (φλόξ) rather than "fire" (πῦρ), though in some contexts he uses the terms interchangeably.[46] Flame is distinct from the element fire: the latter is just the hot and the dry (and prime matter, on some interpretations), while flame – the *burning* thing that we habitually call "fire" – is a "boiling" of the hot and dry (hot–hot–dry, rather than just hot–dry). As Aristotle states in *GC* II 4, flame is fire "most of all" (μάλιστα, 331b25). Thus the point in *GC* II 3 is not that elemental fire is not really fire, but rather that what we call "fire" is the everyday body that consists of elemental fire most of all, that is, everyday fire or flame.[47]

And if everyday fire is the fire of 330b24, it must also be the mixed fire of 330b21–22 rather than one of the simple things. This is because for Aristotle, everyday fire is the burning of a mixture of two other elements: it

44 That is, (a) ἁπλᾶ (b) τοιαῦτα (c) οὐ ταὐτά is parallel to (a) τι τῷ πυρὶ ὅμοιον (b) πυροειδές (c) οὐ πῦρ.

45 Crowley 2013: 175.

46 See, e.g., *Juv.* 5, 469b21–470a5. That he means "everyday fire" here is clear from the fact that he speaks of a lamp's flame (ἡ τοῦ λύχνου φλόξ, 469b33). At *Meteor.* II 8, 366a2–3, by contrast, he distinguishes fire from flame: πῦρ *becomes* φλόξ when accompanied by wind.

47 That is, if the everyday bodies are mixtures of all elemental bodies (see note 42), what is most properly called fire is what is composed of more elemental fire than of anything else (i.e., it is "fire most of all").

is "burning smoke, and smoke is out of air and earth" (*GC* II 4, 331b25–26). For these reasons, the answer to (3) must be that the fiery and airy things of 330b23–25 are the simple elemental bodies, while the fire and air of 330b21–30 are the everyday mixed bodies.

4.6 Proper Places and Qualities of the Simple Bodies

This brings us to the last section of the chapter (*GC* II 3, 330b30–331a6). I will consider the following questions: (1) How do the claims of 330b30–331a6 either support or follow from the chapter's main theses? and (2) Why does Aristotle associate just one primary differentia with each of the bodies, and what governs his choice of which to associate with which?

Before these questions can be answered, we must decide what Aristotle means by the four "simple bodies" at 330b30–31. Since on my reading Aristotle has just stated (330b21–30) that the everyday bodies (the apparent simple bodies of 330b2) are not really simple but mixed and that elemental bodies are "the simple things," the four simple bodies at 330b30–31 ought to be the elemental bodies. This identification is not threatened by the fact that he goes on to say that the ones in the middle regions (air and water) are more mixed than those at the extreme locations (fire and earth) (330b33–331a1), since this just means that the elemental bodies mix with *one another* in the middle regions (e.g., elemental air and water are mixed together in vapor) and not that the elemental bodies are *themselves* mixtures (e.g., that elemental air is itself a mixture and, even more problematically, that it is somehow *more* of a mixture than elemental fire).

(1) I take it that this section of the text is meant to support the thesis that there are four simple bodies (thesis (i)) by appealing to the fact that the four are neatly distributed among the sublunary regions: two each belong to each of the two sublunary regions (the center and the periphery) (330b30–32).[48] And Aristotle's claim that the elemental bodies associated with opposed pairings of differentiae are opposed to one another (331a1–3) seems to follow from the thesis that each elemental body is associated with two primary differentiae (thesis (iii)). But what are we to make of the chapter's last claim, namely that each elemental body is, strictly, of just one of its two differentiae?

[48] The reason Aristotle gives for this distribution is that two of the bodies naturally move toward the periphery while two naturally move toward the center (330b32–33).

(2) Perhaps Aristotle aims to establish that there is another way in which the four differentiae yield four simple bodies (that is, besides the way he established at 330a30–b7). In the first part of the chapter, he showed that each of the four pairings of differentiae matches one-to-one with each of the four bodies; now he shows that each of the four individual differentiae matches one-to-one with each of the four bodies. That this may be his reason is suggested by his way of introducing the proposal: "since they are four (τέτταρα ὄντα), each is strictly of one affection" (331a3–4). This is not an especially compelling reason to make this last move; but it is a reason. Perhaps more baffling is why Aristotle makes the specific assignments he does – and especially why he associates water with cold rather than with moist. Aside from the fact that this is counter-intuitive, it conflicts with his *Meteor.* IV 4 statement that "of the elements earth is especially representative of the dry, water of the moist" (382a3–4). (He repeats this claim at *Meteor.* IV 5, 382b3.)

I offer the following as a possible reconstruction of the chain of reasoning. Aristotle has, in the previous chapter, established that the moist is that which "cannot be determined by a limit of its own," while the dry is "that which is easily determined by a limit of its own" (*GC* II 2, 329b29–31). Of all four elemental bodies, air best exemplifies the former feature and earth the latter. Hence if these features of the humidity qualities (moist and dry) govern Aristotle's choice of which quality to associate with which elemental body, clearly air must be associated with the moist and earth with the dry. The temperature qualities (hot and cold) will then have to be associated with the two remaining elemental bodies, fire and water. In *GC* II 2, Aristotle argues that the hot aggregates things of the same kind, while the cold aggregates indiscriminately (329b25–29). If these features of the temperature qualities govern Aristotle's choice of which quality to associate with which body, then since he states in *GC* II 2 that fire aggregates things of the same kind (329b26–28), clearly fire must be associated with the hot, and so water with the cold. And it is perhaps not quite so counter-intuitive to associate water with the cold rather than the moist when we are considering water *qua* contrary of fire (331a1–3), and fire has been associated with the hot. I see plenty of problems with this story;[49] nevertheless, it is a possible rationale.

[49] For one, I have emphasized the moist's feature of not being bounded by its own boundary (ἀόριστον οἰκείῳ ὅρῳ, 329b29–30), but not its feature of being easily bounded (or "easily delimited," εὐόριστον, 329b30); had I emphasized the latter, Aristotle's *GC* II 8, 334b34–335a1 statement that water is the most easily bounded of the bodies would need to be explained.

4.7 Conclusion

We have seen that Aristotle defends three central theses in *GC* II 3: (i) that there are four primary or elemental bodies; (ii) that these are earth, air, fire, and water; and (iii) that they are systematically associated with the primary differentiae. Since these associations at the elemental level explain much at the higher levels, it would be disappointing if, in *GC* II 3, Aristotle had either grounded them in little more than stipulation or relied entirely on theory while neglecting the empirical data (the φαινόμενα). I hope to have shown, among other things, that by considering what can be observed about the elemental bodies' everyday counterparts, Aristotle offers empirical reasons for assigning particular qualities to particular elements. He also gives a variety of justifications for (i) and (ii), including appeals to mathematical and metaphysical principles (330a30–b1) and to his predecessors' views (330b7–21). Brief though *GC* II 3 may be, Aristotle accomplishes a great deal in this chapter.

5

On Generation and Corruption *II 4*

Margaret Scharle

5.1 Introduction

GC II 4 has two aims: to determine whether each of the elements can come to be out of each of the others and to describe the manner of their change into one another. After arguing that each of the elements can come to be out of each of the three others, the bulk of the chapter presents and compares the speed and ease of three mechanisms by which the elements change into one another. This much is clear, but scholarly disputes lie in the details. Some have thought these three mechanisms are narrowly focused on individual changes, that is, the three ways an element can change into each of the others.[1] On my interpretation, the three mechanisms operate as complete cycles of change by which all four elements come to be, and Aristotle's interest in speed and ease should be interpreted as an interest in how quickly and easily all four elements can be generated out of one another. I understand the first mechanism as a mechanism whereby all four elements come to be singly and consecutively, each coming to be out of one other to form a cycle; by contrast, I understand the second and third mechanisms as processes whereby the elements come to be in cycles pairwise. My interpretation builds on passages other scholars and translators have deemed textual oddities or mistakes and has the additional advantage of showing that Aristotle's interest in the relative

* I thank the participants in the Spetses workshop, especially Andreas Anagnostopoulos for written comments and Sean Kelsey for exceptionally detailed comments on a penultimate draft. I also thank Andrea Falcon for his editorial leadership, and the participants at the Northwest Ancient Philosophy Workshop, and especially Brad Berman for written comments on an early draft. I thank Mary Krizan for organizing our panel on *GC* II at the 2020 Society for Ancient Greek Philosophy meeting. I thank Walter Englert for comments on my translation and Kellyn Yvonne Standley for editorial assistance and Jake Standley for the graphics.

[1] See, for example, Krizan 2013 and Gill 1989.

speed and ease of cycles of elemental transformation lays the groundwork for his metaphysical and scientific projects in *GC* II 10 and the *Meteorology*.

5.2 All Simple Bodies by Nature Change into One Another

GC II 4 opens by contextualizing its task with respect to *DC* III 6, where Aristotle establishes that "the elements of bodies" are "subject to destruction and generation" (304b23–305a14) and are "generated from one another" (305a14–32). Although *GC* II 4 does not rehearse the argument from *DC* III 6, it nonetheless tacitly revisits one of *DC* III 6's premises: that the elements do, in fact, come to be. *DC* III 6's argument begins with the claim that "we see" all four elements "dissolving" (304b26–28). In the *interim*, however, Aristotle has become much more cautious: *GC* I 3 maintains that air is not easily perceptible (*GC* I 3, 318b28–32), thereby casting doubt on the empirical evidence cited in *DC* III 6 for its analysis or dissolution. Perhaps this is why *GC* II 4 offers a new argument based on a much more reliable, albeit indirect, empirical observation. We see things – just regular things, not the elements in particular – altering (*GC* I 1, 314b13–15). Given points established previously in *GC* – namely, that alteration is a change of tangible qualities (*GC* I 4, 319b10–14), and given *that* the elements are the sources of all tangible qualities (*GC* II 2) – then if the elements did not come to be, there would be no change in tangible qualities and there would be no alteration at all (cf. *GC* I 1, 314b10–26). Thus, while the argument in *DC* III 6 begins with the empirical observation of the elements themselves changing, this new argument in *GC* relies on the less tentative empirical evidence of other, non-elemental things altering.

Although *GC* II 4 replaces the observation recorded in *DC* III 6 with a new empirical premise, it mentions neither the context nor the content of the other *DC* III arguments regarding elemental change. For example, in *DC* III 7 Aristotle objects to the Platonic theory of elemental transformation, on which the element earth alone does not transform into or out of any other element (306a2–6), whereas here in *GC* II 4 he does not single out the case of earth, but instead investigates all four elements evenhandedly: "[O]ne must say what is the manner of their change into one another, and whether every one of them is capable of coming to be out of every one, or whether some are capable of this, others incapable" (*GC* II 4, 331a10–12). Again, while in *DC* III 6 Aristotle has shown that each element comes to be out of *at least one* other element, here in

GC II 4 he is concerned with the question of "whether every one of them is capable of coming to be out of every one" (331a11–12); more to the point, he investigates this question from the fresh perspective afforded by *GC* II 2's introduction of the pairs hot/cold and wet/dry, which serve as the differentiae of the four elements:[2]

> It is clear, then, that all simple bodies by nature change into one another. For generation is into contraries and out of contraries, and all elements have contrariety toward one another because of their differentiae being contraries; for in the case of some both differentiae are contraries, for example, in the case of fire and water (for the former is dry and hot, the latter wet and cold), whereas in the case of others only one is, for example, in the case of air and water (for the former is wet and hot, the latter wet and cold). (*GC* II 4, 331a12–20)

Here Aristotle offers the following argument:

1. "[G]eneration is into contraries and out of contraries." (Premise)
2. "[F]or in the case of some [elements] both differentiae are contraries [. . .] whereas in the case of others only one is." (Premise)
3. "[A]ll elements have contrariety toward one another." (2)
4. "Every one of them is of a nature to come to be out of every one." (1, 3)

Aristotle begins with the nature of generation in [1], and then [2] notes that for any two elements, either one or both of their differentiae are contraries. Premise [3] generalizes over the cases presented in [2] to show that every element has at least one contrariety toward every other, which, via [1], yields claim [4], namely that every element is of a nature to come to be out of every other one (that is, each element is of a nature to come to be out of each of the other three).

Aristotle then shifts his focus in the argument that follows:

> Therefore, considered generally, it is clear *that* every one of them is of a nature to come to be out of every one, though as to *how* [they come to be out of one another], considered individually, it is not difficult to see even now. (*GC* II 4, 331a20–23)

Aristotle clearly shifts from considering elemental coming to be "generally" (καθόλου) to "individually" (καθ' ἕκαστον). At the same time, he shifts

[2] *GC* II 4's one case of directly observable elemental change at 331b26 is appealed to not as a fundamental premise, but simply in confirmation of a result established previously.

from the fact *that* (ὅτι at 331a20) each of them can come to be out of each of the other three – a fact he was able to establish by considering the elements in general – to the consideration of *how* (πως at 331a21) *the four elements actually do that.* But to show this requires another shift: Aristotle can no longer simply generalize over individual changes but must consider the mechanisms by which *all come to be out of all.* The shift from the singular "every one" (πᾶν) to the strengthened plural form "all" or "all together" (ἅπαντα) can be understood to signal Aristotle's aim to examine the mechanisms by which all four elements, taken all together, come to be in coordinated complexes.

Whether my interpretation of the mechanisms is correct comes down to whether Aristotle's descriptions of each mechanism bear it out. As we shall see, many commentators assume the three mechanisms are narrowly focused on single changes of one element into one other. I argue that in presenting each mechanism holistically, as a mechanism for generating all four elements, Aristotle considers whether it is quicker and easier for each of the four elements to come to be via a one-contrary change (the first mechanism), or whether it is quicker and easier for the four elements to come to be into and out of the others as pairs operating according to either the second or third mechanism.

5.3 First Mechanism

The first mechanism is a single cycle of four, one-contrary changes by which each element comes to be the next:

> For as many as have counterparts relative to one another, the change of these will be quick, but as many as do not have them, it will be slow, because it is easier for one to change than for many. For example, out of fire there will be air when one quality changes (for if the former was hot and dry, the latter hot and wet, the result if dry should be overcome by wet will be air); again, out of air there will be water if hot should be overcome by cold (for the former is hot and wet, the latter cold and wet, so that when hot changes there will be water). Also, in the same manner, out of water earth and out of earth fire; for both have counterparts relative to both. For water is wet and cold, and earth cold and dry, so that when wet is overcome there will be earth. And again, since fire is dry and hot, and earth cold and dry, if cold is destroyed, fire will be out of earth. Therefore, it is clear both that generation for the simple bodies will be in a circle and that this manner of change is easiest because the counterparts are present in consecutive simple bodies. (*GC* II 4 331a23–b4)

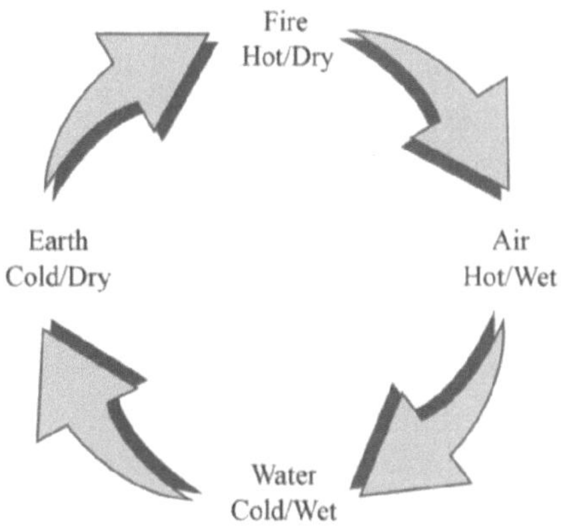

Figure 5.1 First mechanism example cycle

Here Aristotle offers the example cycle in Figure 5.1. This cycle proceeds solely via one-quality changes. In this description, as we shall find in the description of the other two cycles, the mechanism produces the four elements via a number of changes of the same sort (in this case, of one-quality changes); the consideration of each of these changes in turn is highlighted by the expressions "and again," "also, in the same manner," and "and again." Note that it is not until after he has described the entire cycle that Aristotle concludes: "this manner of change is easiest."

While this much is clear, the interpreters of *GC* disagree on several important details. First, while they have not addressed the issue head on, it is unclear whether "this manner of change," which Aristotle claims to be "easiest," refers to the entire cycle by which all four elements come to be or, more narrowly, to any of the single, one-contrary changes comprising that cycle. While Gill and Krizan assume that "easiest" refers to a single, one-contrary change, since they both understand all three mechanisms to focus on the ways a single element changes into each of the three others, I agree with Joachim, who takes "easiest" to characterize the entire cycle, which in fact is the more natural, immediate referent of the expression "this manner of change."[3]

Some support for this interpretation can be found in Aristotle's interest in successive changes: he opens his discussion with the principle, "*as many*

[3] Gill 1989: 70; Joachim 1922: 221. Krizan 2013: 197. My interpretation fits well with Anagnostopoulos' reading of changes "out of one another" in his contribution to this volume.

as (ὅσα) have counterparts relative to one another, the change of these will be quick" (331a23–25), thereby suggesting that his characterization of a single one-contrary change as quick should be seen in the context of many such changes, which he goes on to enumerate. Aristotle then uses the comparative form "easier" (331a25) to explain why a two-contrary change is slow: "because it is easier for one thing to change than for many." It would be odd to move from the comparative "easier" to the superlative "easiest" (331b3) to describe a single one-contrary change given that the comparative already compared one with "many," and he introduces no further comparators in the interim. Instead, in the interim, Aristotle describes the entire series of changes forming a cycle. Aristotle concludes his description of the first mechanism by stating the reason why "this manner of change" is "easiest": "easiest because the counterparts are present in consecutive simple bodies," where the word "consecutive," like "as many as" at 331a23, gestures at the entire series, as opposed to a single change from one element into the next, thereby suggesting that what Aristotle is here claiming to be "easiest" is the entire cycle by which all four elements come to be.

Scholars also disagree about whether Aristotle intends the example given to be the only cycle operating according to this first mechanism. For just as the one-contrary changes from fire to air, air to water, water to earth, and earth to fire form a single cycle wherein all four elements come to be, the same is apparently true of the "reverse" changes – fire to earth, earth to water, water to air, and air to fire – as shown in Figure 5.2.[4]

I call this cycle the "unidirectional reverse" of the example cycle Aristotle offers. In support of recognizing it as a cycle of the first mechanism, note that Aristotle presents the example cycle *as* a mere example (flagged by "for example" (οἷον) at 331a25), and that the principle it illustrates is equally well met by its unidirectional reverse: "for as many as have counterparts relative to one another, the change of these will be quick" (331a23–24). This reverse cycle, like the example cycle, operates exclusively via quick, one-contrary changes. Thus, I see no reason not to classify both the example and its unidirectional reverse as operating according to the first mechanism. And, as I argue below, by the end of the chapter we will find positive reason to include it.

[4] Gill 1989: 71 and Averroes (in Kurland 1958: 129–130) maintain that the reverse operates according to this mechanism, while Krizan 2013: 198–200 disagrees.

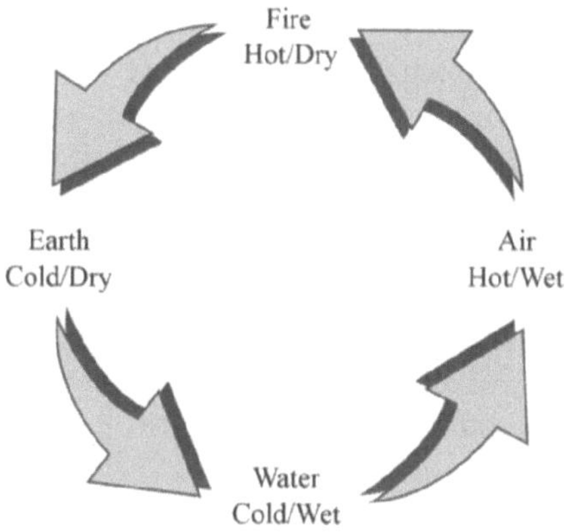

Figure 5.2 First mechanism unidirectional reverse cycle

5.4 Second Mechanism

In this section, I will argue that while the first mechanism is a process whereby each element comes to be the next in a single cycle comprised of four, one-contrary changes, the second and third mechanisms involve a *pair* of elements coming to be into and out of another pair. The second mechanism involves the pair fire and air coming to be into and out of the pair water and earth, while the third involves the pair fire and water coming to be into and out of the pair earth and air.

Let us begin with Aristotle's description of the second mechanism:

> However, though it is possible for water to come to be out of fire and earth to come to be out of air, and again, for air and fire to come to be out of water and earth, it is more difficult because the change is of more things. For it is necessary, if fire will be out of water, that both cold and wet be destroyed, and again, if air will be out of earth, that both cold and dry be destroyed. And similarly, too, if water and earth will be out of fire and air, it is necessary that both things change. So this generation takes longer. (*GC* II 4 331b4–11)

The scholarly controversy concerning this passage centers on whether the change in both contraries described here as necessary, for example, for water to come to be out of fire, could happen simultaneously or whether the change requires stepwise progression, for example, fire first becoming wet, then cooling into water, perhaps via a hidden interim element (in this case, air). Many scholars agree that the change must proceed stepwise, on the grounds that otherwise Aristotle would not claim that the two-contrary change "takes longer" than the one-contrary change of the first mechanism.[5] Others disagree, maintaining that if the two-contrary change is just two one-contrary changes in disguise, the second mechanism simply reduces to the first, making its introduction superfluous.[6]

My novel reading of the passage will show not only that the stepwise interpretation is unnecessary (albeit compatible with my reading) but also will show that the second mechanism is irreducible to the first.[7] On my view, the scholarly debate is centered on the wrong issue: although scholars think that Aristotle only seeks to compare the *speed* of a *single* two-contrary change to that of a *single* one-contrary change, the text in fact suggests that he aims to compare the speed *and ease* of *the entire four-change cycle* of the first mechanism with the speed and ease of the entire circuit of the second mechanism. Just as Aristotle may be interpreted as closing his discussion of the first mechanism by declaring that the entire cycle is "easiest" (331b3), so too here, at the conclusion of the second mechanism, when Aristotle says that "this generation takes longer" (331b9), we may understand the expression "this generation" to refer, not to a single two-contrary change, but rather to the entire circuit in which all four elements come to be.

To understand what Aristotle means by "this generation" in this passage and why he thinks it "takes longer," let us work through the text line by line. He introduces the second mechanism by conjoining two changes: "it is possible for [1] water to come to be out of fire and [2] earth to come to be out of air" (331b4–6). So conjoined, we can diagram these changes as shown in Figure 5.3.

Next, after introducing in tandem these two changes,[8] [1] and [2], Aristotle immediately presents two more changes: "and again (πάλιν), [it is possible] for air and fire to come to be out of water and earth" (331b4–5). The word πάλιν might indicate backward movement, thereby

5 See Krizan 2013: 200-202; Gill 1989: 74–75; Furth 1988: 224; and Lewis 1996: 39–42. Gill 1989: 77, and Lewis 1996: 23 think this passage threatens the traditional view of prime matter: if two-contrary changes must proceed stepwise, then a contrary can underlie the change, obviating the need for prime matter. Lewis 2008: 129n14 argues against the stepwise interpretation, while Gannagé 2005: 67 maintains one-contrary changes equally raise the specter of prime matter.

6 See Lewis 2008: 129n14.

7 My reading of *GC* II 5 is thereby neutral on Aristotle's commitment to prime matter. (See also Scharle 2009.) Yet the fact that I show the second mechanism not to require the stepwise interpretation undermines one of the best arguments against Aristotle's commitment. For more on prime matter, see the coda to this volume's introduction.

8 While the change from fire to water requires both becoming cold and becoming wet, the change from air to earth requires both becoming cold and becoming dry; this suggests the possibility that the cooling required in both changes is simultaneous, which, as we shall see, would make the change faster.

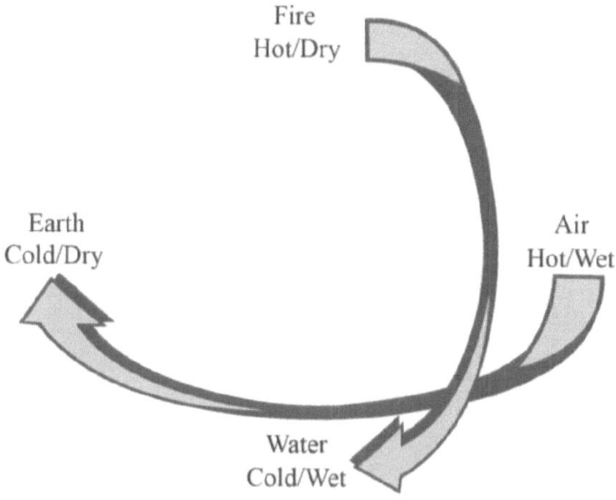

Figure 5.3 Second mechanism changes [1] and [2]

highlighting the fact that these next two changes – [3] air out of earth, and [4] fire out of water – "reverse" the first two changes. Regardless, the fact that changes [3] and [4] *do* reverse changes [2] and [1] (respectively) may be seen from the fact that [3] is out of what [2] was into and into what [2] was out of, while [4] is out of what [1] was into and into what [1] was out of. In short, changes [2] and [3] form a cycle, from air to earth and back to air, as changes [1] and [4] form another cycle, from fire to water and back to fire.

Note that the syntax of the passage has puzzled commentators, for "ἐξ ὕδατος καὶ γῆς ἀέρα καὶ πῦρ" presents the respective products of changes [3] and [4] in chiastic order: since it is fire, not air, that comes to be out of water, and it is air, not fire, that comes to be out of earth, we would expect Aristotle to claim "(*fire and air*) out of (water and earth)" instead of the chiastic order he states here, "(*air and fire*) out of (water and earth)." While Gill's translation retains the textual order "air and fire," formally noting it should be understood chiastically, Joachim and Williams dispense with the chiasmus, rendering the passage "fire and air."[9] On my reading, by contrast, the chiastic order is in fact essential to understanding the mechanism Aristotle is describing. Not only does my translation retain the order, but my interpretation also builds on it, for it suggests that the second mechanism, unlike the first, essentially involves *pairs of elements.*

[9] See Williams 1982 and Joachim in Barnes 1984, and Gill 1989: 70. Rashed 2005: 60 and Forster-Furley 1955: 281 retain the textual order as I do.

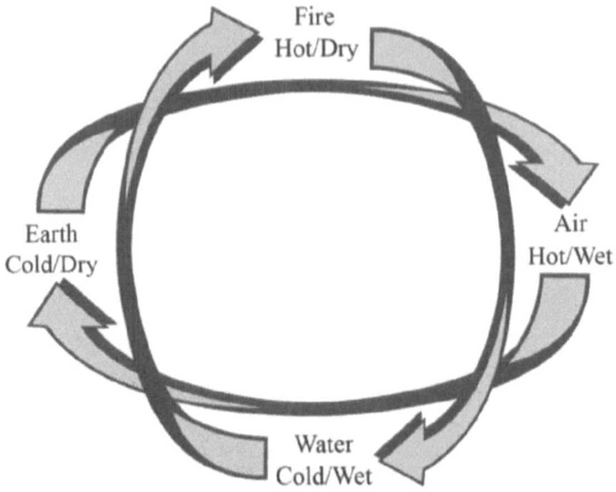

Figure 5.4 Second mechanism entwined cycles

Recall that Aristotle introduces changes [3] and [4] as the coming to be of a *pair* of elements out of another *pair*: "and again, air *and* fire to come to be out of water *and* earth" (331b4–5). If we understand changes [1] and [2] as integrated with one another as in the diagram above, changes [3] and [4] may be understood as integrated as well, reversing changes [1] and [2] and thereby completing two integrated cycles in which each of the four elements comes to be. Further, we can interpret Aristotle's presentation of them in chiastic order as suggesting that the two cycles – the one comprised of changes [1] and [4] and the one comprised of changes [2] and [3] – are entwined. If these two cycles are entwined, then changes [3] and [4] are to be understood as completing the cycles begun with changes [1] and [2], in which case changes [3] and [4] may be described equally well as *either* "the coming to be of *fire and air*" *or* "the coming to be of *air and fire*." So understood, the entire circuit of the second mechanism can be depicted as in Figure 5.4.

After laying out this pair of entwined cycles, Aristotle goes on to observe:

> [However,] it is more difficult because the change is of more things. For it is necessary, if fire will be out of water, that both cold and wet be destroyed, and again, if air will be out of earth, that both cold and dry be destroyed. And similarly, too, if water and earth will be out of fire and air, it is necessary that both things change. So this generation takes longer. (*GC* II 4, 331b6–11)

This passage continues to play with chiastic order, which we may interpret as reinforcing the point that the two cycles here described are entangled with one another. This time, Aristotle presents change [4] before change [3], reversing the first chiasmus, before conjoining changes [1] and [2]. The second mechanism description can be summarized as follows, where bold highlights Aristotle's use of chiastic order:

GC II 4, 331b4–6:
[1] water out of **fire**.
[2] earth out of **air**.
[3] and [4]: (**air and fire**) out of (water and earth).
GC II 4, 331b7–11:
[4] **fire** out of water.
[3] **air** out of earth.
[1] and [2]: (water and earth) out of (**fire and air**).

This summary shows Aristotle offering a chiastic version of the initial presentation: While *GC* II 4, 331b4–6 begins by mentioning first the change out of fire, then the change out of air, and then conjoins them in chiastic order (air and fire), the text immediately goes on at *GC* II 4 331b7–11 to reverse the chiastic order of the initial conjunction by taking fire, then air as the products of these changes and then conjoining them as (fire and air). Putting together the conjunctions, we get:

(**Air and fire**) out of (water and earth) out of (**fire and air**).

If the pair (air and fire) operates as a conjunction, then we might take it as equivalent to the pair (fire and air). And, if the two pairs of elements are equivalent, then we can understand the two pairs of changes – the pair comprised of changes [1] and [4] and the pair comprised of changes [2] and [3] – to begin and end in *the same pair of elements*. Moreover, given Aristotle's later clarification – namely, that "we say that generation has revolved in a circle because it comes back to its beginning" (*GC* II 10, 337a5–6) – if the pairs of changes described here begin from and end in the same pair of elements, we may understand these pairs of changes as constituting a complex cycle in the course of which all four elements come to be. So understood, we may interpret Aristotle's complicated and otherwise confusing use of chiasmus as a device for signaling the equivalence of the pair (air and fire) with the pair (fire and air). The result would be that the second mechanism is in fact a complicated two-phase cycle in which one pair of elements (viz. air and fire, or [equivalently] fire and air) comes to be into and out of another pair of elements (water and earth).

Now let us consider whether the foregoing interpretation can explain Aristotle's claim that the second mechanism takes longer than the first

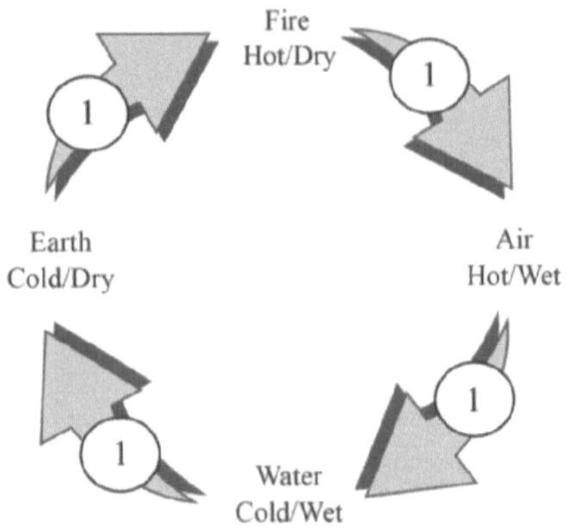

Figure 5.5 Example of first mechanism cycle

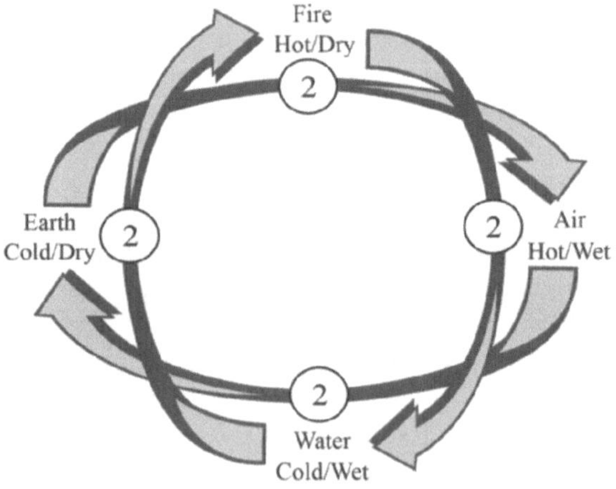

Figure 5.6 Second mechanism entwined cycles

mechanism. Note that the explanation for the second mechanism taking longer relies on its being more difficult, which is explained by "the change is of more things" (331b6–7). The "more things" he goes on to enumerate are the four contraries to be destroyed in the second two legs of both cycles, and "similarly, too" for the first two legs, thus bringing the grand total to eight (331b7–11). We are now in a position to see why, in the second mechanism, the change is "of more things" than in the first: while the paired cycles of the second mechanism require the destruction of a total of eight qualities, a first mechanism cycle requires the destruction of only four (Figures 5.5 and 5.6).

I have identified the "more things" that make the second mechanism cycles more difficult, but I still need to explain why the cycles of the second mechanism "take longer." I appear to face the same challenge that has led other scholars to adopt the stepwise interpretation: if the second mechanism's two-contrary changes can happen simultaneously, and two simultaneous changes take the same amount of time as just one of them would alone, then the four two-contrary, simultaneous changes that compose the second mechanism would take the same amount of time as the four successive one-contrary changes of the first mechanism. Even worse, if changes [1] and [2] of the second mechanism occur simultaneously, and changes [3] and [4] occur simultaneously, the second mechanism would take *half* the amount of time as the first mechanism. If this is accurate, I need the stepwise interpretation both to explain two-contrary changes *and* to explain why the legs of the entwined cycles operate successively (or at least partially successively).[10] Both explanations are needed to account for why the second mechanism takes longer. My interpretation would, at least, have one advantage over its other advocates: while others face the objection that the stepwise interpretation makes Aristotle's introduction of the second mechanism superfluous, my interpretation as two entwined cycles makes the second mechanism irreducible to the first.

In fact, however, I do not need the stepwise interpretation to account for the second mechanism's slower speed. While the stepwise interpretation is compatible with my reading, I see no reason to adopt it. For Aristotle, the cycle's speed depends upon its ease: a change is "slow, because it is easier for one to change than for many" (331a25–26), and a change is "more difficult because the change is of more things" (331b6–7).[11] The order of explanation, then, seems to be that the number of contraries destroyed determines the level of ease and difficulty, which, in turn, determines the speed. As Sarah Broadie determines, Aristotle inserts the measure of ease and difficulty between the *explanans* – the change is of more things – and the *explanandum* – the speed.[12] Thus, we do not need

[10] Change [1] requires cooling and wetting, change [2] requires drying and cooling, change [3] requires heating and wetting, and change [4] requires heating and drying. Partially successive operations would allow simultaneous cooling for changes [1] and [2] and simultaneous heating for changes [3] and [4]. Depending on how the stepwise interpretation of two-contrary changes is fleshed out, it might not allow for such partially successive operations: for example, if change [1] operated by ordered steps cooling then wetting, but change [2] operated by ordered steps drying then cooling, there would be no potential for overlapping cooling in these changes.

[11] The language here of "many" and "more" supports my view that he compares entire cycles, not just the transformations of one element into another.

[12] See Broadie 2004: 144.

the stepwise interpretation to explain the slower speed of the second mechanism. For example, perhaps Aristotle thinks the speed of change depends not only on *how long the change takes*, whether simultaneous or successive, but also on *the number of agents required to cause the change.*[13] More changes can require more agents and more complex coordination among them. Even if Aristotle would acknowledge the possibility of two-contrary changes occurring simultaneously, or even of changes [1] and [2] occurring simultaneously with changes [3] and [4] (respectively) – and I see no reason to rule this out – the occurrence would depend on adequate local conditions. Given that it is difficult for agents to act simultaneously, and without any guarantee of such simultaneity built into the second mechanism, one complete circuit of the second mechanism – that is, completing the two entwined cycles – is (always or for the most part) slower than one complete cycle of the first mechanism. I return to these considerations below.

5.5 Third Mechanism

Aristotle goes on to introduce a third mechanism, which he describes as "easier" than the second mechanism (331b12). While the second mechanism describes pairs of elements coming to be into and out of another pair via two entwined cycles, I will argue that the third mechanism describes a single cycle in which a pair of elements comes to be into and out of another pair as follows:

> So this generation [i.e., the second mechanism] takes longer; however, if one affection of each pair is destroyed, the change though easier is not into one another, but rather (ἀλλ') earth and air will be out of fire and water, and fire and water out of air and earth. For whenever the cold of water and the dry of fire is destroyed, air will be (for the hot of the latter and the wet of the former remain), but whenever the hot of the fire and the wet of the water is destroyed, earth will be, because there will remain the dry of the former and the cold of the latter. And, similarly, fire and water will be out of air and earth, for whenever the hot of air and the dry of earth is destroyed, water will be (for the cold of the one and the wet of the other remain), but whenever the wet of air and the cold of earth is destroyed, fire will be, because the hot of the former and the dry of the latter will remain, which very things are fire. (*GC* II 4 331b11–24)

[13] According to *Phys.* VIII 4, elements require an external source of movement.

Scholars uniformly read this sentence to deny the third mechanism forms a cycle. For example, Krizan dubs third mechanism change "Non-reciprocal change," while Gill names it "Quick Irreversible Change."[14] In my view, this characterization is a mistake.

What is not cyclical is the change from the pair in which "one affection of each . . . is destroyed" to a single element produced. Consider Aristotle's example, "whenever the cold of water and the dry of fire is destroyed, air will be" (331b14–16). *This* change is not reversible to complete a cycle: even though air comes to be out of water and fire, we cannot reverse that process to get water and fire back out of the single element air.[15] Thus, Aristotle denies that the pair (fire and water) form a cycle with air. To summarize:

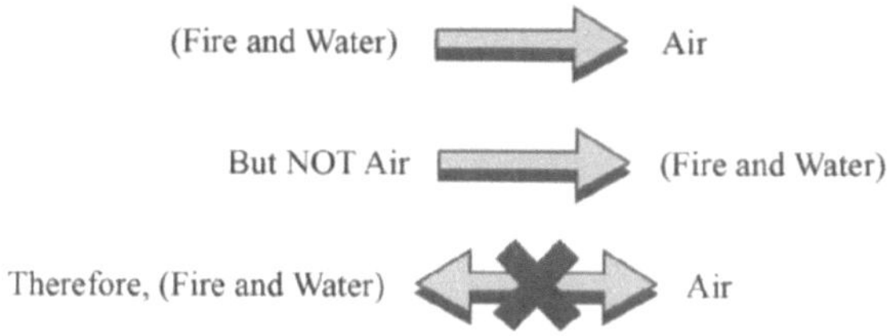

Scholars have not properly attended to the fact that after denying the cyclical relationship between a pair (fire and water) and the single element air, Aristotle immediately inserts a "but rather" (ἀλλ') at 331b13: "but rather earth and air will be out of fire and water, and fire and water out of air and earth." In other words:

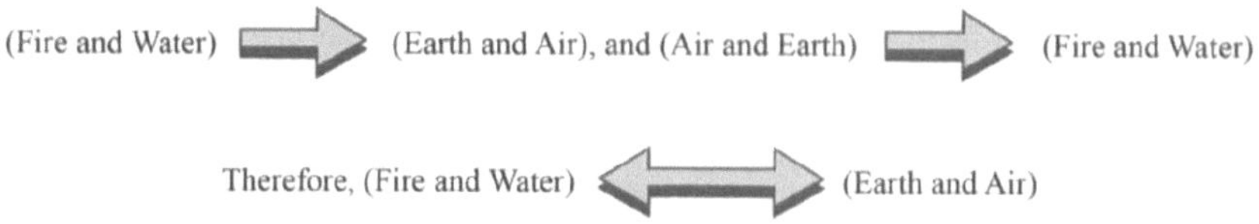

On my reading, the first contrast at 331b13 denies one cyclical relationship, namely, between the pair (fire and water) and the single element air, while the second contrast at 331b13–14 acknowledges (and so endorses) a different cyclical relationship: namely, the cyclical relationship that obtains between the pair (fire and water) and the pair (air *and earth*). On this reading, while air alone does not form a cycle with fire and water, the pair air *and earth* do form such a cycle.

[14] See Krizan 2013: 202 and Gill 1989: 71. [15] Gill 1989: 71n43 agrees.

Like the description of the second mechanism, Aristotle once again uses the syntax of chiasmus to highlight the fact that this mechanism operates by way of pairs of changes: so, "earth and air will be out of fire and water, and fire and water out of air and earth," and again, five lines later, "and, similarly, fire and water will be out of air and earth" (*GC* II 4 331b13–14, 18–19). The chiasmus can be interpreted to show the two pairs are equivalent, so we can understand the two changes – (earth and air) out of (fire and water) out of (air and earth) – to begin and end in *the same pair*, thereby completing a cycle of *pairs* of elements.

This reading enables us to solve a puzzle that has long vexed translators and commentators regarding the translation of καὶ in two of its occurrences at 331b14 and in one occurrence at 331b19: "earth *and* air will be out of fire and water, and fire *and* water out of air and earth [. . .] fire *and* water will be out of air and earth." I preserve the translation of the καὶ as "and" since I interpret the conjunction to be central to recognizing that Aristotle understands the mechanism he is describing here to involve pairs of changes. Gill follows Williams in translating the second καὶ as "or," despite Williams' admission that "the word I have translated 'or' is the normal word for 'and'," but to render it so "with no further adjustment would have been positively misleading."[16] Forster splits the difference to translate "and (alternatively)."[17]

Further justification for my translation and interpretation can be found in the next segment of text where Aristotle explains why the pair fire and water form a cycle with *both* air *and earth*:

> For (γὰρ) whenever the cold of water and the dry of fire is destroyed, air will be (for the hot of the latter and the wet of the former remain), but whenever the hot of the fire and the wet of the water is destroyed, earth will be, because there will remain the dry of the former and the cold of the latter. (*GC* II 4, 331b14–18)

This passage aims to explain *why* fire and water form a cycle with *both* air *and earth*. On my reading, the immediately preceding lines first say that when air comes to be out of fire and water, we do not get fire and water back again out of air, and then concede that we *do* get fire and water out of *both* air *and earth* (*GC* II 4, 331b11–14). Here then we are told the reason why: it is because although it is true that (sometimes) air comes to be out

[16] See Williams 1982: 163. Gill 1989: 71n44 also recognizes that her own translation is problematic: "The phrasing is somewhat perplexing because Aristotle uses the same idiom with a different sense to describe the second mechanism." Krizan 2013: 202 and Joachim in Barnes 1984: 542 translate "and" without offering justification.

[17] See Gill 1989: 71n44, Williams 1982: 163, and Forster-Furley 1955: 283.

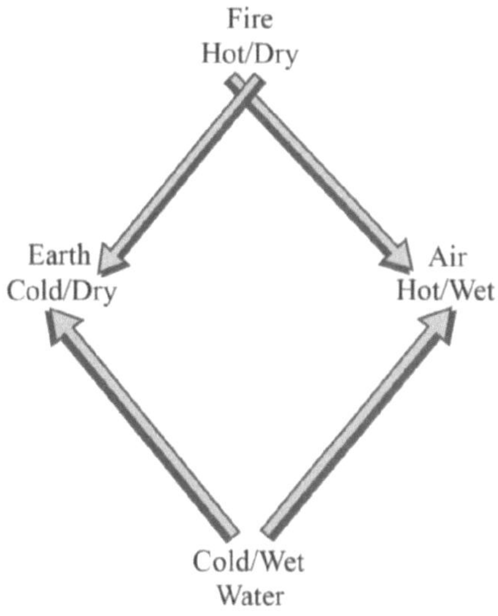

Figure 5.7 Third mechanism first phase

of fire and water, still it also happens, simultaneously, by the very same mechanism, that earth comes to be out of fire and water. The result is that we get *both* (air *and earth*) out of (fire and water). On this reading, what it is for "the cold *of water*" and "the dry *of fire*" to "be destroyed" is for those qualities to remain *in earth*, while what it is for "the wet *of water*" and "the hot *of fire*" to "be destroyed" is for those qualities to remain *in air*, as depicted in Figure 5.7.

The arrows depict what of the originating pair "remains" in the product. The hot of fire and the wet of water remain in air, so the arrow points from those contraries to the contraries that constitute air, and the dry of fire and the cold of water remain in earth, so the arrow points from those contraries to the contraries that constitute earth. Given that this description aims to explain why both air and earth are out of fire and water, we can read it to show this: what it is for the hot of fire to be destroyed *is for it to remain in air*, while what it is for the dry of fire to be destroyed *is for it to remain in earth*. And likewise: what it is for the wet of water to be destroyed *is for it to remain in air*, while what it is for the cold of water to be destroyed *is for it to remain in earth*. Thus, by this third mechanism, the coming to be of air is both necessary and sufficient for the coming to be of earth, which is why it is the pair (air and earth), not just the single element air, that comes to be out of (fire and water).

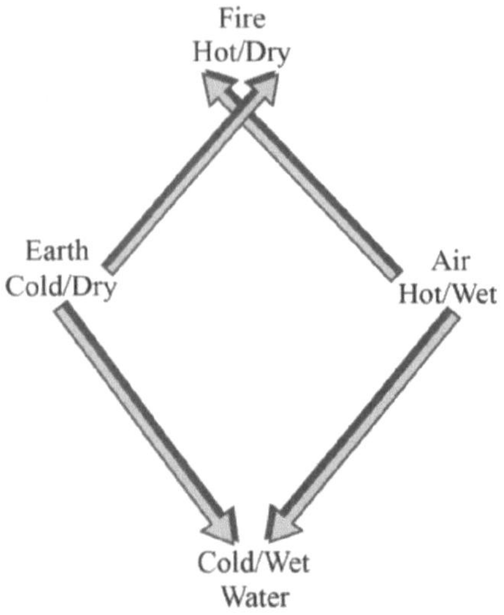

Figure 5.8 Third mechanism second phase

Aristotle then explains the second phase of this third mechanism, which completes the cycle whereby we get the pair (fire and water) back again out of the pair (air and earth) (Figure 5.8):

> And, similarly, fire and water will be out of air and earth, for whenever the hot of air and the dry of earth is destroyed, water will be (for the cold of the one and the wet of the other remain), but whenever the wet of air and the cold of earth is destroyed, fire will be, because the hot of the former and the dry of the latter will remain, which very things are fire. But the generation of fire is confirmed also by perception: for flame is fire most of all, but this is burning smoke, and smoke is out of air and earth. (*GC* II 4, 331b14–26)

Once again, this passage aims to explain why we get back to the original pair, fire and water, from *both* air *and earth*. We can read "the cold of earth is destroyed" to mean that the cold *of earth* remains *in water*, leaving the dry to remain *in fire*. In addition, we can read "the wet of air [. . .] is destroyed" to mean that the wet *of air* remains *in water*, leaving the hot to remain *in fire*.

Having laid out the operations involved in the third mechanism, we are in a position to see why Aristotle claims this mechanism is "easier" than the second (331b12). As I argued above, Aristotle thinks that the ease of a mechanism depends on the number of contraries that must be destroyed, and thinks that the second mechanism takes longer than the first because it

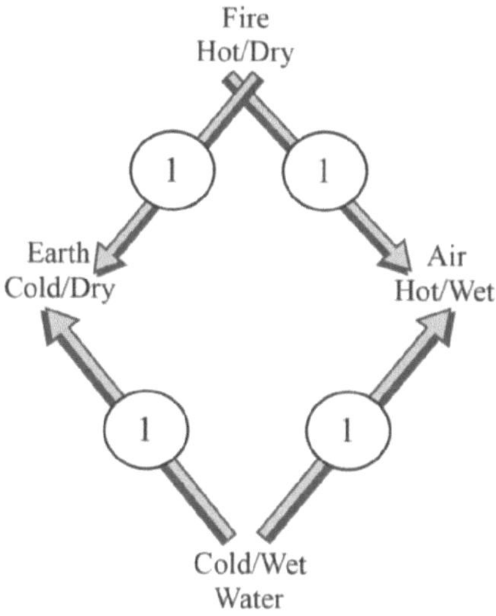

Figure 5.9 Third mechanism first phase

requires the destruction of more contraries. (Note that the mechanism does not guarantee that their destruction will be simultaneous.) To discern the number of contraries that must be destroyed in the third mechanism, let us return to the text to tally them up. The coming to be of each element requires the destruction of two contraries, as, for example, "whenever the cold of water and the dry of fire is destroyed, air will be." So, the coming to be of the pair earth and air out of the pair fire and water requires the destruction of a total of four contraries, and, again, the coming to be of the pair fire and water back out of the pair earth and air requires the destruction of a total of four contraries, for a grand total of eight as depicted in Figures 5.9 and 5.10.

Thus, the entire cycle of the third mechanism requires a total of eight contraries to be destroyed as shown in Figure 5.11.

Now compare the entire cycle of the third mechanism to those of the second mechanism, which also requires a total of eight contraries to be destroyed (Figure 5.12).

We are now in a position to see why the third mechanism is "easier" (331b12) than the second, despite the fact that each mechanism requires the destruction of eight contraries. As I argued above, the number of contraries destroyed determines the ease of a cycle, which, in turn, determines its speed. Yet, if I am correct that in the third mechanism pairs of

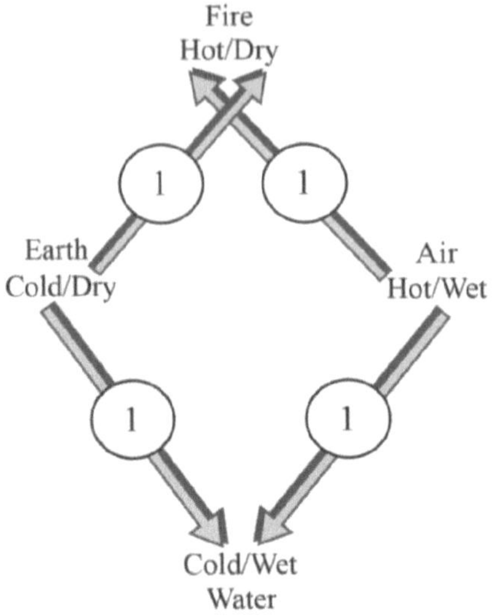

Figure 5.10 Third mechanism second phase

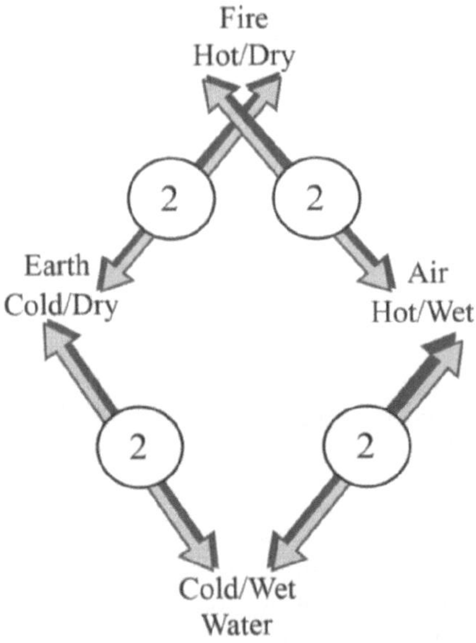

Figure 5.11 Third mechanism cycle both phases

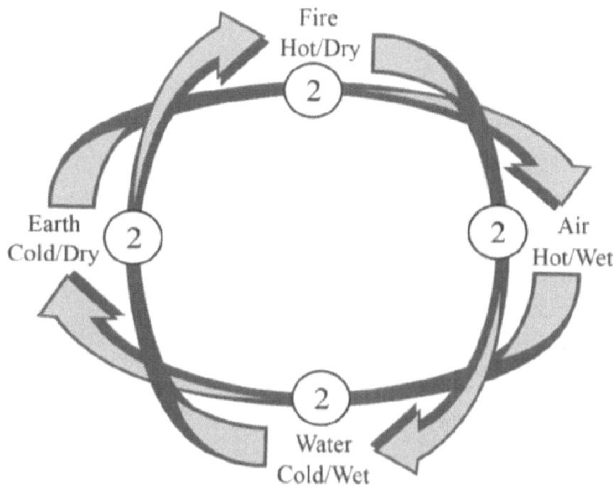

Figure 5.12 Second mechanism entwined cycles

elements come to be simultaneously – this is an essential feature of the third mechanism as described above – whereas this is *not* an essential feature of the second mechanism as described above, then we can see why the third mechanism is easier. The reason is that the destruction of four contraries in each of its two phases is guaranteed to be simultaneous; for (again) the coming to be of earth is necessary and sufficient for the coming to be of air, in as much as the generation of earth requires that the hot of fire and the wet of water must be destroyed, which in turn is simply for them to remain *in air*. So understood, although it is true that the second and third mechanisms require the destruction of the same number of contraries – a grand total of eight in each case – still, the third mechanism guarantees that the four destroyed in each phase of the cycle are destroyed simultaneously. And that is what makes the third mechanism both easier and quicker than the second.[18]

5.6 An Impossible Pairing for the Third Mechanism

Before bringing the chapter to a close, Aristotle considers whether the third mechanism can operate between pairs of "consecutive" elements, which he

[18] While Aristotle's calling the first mechanism "easiest" might be taken to imply that it is easier than the third, perhaps the third mechanism's cycle of pairs also meets the sufficient condition for "easiest" offered at 331b3 – "the counterparts are present in consecutive simple bodies" – since the paired changes of the third mechanism toggle back and forth between consecutive elements.

Table 5.1 *Impossible third mechanism example case*

	Fire	Air
1	**H** D	**H** W
2	**H** D	H **W**
3	H **D**	**H** W
4	H **D**	H **W**

rejects as impossible (331b26–332a1). Scholars have puzzled over what they interpret as a bad argument for this impossibility, but on my reading of the third mechanism, the argument is sound, thereby providing yet more evidence for my interpretation. Aristotle argues,

> However, in respect of the consecutive elements, it is not possible, with the destruction of one contrary in each of the elements, that there occur a change into any of the bodies, because either the same or contrary qualities remain in both. But neither of these options admits of a body's coming to be. For instance, if the dry of fire were destroyed, the wet of air would be destroyed, for in both the heat remains; but if the heat out of each were destroyed, the contraries dry and wet would remain. Likewise also in the other cases: for in all consecutive elements there exists one same and one contrary quality. (*GC* II 4, 331b26–34)

In this passage, Aristotle argues for the impossibility of a pair of consecutive elements producing "any of the bodies" by destroying one contrary in each of the initial pair. Since there are four possible combinations by which for any pair of elements each would lose one contrary, scholars have puzzled over why Aristotle considers only two possible combinations before declaring such an operation impossible.

Consider his example of fire and air, which he claims cannot produce any element by the third mechanism. Table 5.1 enumerates the four possible combinations of one contrary each destroyed, where the destroyed contrary is in bold.

The text considers only two of these combinations, namely [1] and [4], before declaring it impossible for consecutive element pairs to figure in a process of elemental transformation via the third mechanism. But what rules out combinations [2] and [3]? As Williams puts it, [2] and [3] are the only "intelligible" options and "neither of these is impossible": while the remaining dry/wet in [1] and hot/hot in [4] could not form an element, it appears the remaining dry and hot in [2] and hot and wet in [3] would produce fire and air respectively. While Williams declares Aristotle "misled," Joachim and Krizan suggest a way to patch up the argument.[19] On their view, Aristotle must mean that no *third* element can come to be out of the initial two, thereby excluding combinations [2] and [3], which each simply reproduce one of the initial elements.

However, Joachim and Krizan's proposed solution strains the text: Aristotle argues, "It is not possible, with the destruction of one contrary in each of the elements, that a change occurs into *any* of the bodies" (331b26–28), yet Joachim and Krizan suggest that Aristotle excludes combinations [2] and [3] since they produce an initial element, but do not produce a *third* element. But for Aristotle "any" would presumably *include* the fire and air Krizan and Joachim interpret these combinations to produce.[20]

My interpretation of the third mechanism can accommodate the text. It shows why combinations [2] and [3] cannot yield the initial elements fire and air, but it also shows (as we will see) that they cannot occur for the exact reason Aristotle says they cannot occur: "either the same or contrary qualities remain in both." As I go on to show, combinations [2] and [3] leave a single contrary remaining in each of the initial elements; this is impossible because no body can be formed out of just one contrary.

Consider combination [2], in which the hot of fire and the wet of air is destroyed (Figure 5.13).

As I argued above, the destruction of the heat of fire just *is its remaining in air*, and the destruction of the wet of air just is its remaining in water. More generally, destruction of a quality amounts to its remaining in another. So understood, no body has come to be from what remains. For after the proposed destructions, the dry would remain in fire, hot would remain in air, and wet would remain in water, *but there is no fire which is just dry (and not also hot), and there is no air which is just hot (and not also wet), and there is no water that is just wet (and not also cold).*

We can see how this reading fits the text: "either the same or contrary qualities remain in both. But neither of these options admits of a body's coming to be." As the diagram shows, dry, the "contrary quality," remains in fire and hot, the "same quality," remains in air, but a single contrary remaining in each is not enough to admit of a body's coming to be. In other words, to produce fire, it is not enough for hot to remain somewhere.

[19] Williams 1982: 163, Krizan 2013: 204-205, Joachim 1922: 222.

[20] It also seems Krizan 2013 and Joachim 1922 show Aristotle to allow a pair of elements to produce one of the pair in this way. Does this then commit Aristotle to yet another mechanism by which the elements can come to be out of one another? If so, why does he not formally introduce it as another mechanism?

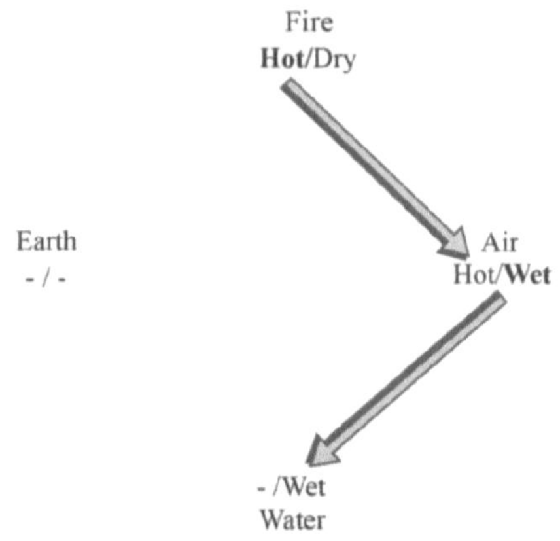

Figure 5.13 Third mechanism impossible pairing

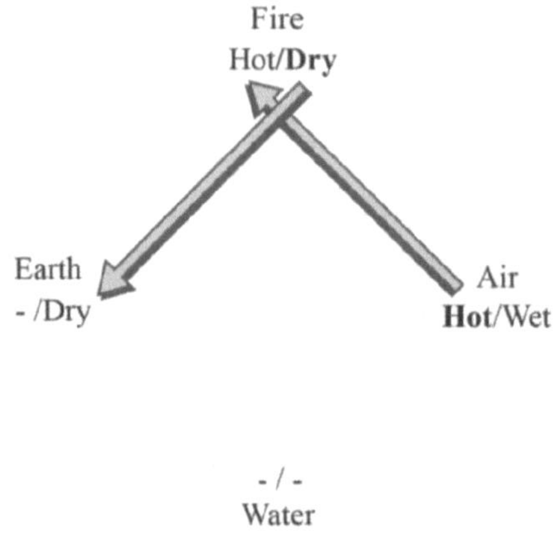

Figure 5.14 Third mechanism impossible pairing

Rather, the hot of air would have to remain *in fire* to form fire. But for the hot to remain in fire would be for the hot *of air* to be destroyed, which *ex hypothesi* it was not. Only the wet of air was destroyed.

Consider combination [3], in which the dry of fire and the hot of air are destroyed (Figure 5.14).

Once again, given that the destruction of the dry of fire just is its remaining in earth, and the destruction of the hot of air just is its remaining in fire, there is not *any* body that has come to be: the wet would remain in air, hot would remain in fire, and dry would remain in earth, but there is no air with just wet, there is no fire with just hot, and there is no earth with just dry. Thus, there is *no body* in which the qualities can remain. Again, we can see how this reading fits the text: "either the same or contrary qualities remain in both. But neither of these options admits of a body's coming to be." As the diagram shows, hot, the "same quality," remains in fire; and wet, the "contrary quality," remains in air. But a single contrary remaining in each is not enough to admit of a body's coming to be. For air to be produced by what remains would require the hot to remain in air. But for the hot to remain in air would be for the hot *of fire* to be destroyed, but it was not. Only the dry of fire was destroyed. Thus, my interpretation of the third mechanism successfully shows consecutive elements cannot produce any body whatsoever via the third mechanism.

5.7 Conclusion

After considering these three mechanisms, Aristotle concludes the chapter with the following words: "so we have established that all come to be out of every one, and in what manner the transition into one another occurs" (331b36–332a2). *GC* II 4 began by establishing all come to be out of every one (331a12–20), but the bulk of the chapter presented three possible mechanisms for how all four elements come to be into one another, either singly in succession according to the first mechanism, or by coming to be in pairs according to the second and third mechanisms (331a21–331b34), dismissing the possibility of consecutive elements operating via the third mechanism.

It is clear how my interpretation holds explanatory power in reading the text. Further, my interpretation demonstrates that the chapter exhausts all possible ways in which the four elements come to be out of one another: while the first mechanism describes the only way all four can come to be out of one another singly, both in the order of the example cycle or its unidirectional reverse, the second and third mechanisms (including the impossible iteration) exhaust all the ways the elements can come to be out of one another as pairs. Aristotle's argument at *GC* II 4 331b26–34 against "consecutive" element pairs operating via the third mechanism implies that the third mechanism is exclusive to non-consecutive pairs of elements. This suggests that the second mechanism can be seen as the exclusive mechanism for "consecutive" element pairs. Thus, while his description of the second mechanism at *GC* II 4 331b12–26 only explicitly considers the pairs (fire and air) and (water and earth), he might also think the other set of consecutive pairs – the unmentioned pairs (air and water) and (earth and fire) – operate via the second mechanism. If so, the chapter exhausts all possible ways the elements can come to be, both singly and in pairs.

Not only does my interpretation of *GC* II 4 – as a chapter aimed at comparing complete cycles of elemental transformation with respect to speed and ease – fit the text by retaining and building on what other scholars and translators deem textual mistakes or oddities; my reading also shows the chapter to set the stage for the role cycles will play in *GC* II 10 and the *Meteorology*. As I argue elsewhere, Aristotle understands the relative speed and ease of such cycles to have a teleological significance:[21] the quicker, easier, and more regular the cycles of elemental transformation are, the better the elements are and the more closely they imitate the divine. While I do not have the space to develop the suggestion, *GC* II 10 and *DC* II can be interpreted to show the quickest, easiest, and most regular cycles of elemental transformation are caused by the sun in its circular course, which is the ultimate cause of heating and cooling, while other elemental transformations are harder, irregular, and lack teleological direction. The *Meteorology* goes on to show the regular cycles that take the sun as their efficient cause to be the ones that most closely approximate the divine and can thereby serve as the *explanantia* of various meteorological phenomena. *GC* II 4's connection to these other scientific texts shows just how essential understanding the mechanisms of elemental transformation is to comprehending Aristotle's natural philosophy more broadly.

[21] Scharle 2008; Scharle 2015*a*; Scharle 2015b.

6

On Generation and Corruption *II 5*

Mary Krizan

6.1 Introduction

GC II 5 contains a long and complicated argument against the view that there is a single primary body. Specifically, Aristotle shows that a single primary body cannot function as the single material or origin of other bodies.[1] First, at 332a6–26, Aristotle argues against positions that he takes to accept a single material from which other bodies are generated that also survives a change into another body. Then, at 332b5–333a15, Aristotle turns to positions that consider a single primary body to be a principle or origin (ἀρχή) of the others.[2] Aristotle shows that the only way to make sense of a body as an origin for the others implies that simple bodies are, instead, origins for one another.

The argument of *GC* II 5 confirms Aristotle's own account of the simple bodies and their changes into one another. In *GC* II 1–4, Aristotle constructed his account of the simple bodies, beginning with principles, then describing primary contraries, simple bodies, and the generation of simple bodies from one another. In *GC* II 5, Aristotle demonstrates the incoherence of an alternative position, which assumes a single body is the matter or origin of the others. In doing so, he responds to a possible objection against his own position.

* I would like to thank Sean Kelsey for detailed comments on a previous draft of this paper, as well as the participants of the *On Generation and Corruption* II reading group for helpful discussion and insights. Funding for parts of this project was provided by the University of Wisconsin, La Crosse, and the National Humanities Center.

[1] Philoponus offers an instructive *diairesis* of Aristotle's take on previous views of simple bodies; see Kupreeva 2005: 28, and Philoponus, *In GC* 237.22–238.29. Philoponus' differentiation of the accounts discussed in *GC* II 5 is a bit different from my proposal; he treats the final account of the chapter as allowing that the single material element is matter "in a way" for the other bodies, whereas I treat the single material element as an origin.

[2] The single body may be thought of as a material cause in the same way that each simple body is, in a way, a material cause of the others within Aristotle's own account.

Although the primary topic of the chapter is a refutation of material monism, it confirms Aristotle's own theory of the origins of simple bodies. Aristotle shows that the simple bodies must be origins for one another, and hence, no single body can be the principle of the others. In doing so, he highlights the problems that result if one assumes there to be a single contrariety or infinite contrarieties. Thus, Aristotle confirms his own account of the contrarieties: the simplest explanation of simple bodies requires four contraries, arranged in two pairings of opposites.

6.2 First Argument (*GC* II 5, 332a6–20)

Aristotle's first argument is directed against accounts that maintain one of the four usual simple bodies – air, water, fire, or earth – as the matter of other bodies. Aristotle assumes, as he believes his predecessors would, that change is into and out of contraries. He then shows that a single simple body cannot be matter for other bodies because it must remain through a change and possess a contrary opposite to that of the body it produces.

This account is problematic, first, because it appears to explain alteration, but not generation. If air is the primary body, then it must be the matter for fire, which is hot. Yet if air remains in the change into fire, fire is simply hot air. Thus, fire is air, along with a contrary: air has altered, and fire has not been generated. The second problem emerges because fire should be able to change back into the primary body, air. Since change requires opposites, and fire is hot, air must be cold. Yet if air is cold and remains as the matter of fire, it must remain cold when fire is generated. Since fire is hot air, and air is cold, fire is simultaneously hot and cold.

A single simple body cannot be the matter for other bodies because it has two inconsistent roles in the explanation of change: it remains as matter within a new body, even though it is opposite to the new property of the body. Accordingly, an explanation of simple bodies must differentiate these two roles: the matter that remains and the bodies that possess contrary qualities cannot be identical. Aristotle remarks that fire and air must be something else that is the same; a logical step in a defense of a single matter is to differentiate the matter from what possesses the contraries.[3]

[3] Lines 332a16–18 initially seem to introduce prime matter (Williams 1982: 214, and Joachim 1922: 224). In this argument, it only indicates that fire and air have some matter in common; this matter could reasonably be one of the contraries or another body.

6.3 Second Argument (*GC* II 5, 332a20–26)

Aristotle has shown that a single body, identified by a particular contrary quality, cannot be the matter for other bodies. Next, he turns to an account that introduces a common matter for bodies such as fire and air. On this account, bodies such as fire and air possess contraries, but are not matter for one another. Rather, a single body can be common matter for all bodies because it does not possess contraries. In 332a20–26, Aristotle shows that a single body without contraries is problematic because the single body must be either all of the bodies or nothing at all.

Aristotle's argument runs as follows:

> Nor is there anything else besides these, such as an intermediate between air and water or air and fire, which is coarser than air and fire, but finer than the others: for air and fire will be this (ἐκεῖνο) with a contrariety. But one of the contraries is a privation; hence, it is not possible for this (ἐκεῖνο) ever to be isolated, as some say regarding the *apeiron* and the surrounder. Thus, in the same way, it is any one of these or it is nothing.

This argument is directed against a view that considers something beyond the simple bodies as the common matter for at least two of them.[4] The argument may be outlined as follows.

(1) Assume an intermediate besides the four elemental bodies: either between air and fire or between air and water.

(2) The intermediate is coarser than air and fire but finer than the others.

(3) Air and fire are the intermediate along with a contrary.

(4) One of a pair of contraries is a privation.

(5) So, the intermediate cannot be isolated.

(6) So, the intermediate is any one of the elemental bodies or nothing.

6.3.1 Aristotle's Description of the Intermediate

Aristotle begins in (1) by considering an intermediate: this intermediate falls either between air and fire or between air and water. The disjunction is not meant to be exclusive or exhaustive; rather, both options offer likely accounts of an intermediate that distinguishes the bearers of contrarieties

[4] Aristotle elsewhere attributes a view of this kind to Anaximander; here, he refers to the single body as the "*apeiron*" and "surrounder."

from their matter.[5] In (2), Aristotle situates the intermediate in the way he thinks his predecessors might; fire and air are rarer than other bodies, and fine and coarse correspond with rare and dense.[6]

In (3), Aristotle introduces a relationship between fire and air, a contrariety, and "this" (ἐκεῖνο), which refers to the intermediate. The subject of the sentence is unclear: it could be (a) "air and fire," or (b) "this." There are thus two ways to render (3):

(3a) **Air and fire** are this, along with a contrariety.[7]
(3b) **This** is air and fire, along with a contrariety.[8]

On (3a), the intermediate is also indeterminate. On this reading, the indeterminate entity is a constituent of air and fire; hence, both are the intermediate along with a contrariety. On (3b), the intermediate is a combination of air and fire along with an additional contrariety; it is a mixture of air and fire and contains a contrariety that both air and fire lack.[9]

Premise (3a) is preferable; *GC* II 1, 328b34–35 offers evidence that the intermediate is indeterminate. There, Aristotle mentions that some accept an intermediate between air and fire that is corporeal and separable as the matter underlying perceptible bodies; later, he introduces difficulties for this view. The intermediate body is rejected because it cannot exist as a body that is *also* separable from all perceptible contraries. *GC* II 5 marks a return to Aristotle's discussion of an intermediate; as in *GC* II 1, it is best understood as indeterminate.[10]

[5] Michael Wedin, in contrast, thinks that Aristotle examines two distinct intermediates: the intermediate between air and water is not the same as the intermediate between air and fire (Wedin 2013: 18–21). Aristotle, in fact, considers the intermediate in different ways throughout the corpus: he identifies an intermediate between water and air in *DC* III 5, 303b9–12, an intermediate between fire and air in *GC* II 1, 328b34–35, and an intermediate between water and fire in *Phys.* I 6, 189b1–2. See also *Phys.* III 4, 203a17–18; *Phys.* III 5, 205a27; and *Phys.* I 4, 187a13–15. Williams 1982: 164 comments that Aristotle is uncertain whether Anaximander's "infinite" falls between air and water or air and fire.

[6] Aristotle takes his predecessors to associate fine and coarse with rare and dense, and sometimes, great and small. See *DC* III 5, 303b19–30. On Aristotle's own view in *GC* II 2, fine and coarse belong to moisture and dryness; one would thus expect fire to be coarser than air and water to be finer than fire.

[7] Joachim 1922: 224–225 accepts this reading.

[8] Wedin 2013: 22, 25–26. The translation of Forster agrees with this decision: it reads, "the intermediate will be Air and Fire with the addition of a pair of contraries" (Forster-Furley 1955: 287).

[9] Graham supposes that there is room for both interpretations – Aristotle considers the *apeiron*, in different places, to be an intermediate or a mixture from which other bodies are extracted (Graham 2010: 67).

[10] *Phys.* I 4, 187a12–23 may present an objection to (3a). There Aristotle seems to consider Anaximander among the previous thinkers who produce other entities from a mixture. The passage does not definitively show that Aristotle commits Anaximander to a mixture; it may

Furthermore, (3a) makes better sense of Aristotle's argument in the context of *GC* II 5. Earlier, Aristotle argued that a single body with a particular contrary cannot function as the matter for other bodies. Now, Aristotle examines a third thing which is different from both bodies but at the same time is also their common matter. The point of the argument is to show that the third thing cannot be common matter because it cannot be isolated from the contraries. Aristotle assumes that the intermediate can be isolated because it lacks contraries, and then shows the problems that result from such an assumption. If the intermediate were already a mixture of contraries, Aristotle would not need an additional argument to show that it cannot be isolated.

6.3.2 *Aristotle Against the Intermediate*

In (3), Aristotle claimed that fire and air are the intermediate along with a contrariety. Call the intermediate "*x*"; given the contraries introduced in the 332a6–20, fire and air may be identified as follows:

Fire: x + Hot
Air: x + Cold.

Here, both fire and air consist of the intermediate along with a contrariety, and both possess one of the two opposed qualities in the contrariety hot-cold. This picture avoids the problem found in 332a6–20; air no longer contains inconsistent qualities because the intermediate is matter for both air and fire. Bodies possess contrariety; the common matter does not.

Premise (4) explains why the intermediate is problematic as the single matter of other bodies. According to Aristotle, one of a pair of opposites is a privation: within each contrariety, one is a positive characteristic or state, whereas the other is the privation of the positive characteristic. A privation, in this sense, is not the complete lack or absence of any property: bodies that lack heat have coldness, and bodies that lack lightness have heaviness.[11] Thus, the lack of a contrary is just the possession of its opposite.

Accordingly, there are two ways in which one might attempt to isolate the intermediate. First, the intermediate might be isolated because it lacks the *specific* contraries that belong to air and fire. Fire is hot; so, if the intermediate lacks heat, it is cold. Likewise, air is cold; so, if the

simply mean that Anaximander is like Empedocles and Anaxagoras because he appeals to separation, unlike previous thinkers whom only appeal to condensation and rarefaction.

11 See *Meta.* XI 11, 1068a5–7; *Phys.* V 1, 225b3–5; *DC* II 3, 286a25–28; *Meta.* I 4.

intermediate lacks cold, it is hot. By lacking the specific properties of fire and air, the intermediate possesses both opposites. The intermediate is hence inconsistent because it is both hot and cold: it is, in a way, both fire and air. The problem generalizes: for as many specific contraries as there are, isolating the intermediate from any quality implies that it possesses the opposite. The intermediate becomes a maximally inconsistent object; it contains all the contraries. This explains the first half of Aristotle's conclusion in (6): as a material that contains all the contraries, the intermediate is all the bodies.[12]

Second, the intermediate might be isolated because it lacks the *contrarieties* that identify fire and air: the intermediate is neither hot nor cold nor any intermediate of the two. This also avoids the problem in 332a6–20; now, the intermediate – which has no qualities – is common to air and fire. Yet without any contraries or qualities whatsoever, the intermediate becomes vacuous: if it has no properties whatsoever, it cannot be identified. So, Aristotle concludes, in the second half of (6), that it is nothing.

In either case, the intermediate, as described, cannot be matter for other bodies. If it is isolated from all contrarieties, then it turns out to be nothing; if it is isolated from each of the specific contraries that other bodies possess, then it is everything. Neither option explains why such an entity remains as matter in other bodies.

6.4 A Single Body as Origin (*GC* II 5, 332b5–333a15)

In the first half of *GC* II 5, Aristotle has argued against positions that suppose a single body or entity to be the matter of the others. In the second half of the chapter, he turns to accounts that make a single body the principle of the others. As a principle, a single body need not literally remain in the bodies that it produces. Rather, it has a role in producing the other simple bodies as an origin or starting point from which the remaining bodies proceed.[13]

At 332b5–14, Aristotle assumes a linear arrangement of bodies from highest to lowest: fire is highest, followed by air and water, with earth at

[12] It will not help to suppose that the intermediate is a mixture of contraries, or a literal intermediate of heat and cold: intermediates are inconsistent with extremes, so even if fire is an intermediate of heat, it cannot also be hot (in the extreme sense).

[13] Here the ἀρχή is like the origin or starting point of a line, as described in *Meta.* V 1, 1012b34–1013a1. The origin need not be a productive cause or a constituent of its product.

the bottom.[14] Fire and earth are at the extremes, whereas air and water are in the middle.[15] If one of the bodies at the top or bottom is the origin, then the remaining bodies follow from it along a line in one direction. If one of the bodies in the middle is the origin, then the remaining bodies follow from it along a line in two contrary directions. Bodies must be arranged in a line in order for one to be the origin of the others; if they are arranged in a circle, as on Aristotle's own account in *GC* II 4, then no single body is the origin of the rest.

6.4.1 Extremities as Principles

Aristotle first shows that one of the bodies at the extremities, either fire or earth, cannot be the origin.[16] Aristotle states that this account is the same as saying that everything is from (ἐκ) fire or earth, and moreover, makes everything fire or earth. Aristotle does not discuss the view in further detail; he thinks that by making everything fire or earth, the account is subject to the same objections already raised against a single body as matter in 320a6–20.

Here is the problem. Assume that fire, which is hot, is the origin of other bodies. Since change occurs between contraries, a change from fire into its neighbor, air, produces a body that is cold. Since fire and air have no other principle in common, air is nothing more than cold fire. The origin has become the matter, and fire is a body with inconsistent contraries.[17]

The problem is not solved by introducing a second contrariety common to both. If fire and air have heat in common but are differentiated by dry and moist, respectively, then change can occur between contraries because fire is dry, air is wet, and heat remains; fire is the origin of air. When air changes into water, there are two options: *either* there is a change between the opposites hot and cold with wet remaining; *or* there is a change between the opposites wet and dry, with heat remaining. The first option

[14] The same linear arrangement of bodies occurs in *DC* IV. It is not novel to Aristotle. See Kahn 1960: 121–122.

[15] Here "the middle" indicates what falls between the highest and lowest bodies, rather than the center, which would be earth.

[16] Williams comments that Aristotle may have the view of Heraclitus in mind (Williams 1982: 165). See also *Phys.* III 5, 205a3–4 and *DC* III 1, 298b30–33.

[17] If we deny that something must remain in the change from fire to air, and that what is hot can become cold (and *vice versa*) without anything remaining, then fire is no longer the principle; fire and air are principles of one another.

implies that fire is not the origin of water; rather, air is the origin.[18] The second option is problematic because it fails to produce a body, water, which is different from fire.

One might maintain that fire is the origin of other bodies by suggesting that other bodies are various states of fire. If fire is the origin and other bodies are generated through condensation and rarefaction, then perhaps other bodies are condensed states of fire.[19] Such an account does not differentiate simple bodies; instead, it shows that a single matter is subject to different configurations. The result is either a single material differentiated by quantity (the result of fire condensing) or a single material with void interspersed. As Aristotle points out in *DC* IV 2, neither of these views differentiates simple bodies because they cannot explain the motions of intermediate bodies such as air and water.

6.4.2 The Intermediates

Since a single body at the extremities cannot be the origin of the others, the remaining option is that a single body in the middle – either air or water – can be the origin of the others. Unlike fire and earth, air and water can change into both higher and lower bodies. Since air and water can each change into two different bodies, an explanation of air or water as the origin of other bodies requires at least two contrarieties. The introduction of an additional contrariety explains why air and fire have something in common and why fire is something more than hot air.

Aristotle begins by setting out the problem to which he will return later:

> On the other hand, it is not an intermediate, as some suppose air changes into water, and water changes into air and earth, but the extremities do not further change into one another: for change must stop and must not proceed straight out to infinity in both directions. For in that case, infinite contrarieties will belong to one of them. (*GC* II 5, 332b10–14)

Here, Aristotle offers his initial reasoning against the assumption of an intermediate as the origin of other bodies. For an intermediate to be the origin of other bodies, two conditions must be met: the extremities must not be able to change into one another, and change must not continue

[18] As Aristotle shows later, the introduction of two contrarieties results in four simple bodies, arranged such that the extremities can change into one another. Accordingly, each of the bodies is an origin for each of the others.

[19] Cherniss makes a similar claim: each simple body is the single element in various phases (Cherniss 1935: 123). Aristotle seems to have in mind theories that generate all things through condensation and rarefaction; see also *DC* III 5.

straight out to infinity in both directions. In the remainder of the chapter, Aristotle confirms that it is not possible to meet both conditions. He also explains why infinite contrarieties cannot explain a plurality of bodies.

First, Aristotle shows that the assumption of two contrarieties implies that no single body is the origin of the others. In 332a15–30, he introduces an account of the four simple bodies. Rather than appealing to his own primary contraries, Aristotle introduces the contrarieties wet-dry and white-black to explain the simple bodies.[20] He arranges them as follows:

Fire: Dry – Black
Air: White – Dry
Water: Moist – White
Earth: Black – Moist.[21]

On this model, each of the bodies can be transformed into each of the others. Hence, each body is an origin for each of the others, and one single body is not the origin. Aristotle still assumes that change is into and out of contraries; any two bodies have at least one pairing of opposites. Even if one denies that non-adjacent bodies such as air and earth can directly change into one another, there remain two possible ways to generate each of the simple bodies from its neighbors; earth and fire can generate one another due to their opposites.[22] No single body is a principle of the others; they are all in some way principles of one another.

Aristotle's model of simple bodies in 332a15–30 shows that the assumption of two contrarieties can explain why bodies are origins of one another but cannot explain why one body is the origin of all others. Given the relationships between contrarieties, even the bodies that are at the extremities on a linear arrangement – fire and earth – can directly change into one another because they have contrariety. Hence, the introduction of two contrarieties cannot explain why air or water would be an origin

[20] This account is defended in *GC* II 2 and applied within the account of elemental transformations in *GC* II 4. In *GC* II 5, the contraries dry, wet, white, and black take the place of *GC* II 4's hot, cold, wet, and dry, respectively.

[21] Aristotle's choice of contrarieties shows that his argument succeeds even if one does not select heat, cold, moisture, and dryness as primary. His selection of wet-dry and white-black suggests that he likely has Milesian thinkers in mind. Kahn 1960: 160–162 notes that hot-cold, moist-dry, and bright-dark (or white-black) were among the common Milesian pairings. Aristotle may (as Cherniss 1935: 124 claims) be offering an argument against Anaximenes.

[22] It is unclear whether Aristotle's account of simple bodies in *GC* II 4 allows for direct changes between non-adjacent bodies; the answer partially depends on the correct interpretation of elemental structure. Gill 1989: 74–75 admits that the process takes longer because the element (e.g., fire) must pass through air before becoming water. I agree with this assessment.

productive of changes that occur straight out in two directions and come to a stop; it cannot deny the possibility of extremities that directly change into one another.[23]

6.4.3 *Infinite Contrarieties*

Aristotle's model implies that *no* finite assignment of contraries permits one of the intermediates to be an origin of the others. If an additional contrariety were added, the outcome would be similar: now, there would be eight bodies, each would have contrariety with each of the others, and direct changes between the extremities would remain possible. In general, any finite addition of contrarieties leaves open the possibility of direct changes between the extremities.[24] So, the only way to maintain one body as the origin of the others is to allow for the infinite addition of contrarieties. Aristotle finds this view to be problematic because he thinks it implies that all simple bodies are one.

6.4.3.1 *Adding Contrarieties*

In 332b32–333a5, Aristotle sketches the results of adding bodies and contrarieties to the previous picture. First, assume that there is a body above fire: call it *X*. Second, assume that *X* is not identical with air, water, or earth, as previously introduced. Third, *X* is distinguishable from fire as well as air, water, and earth because it has a different contrary: it possesses *J*, whereas the remaining elements possess *K*. Thus, the introduction of a new body introduces a new *contrariety*: in addition to hot-cold and

[23] It is unclear whether Aristotle thinks that the account in question must deny any changes between fire and earth whatsoever or only direct changes (those that do not proceed through another simple body). I take it that Aristotle only needs to deny direct changes between fire and earth (or whatever the extremities are); an indirect change between fire and earth requires a return to the point of origin. Cherniss accepts this interpretation, noting that Aristotle may have Anaximenes' oscillatory theory of change in mind (Cherniss 1935: 124; cf. Verdenius-Waszink 1966: 55). Williams thinks that Aristotle must deny any change, direct or indirect, between fire and earth (Williams 1982: 165). Which view is to be accepted depends, in part, on one understands Aristotle's claim at 332b33 that when fire becomes the next higher body it does not return (ἀνακάμψει); other uses of the verb (*GC* II 10, 337a6; *GC* II 11, 338a5 and 338b4) take "returning" to be the completion of a cycle. Whether the change of fire back into air counts as a "return" is unclear; at minimum, Aristotle must deny that bodies at the extremities can change into one another.

[24] Interestingly, Aristotle does deny the possibility of the extremities to (directly) change into one another in *DC* IV; whether they can change back into intermediates at all is a point of contention surrounding a passage at *DC* IV 3, 310b11–15.

wet-dry, there is a new contrariety, *J–K. X* is different from fire: fire is dry, black, and *K*, whereas *X* is dry, black, and *J*.

Next, Aristotle shows that this pattern can be repeated. If *X* can change into yet another body, *Y*, then another contrariety must belong to both *X* and fire; call this contrariety *L–M*. Following the pattern, one of the contraries–*M*–belongs to *X*, fire, and all of the simple bodies below fire. *Y* is different from *X* because it has *L*. The introduction of the first two bodies and their contrarieties is represented in Table 6.1.

Table 6.1 *Adding two bodies and contrarieties*

Bodies (top-bottom)	*Contraries*[a]			
Y	*L*	*J*	Hot	Dry
X	*M*	*J*	Hot	Dry
Fire	*M*	*K*	Hot	Dry
Air	*M*	*K*	Hot	Wet
Water	*M*	*K*	Cold	Wet
Earth	*M*	*K*	Cold	Dry

[a]I used Aristotle's primary contraries, as assigned to the simple bodies in *GC* II 3–4, because they are more familiar. The same picture would result if one were to select different contraries.

Commentators have pointed out that Aristotle's proposed introduction of new contrarieties and bodies faces an immediate difficulty: since two contrarieties produce four bodies, a third contrariety would produce eight bodies, and a fourth would bring the total to 16. In general, the total number of bodies should be 2^n, where n is the number of contrarieties.

Aristotle must realize that adding a single contrariety implies the addition of more than one body; otherwise, the picture above would demonstrate that one of the bodies *can* be the origin of the others. In the simple case in which only *X* and the contrariety *J–K* are added to the original four, *X* and earth are opposites due to multiple contrarieties. If bodies cannot directly change into one another if they are opposite in virtue of multiple contrarieties, *X* and earth cannot, on the proposed picture, directly change into one another. In the absence of additional bodies beyond *X*, it is unclear why Aristotle must entertain the infinite addition of contrarieties.

The new contrariety *J–K* must ultimately result in eight bodies, not five; moreover, there is a perfectly natural arrangement for the new bodies.[25] Within his own arrangement of the simple bodies in *GC* II 4, Aristotle notes that the easiest and fastest manner of transformation is one in which two bodies are different due to a single contrariety; these bodies are also adjacent on a linear arrangement from highest to lowest.[26] There is no reason to suppose that this pattern should be different in *GC* II 5, or that it would change as bodies are added on.

Assuming that adjacent bodies are only opposites along one contrariety, as reflected in Aristotle's own theory and in 332a15–30, the three additional bodies are arranged as follows. *X[4]* is below earth; it is *J*, cold, and dry. *X[2]* is immediately below air; it is *J*, hot, and wet. Finally, *X[3]* is immediately above water and below *X[2]*; it is *J*, hot, and wet. This arrangement motivates the addition of infinite contrarieties; now, *X* and *X[4]* can directly change into one another, just as fire and earth could previously. The arrangement of all eight bodies is represented in Table 6.2.

Table 6.2 *The arrangement of bodies*

X [1]	Hot	Dry	*J*
Fire	Hot	Dry	*K*
Air	Hot	Wet	*K*
X [2]	Hot	Wet	*J*
X [3]	Cold	Wet	*J*
Water	Cold	Wet	*K*
Earth	Cold	Dry	*K*
X [4]	Cold	Dry	*J*

The proposed assignment of contrarieties on this picture is consistent with the standard assumptions that heat travels upward, cold travels downward, and the middle part of the sublunary region is moist. Adding *Y[1]–Y[8]* with the contrariety *L–M* produces a similar result.

[25] Williams 1982: 167 comments that there is "no unique system of ordering," although he admits that the theory to which Aristotle objects *seems* to have one. He wonders why the element that is (in my terminology) *J*, cold, and dry should be below earth rather than below fire. In this case, however, *X[4]* would have two contrarieties with fire; it would be akin to placing water under fire and above air in Aristotle's actual scheme.

[26] The linear arrangement in *DC* IV is the same, although Aristotle does not explicitly invoke hot-cold and moist-dry.

6.4.3.2 *Infinite Contrarieties*

In 333a5–15, Aristotle turns to his final argument against the view that a single body can be the origin of the others. If a single body is the origin of the others, then the contrarieties cannot be finite: any finite arrangement of contrarieties implies that the extremities can change directly into one another. The remaining option is that the contrarieties are added infinitely; a single body is the origin, and the remaining bodies stretch out infinitely in both directions from the origin.

Aristotle's argument reads:

> (1) Similarly, with continual addition a contrariety belongs to the previous bodies, so that if infinite, infinite contrarieties will also belong to one thing. (2) But if this is the case, then there will be neither definition nor generation: for it is necessary, if they come to be from one another, to proceed through so many contrarieties, and still more, so that there will never be change into certain things, for example, if the intermediates were infinite (but this is necessary, if indeed the elements are infinite); furthermore, fire will not come to be from air, if the contrarieties are infinite. (3) And all of them will be one: for all of them above *F* must belong to those below, and these to those above, so that all of them will be one. (*GC* II 5, 333a5–15)

This passage concludes that the infinite addition of contrarieties results in a single thing and the impossibility of change. Accordingly, one body is not the origin of the others; one body is everything, that is, all the simple bodies.

In (1), Aristotle confirms the result of adding contrarieties. The sketch in 332b32–333a5 demonstrates that when a new contrariety is added, it belongs to all previous bodies. The addition of *X* means that air, water, and earth now have three contrarieties, not two. If there are four contrarieties, then any selected body has four contrarieties; if there are infinite contrarieties, then any selected body has infinite contrarieties.[27]

In (2), Aristotle shows that if any selected body has infinite contrarieties, then definition and generation do not occur. Aristotle claims that generation will not occur because any change will require proceeding through "so many contrarieties, and still more." It is not initially clear why this must be true. When air changes into fire – on Aristotle's own theory – there is a change from moist to dry, but there is not a change from hot to

[27] I take having a contrary within a contrariety as sufficient for having a contrariety: hot-cold belongs to fire because fire is hot.

cold. Thus, one might suspect that the change only proceeds through one contrariety; there is not a change along the other contrariety, hot-cold.

Aristotle must think that it is necessary to proceed through the additional contrarieties in each instance of generation, even if the contraries do not change. In the change from air to fire, wet and dry are gone through in virtue of the change. Hot and cold are also gone through, but in a different way, because one of the contraries remains the same. The change from air to fire is different from the change from air to earth; in the second case, both contrarieties are involved because two contraries change, whereas in the first case, both contrarieties are involved because one changes and the other remains the same.[28]

Accordingly, each case of generation must proceed through as many contrarieties as there are. If there are three contrarieties, then a change from one body into its immediate neighbor requires that there is change along one contrariety while two remain the same. If there are infinite contrarieties, then it is no longer possible to go through all the contrarieties: whether a contrary changes or stays the same, there is always another contrariety beyond it.[29]

For Aristotle, this result has two consequences. First, Aristotle claims that there will never be change into certain things because the intermediates are infinite if the elements are infinite. As demonstrated above, the addition of each contrariety introduces additional bodies in the middle, between air and water. So, if the elements or bodies are infinite, there are indefinitely many bodies that exist between air and water. This result rules out the possibility of change into certain things; for air to become water, it must pass through the glut of intermediate bodies. Second, Aristotle claims that even fire will not come to be from air. Although air and fire remain immediate neighbors, a change from air to fire requires going through the contrariety that changes, wet-dry, as well as going through the infinite contraries that do not change into their opposites.

In (2), Aristotle also claims that the introduction of infinite contrarieties does away with definition. Simple bodies are perhaps defined by listing their qualities: air is hot and wet, and fire is hot and dry. If there are

[28] Williams, in contrast, suggests that it is necessary to proceed through infinite contrarieties because in each case of change, one thing (e.g., something hot and wet) ceases to exist and another thing (e.g., something hot and dry) comes into existence (Williams 1982: 167–169). This account implies that all contraries somehow go out of existence in the change from one body to another – a view frequently found in defenders of traditional prime matter.

[29] Aristotle highlights the problems for going through the infinite at *Phys.* III 5, 204b9; *Phys.* VIII 8, 263a6; *Meta.* II 2, 994a31.

infinite contrarieties, it becomes impossible to complete the task; for each contrariety, there is yet another, and the body in question possesses one of the two contraries. Accordingly, an attempt to define any body would require infinite predicates, which Aristotle believes does away with definition.[30]

Aristotle concludes in (3) that all of them will become one. In support of the conclusion, he claims that all those that belong to the bodies above fire will belong to the bodies below fire (and *vice versa*). There are two ways of understanding what Aristotle thinks belongs to the bodies above and below fire; neither one immediately makes sense of Aristotle's claim that everything will become one.

If one understands "all of them" to be the contrarieties, then Aristotle appears to be confusing contrarieties and contraries. Even if any selected body has infinite *contrarieties*, it does not follow that the bodies above and below fire have identical *contraries*. As Joachim points out, earth, air, fire, and water share the contrarieties hot-cold and wet-dry, yet fail to be identical. If instead one understands "all of them" to be the contraries, then the conclusion that all are one follows. If indeed all the contraries that belong to bodies above fire also belong to those below, then the bodies above and below fire are identical. Nonetheless, it remains unclear why Aristotle would think that the bodies above and below fire have the same contraries.

In my view, it is better to understand "all of them" at 333a13 to be the contrarieties, rather than the contraries. This makes the most sense in the context of *GC* II 5, as Aristotle has been discussing contrarieties, not particular contraries. Furthermore, it is true given Aristotle's previous arguments; any body one selects will have infinite contrarieties, so all the bodies above and below fire have the same contrarieties. Perhaps Aristotle is confusing contrarieties and contraries; the problem, however, is not as drastic as Joachim takes it to be.

Joachim is right that the possession of two contrarieties by four bodies does not make the four bodies identical. There is, however, something quite different that happens in the case of infinite contrarieties. If there are infinite contrarieties, then at least one body has infinite contraries, and since it has infinite contraries, everything is identical with it. Aristotle does

[30] See *APo* I 22, 83b5–8. Here Aristotle shows that since it is impossible to proceed through infinite things in thought, there cannot be infinite genera and species proceeding upward or infinite substances proceeding downward; the result would be infinite predicates, which does away with definition.

not need to infer from the fact that the bodies above and below fire have the same contrarieties that they are identical. Rather, he only needs to infer that fire is the nexus, and that contrarieties are not used up in the process of adding a new body.[31]

Here is the thought. As before, we can continue adding bodies both above and below by adding contrarieties. Each body has one difference with each of its neighbors, but otherwise shares the same contraries. For simplicity, consider each pair of contraries as a positive and negative number: 1 and -1 is a contrariety. Beginning with a single contrariety, assume fire and air: fire is [1], and air is [-1]. A second contrariety adds bodies both above and below; *X1* is [1, -2], *X2* is [-1, -2], whereas fire is [1, 2] and air is [-1, 2]. A third contrariety introduces *Y1* above and *Y2* below; *Y1* is [1, -2, -3]; *Y2* is [-1, -2, and -3]; the remaining bodies, *X1*, *X2*, air, and fire all gain the contrary 3. This picture is represented in Table 6.3.

Table 6.3 *Contrarieties represented as numbers*

Y1	1 −2 −3
X1	1 −2 3
Fire	1 2 3
Air	−1 2 3
X2	−1 −2 3
Y2	−1 −2 −3

As contrarieties are added, each individual body gains additional contraries; so, in the case of infinite contrarieties, fire (and indeed every other body) also possesses infinite *contraries.* The picture above attempts to display this result numerically. For Aristotle, this gets the desired result: at least one body that is infinite with respect to its qualities.

If there is a body with infinite qualities, then the remaining bodies must be identical with it. If indeed its qualities are infinite, then there is no distinction between that body and any other body. One might suppose – given the picture above – that everything becomes fire. In fact, the picture precludes any single body from being the single material or the origin of the rest: with infinite qualities, it is no longer possible to differentiate the body as any material.

[31] Philoponus entertains a position of this sort: since the body above *F* does not change back into *F*, the contrariety has been destroyed; hence, all the bodies eventually become the same (Philoponus, *In GC* 257.9–21).

One may still object that Aristotle is confusing contrarieties and contraries; this remaining objection is closer to Williams' contention that Aristotle failed to recognize different sizes of infinite sets. On the model above, using numbers, fire would be like the set of natural numbers; all the contraries (both positive and privation) would be like the set of positive and negative integers. Although the set of positive and negative integers (i.e., all contraries) is different from the set of natural numbers (i.e., the contraries of fire), the cardinality is the same; the fact that the two sets can be put into one-to-one correspondence shows that Aristotle's reasoning is not entirely suspect, even by contemporary mathematical standards.

Aristotle, of course, is not interested in set theory, and does not consider the contraries and contrarieties to be numbers. Rather, he is attempting to apply a notion of the infinite – which is a quantitative concept – to the problem of adding qualities. On the theory of infinity that Aristotle rejects, the infinite contains everything; this would presumably include the privative qualities as well as the positive ones. For Aristotle, the infinite always has something outside of it, and the idea of completing an infinite series of qualities is absurd. Since the infinite series is absurd and every finite series permits cyclical change, there is no single origin for the other bodies.

6.5 Aristotle's Positive Account in *GC* II 5

In *GC* II 5, Aristotle has shown that a single body cannot be either the matter or the origin of other bodies. Although the arguments of the chapter are largely negative, they provide insight into Aristotle's own theory of the material elements. Specifically, they confirm Aristotle's theory that there is no single origin of the material elements because they can change into one another. In addition, they provide parameters for unpacking Aristotle's own theory of fundamental matter.

In the second half of the chapter (332b5–333a15), Aristotle shows that a single body cannot be an origin or principle of the other bodies. A body at the extremities cannot be the origin of the others because it would be matter for the other bodies, which implies that other bodies cannot be differentiated from it. If a body in the middle is an origin of the others, then the contrarieties cannot be infinite; infinite contrarieties imply that all the bodies are one. If the contrarieties are finite, then no single body is an origin of the others; rather, simple bodies are origins of one another. The only way to make sense of multiple bodies is to maintain that they are origins of one another, as Aristotle showed in *GC* II 4.

Aristotle's two arguments against a single matter for other bodies at 332a6–20 and 332a20–26 imply that if some matter remains within generated bodies, such matter cannot possess a single quality, all qualities, or no qualities. First, the argument at 332a6–20 shows that if matter remains in a generated body, it cannot itself be a body that possesses some quality. If air is cold and produces fire, which is hot, then air cannot be the matter that remains in fire; rather, there must be "something else that is the same, and some other common matter" (332a16–18).

The argument at 332a20–26 shows that if matter remains in a generated body, it cannot be an entity that possesses all qualities or lacks all qualities. If the matter possesses no qualities and is isolated from the contrarieties, then it is a vacuous non-being. If the matter possesses the opposite of every *contrary* and thus contains all contraries, then it is everything, and is an inconsistent object.

The first argument (332a6–20) implies that a correct theory of matter must not take one of the simple bodies to be the matter of all the others. The second argument is less decisive (332a20–26). The argument might imply that no single *body* beyond the four can be the matter of the others; in this case, it remains possible that an incorporeal entity could remain as matter. Given that Aristotle does not explicitly call the intermediate a body in 332a20–26, one might alternatively take the argument to imply that there can be no *matter* beyond the four simple bodies. If this is the case, then the simple bodies are also the simplest kind of *matter*.

The correct interpretation depends, in part, on how one reads a passage just a few lines later (332a34–b5). Following his discussion of the intermediate, Aristotle reminds the reader that the simple bodies can change into one another. Then, he introduces part of his own theory; he writes:

> If, then, the contrariety according to which they change is one, they must be two: *for the matter is the intermediate which is imperceptible and inseparable.* Since they are observed to be more [than two], there must be at least two [contrarieties according to which they change]. But the contrarieties being two, they cannot be three, but four, as they are observed: for the pairings are this many: for although there are six combinations, two cannot come to be because they are contrary to one another. (*GC* II 5, 332a34–b5)[32]

The immediate issue in this passage is the intermediate; according to this passage, it is matter, and is imperceptible and inseparable. It is unclear

[32] This passage is extremely compressed; in my translation, I have filled in what I take to be reasonable meanings. In each case, I agree with Gill's assessment (Gill 1989: 248–249).

whether the intermediate, as introduced in this passage, is an entity with a metaphysical status distinct from that of the simple bodies.

Traditional interpretations take this passage to imply Aristotle's commitment to prime matter.[33] On traditional interpretations, 332a16–18 introduces a matter common to two simple bodies, something Aristotle accepts. Furthermore, 332a20–26 show that an intermediate is problematic if it is a *body*; 332a34–b5, in turn, offers a corrective to the failed account of the intermediate.[34] On this interpretation, it is true that the intermediate is both everything and nothing: it is everything *potentially* but nothing *actually*. Since it is potentially everything but nothing in itself, this intermediate is inseparable from the simple bodies and remains imperceptible. There is a common matter for bodies, but it is not a body itself.

Interpretations that reject Aristotle's commitment to prime matter offer an alternative; on such interpretations, the purpose of 332a20–26 is to show that there cannot be *anything* material beyond the four simple bodies. Thus understood, 332a34–b5 introduces the real primary materials, which are the four simple bodies. The intermediate introduced here may be a reference back to the intermediate previously rejected in 332a20–26; it is problematic precisely because it is imperceptible and unable to be isolated.[35] It may also be a more general reference to an incorrect theory that supposes a third entity distinct from the two simple bodies it supposedly underlies.[36]

Both interpretations fail to consider that the primary issue throughout *GC* II 5 is the status of contrarieties in the generation of simple bodies. At 332a6–20, Aristotle shows that a single contrariety cannot produce one body from another; there must be something common to the contraries, which ultimately belong to two bodies. Lines 332a20–26 highlight the

[33] I treat as "traditional interpretations" all those that require prime matter as a persisting substratum and a constituent of simple bodies. Williams 1982: 214 and Dancy 1978: 389–390 find a commitment to prime matter in *GC* II 5, 332a17–20 and 332a35–b1.

[34] Williams 1982: 214.

[35] Gill 1989: 248–249 argues that the intermediate remains that of Anaximander. Aristotle's reference to matter as an "intermediate" is indeed puzzling. Aristotle admits elsewhere that "intermediate" admits a variety of meanings. Aristotle variously uses "intermediate" in referring to: (i) the center of the universe, which is one of the extremities; (ii) the intermediate region between air and fire, where air and water are located; and (iii) an intermediate between two opposed contraries, such as the gray that falls between white and black. The intermediate in this context does not seem to be any of these.

[36] See Charlton 1970: 136. Charlton accepts an account of elemental change according to which generation is distinct from alteration precisely because there is no persisting substratum. Charlton 1983: 202, 207–208 offers a further defense of this reading of *GC* II 5. This account is also defended in Broadie 2004: 123–150.

problem for making matter entirely different from all of the contraries or something that contains all of the contraries. Hence, matter – whatever it is – must be different from the pair of opposites that are changing, cannot be entirely isolated from all contraries, and cannot contain all of the contraries.

This leaves an option intermediate between prime matter and the simple bodies: one of the contraries is the matter in each case of elemental generation. This interpretation accepts that Aristotle needs there to be a matter common to air and fire, as he states at 332a16–18; it is neutral on whether Aristotle wishes to reject a single material or just a single body in 332a20–26. There is no body beyond the four mentioned; in each case of change, there is a different sort of matter involved.[37]

On Aristotle's own theory, fire and air do have a third thing in common. In a change from fire to air, there is a change in the contrariety dry-wet; heat is common to both bodies. Heat is not itself a body; it is a characteristic of bodies but does not exist in separation from bodies. It is also not, on its own, perceptible.[38] Given that heat never exists in separation from a body, and bodies require at least two contraries, one never perceives heat as such; rather, heat is always perceived in conjunction with dryness, moisture, or some combination thereof. Accordingly, heat – not prime matter – is the matter in common for fire and air; in each case of a simple body and its immediate neighbor, a similarly shared contrary can be found.[39]

This interpretation can explain why Aristotle goes to the trouble of introducing two contrarieties in 332a32–b5. The intermediate that is inseparable and imperceptible is Aristotle's own, but it requires a further qualification; the further qualification is possible because there are observed to be more than two simple bodies, and hence, there are two contrarieties. The two contrarieties, in turn, provide the simplest coherent

[37] A view of this kind is introduced in Lewis 1996: 15–26. Although Gill and Furth accept that one of the primary contraries can be the substratum for a change from one simple body into the next, both deny that this is "matter." See Furth 1988: 221–227 and Gill 1989: 41–82, 243–252.

[38] Broadie 2004: 142 makes the point that the primary contraries, as constituents of simple bodies, are not perceptible at this level.

[39] I agree with Charlton 1970 that Aristotle's purpose is to introduce four distinct kinds of matter, but disagree that this matter must be identical with the simple bodies. On four kinds of matter, albeit with a reference to common matter as well, see *DC* IV 5, 312a20–23. One might deny that primary contraries can function as matter, given Aristotle's comment at *GC* I 4, 319b21–24 that generation does not occur if one contrary gains an opposite as an affect and remains the same in the new body. Nothing in *GC* II 5 demonstrates that contraries as matter violate the conditions of *GC* I 4, as the text does not consider how contraries relate to one another or to simple bodies.

explanation for why multiple simple bodies come to be.[40] Neither a single contrariety nor infinite contrarieties offer a coherent explanation; thus, the contrarieties must be finite. Aristotle shows that he can coherently explain the simple bodies appealing to just two contrarieties. The argument of *GC* II 5 thus confirms not only Aristotle's commitment to four simple bodies that are origins of one another, but also his commitment to two primary contrarieties.

40 This thought is echoed in Solmsen 1960: 348, who contrasts Aristotle's contrarieties with those of Anaxagoras.

7

On Generation and Corruption *II 6*

Christopher Frey

7.1 Introduction

GC II 6 is an extended critique of Empedocles' natural philosophy. This is not the only time we witness such an attack. Though Aristotle is more complementary to Empedocles than he is to most of his predecessors, Empedocles remains a frequent target of his ire.[1] Indeed, he argues that Empedocles' account is fundamentally flawed earlier in *GC* itself. In this prior discussion, he concludes that Empedocles' view makes alteration (ἀλλοίωσις) impossible.[2] If sound, this would, on its own, be sufficient grounds to reject Empedocles' view. But there is still value in the additional arguments Aristotle levies against him in *GC* II 6. There are two reasons why this is so (beyond the intrinsic interest of the arguments themselves).

First, Aristotle's criticism in *GC* II 6 belongs to a larger project that he initiates at the beginning of *GC* II. He describes this project as an inquiry into the principles and elements of bodies.[3] More specifically, he is "searching for the principles of perceptible body, and this body is tangible"

* Unless otherwise noted, translations of Aristotle outside of *GC* II are modifications of the revised Oxford translation. Translations of Empedocles are from Laks and Most 2016 and Philoponus from Kupreeva 2005.

[1] Notable discussions of Empedocles' views outside of *GC* include *Phys.* VIII 1, 252a7–b6; *DC* II 13, 295a14–b7; *DA* I 4, 408a19–29; *Sens.* 1, 437b11–438a4; *Resp.* 7; *PA* I 1, 640a19–33; *Meta.* I 4, 985a21–b5; *Meta.* I 8, 989a18–30; and *Meta.* III 4, 1000a24–b20.

[2] *GC* I 1, 314b15–26; cf. *GC* II 4, 331a7–10, and *Meta.* I 8, 989a26–30. According to Aristotle, Empedocles accepts two claims: (i) the affections (πάθη) involved in alteration are differentiae of the elements and (ii) if a composite body with a given affection alters so as to have a different affection, then an element the composite body comprises undergoes the same alteration. It follows from these two claims that the elements can come to be from each other, which, as we will discuss in more detail, violates one of Empedocles' principal commitments.

[3] In this context, he uses the expressions "principle" and "element" interchangeably, for "it is an excellent thing to say that the primary are principles and elements – the things from which, by their changing either by combination and separation or by some other kind of change, generation and perishing result" (*GC* II 1, 329a5–8).

(*GC* II 2, 329b6–7). For when we consider "all the naturally constituted substances, their generation and perishing are not without the perceptible bodies" (*GC* II 1, 328b32–33). This focus allows Aristotle to critique the heart of Empedocles' pluralistic natural philosophy since Empedocles' views about the elemental status of tangible bodies form the foundation upon which the rest of his account rests. Indeed, *GC* II 6 is the longest and most comprehensive discussion of Empedocles' views in the whole Aristotelian corpus.

Second, though Aristotle never wavers from criticizing Empedocles in this chapter, his arguments are not entirely negative. We can also view them as collectively constituting a sustained defense of the superiority of his preferred account. Aristotle's natural philosophy centers on his novel understanding of nature (φύσις) as an internal principle and cause of movement and rest (*Phys.* II 1, 192b21–33). The phenomena that Empedocles is unable either to explain adequately or to endorse consistently are precisely the phenomena that Aristotle's conception of nature is well-suited to explain.

Aristotle's criticism is structured around four theses that he attributes to Empedocles:

(EMP1) There are four kinds of elemental body – earth, water, air, and fire – and instances of these kinds do not change into one another.
(EMP2) Everything that comes to be by nature does so by chance.
(EMP3) Love and Strife are the principles of natural bodily movement.
(EMP4) Soul is, or is constituted by, one or more of the four elemental bodies.

I will follow Aristotle's lead and discuss his criticism of these four theses in turn.

7.2 Comparability and Change

Aristotle begins his attack on the first major thesis of Empedocles' natural philosophy by describing the reaction one ought to have toward its advocates. He says:

> One may well marvel at those who claim that the elements of bodies are more than one so that they do not change into one another, as Empedocles does, and ask how it is admissible for them to assert that the elements are comparable. Yet Empedocles says, "*For these are all equal.*" (*GC* II 6, 333a16–20)

This sense of puzzlement is warranted, he argues, because anyone who advances a broadly Empedoclean account commits themselves to an ultimately inconsistent triad of claims.

(E1) There is more than one kind of elemental body.
(E2) Instances of distinct kinds of elemental body do not change into one another.
(E3) Instances of distinct kinds of elemental body are comparable.

Empedocles endorses all three claims explicitly. That there is more than one kind of elemental body is *the* central tenet of his pluralist natural philosophy. He countenances four basic kinds of body which he calls "roots" (ῥιζώματα): "Fire, water, earth, and the immense height of air."[4]

Equally central to his account is the claim that these roots do not change into one another. What we readily classify as cases of coming to be and ceasing to be in nature are, for Empedocles, nothing but the mixing and the separation of these elements. But the elements themselves are eternal. That is, there is no process, including mixing, that can change the elements' essential, intrinsic features. So "it is impossible for water to come into being from fire or earth from water" (*GC* I 1, 314b25–26) and, in general, "none of the elements comes to be from any other" (*GC* I 1, 315a4–5).[5]

Empedocles asserts that these four unchanging elements are comparable in the passage Aristotle quotes in our chapter's opening lines. The complete passage is as follows:

> For these [*sc.* the four roots] are all equal and identical in age,
> But each one presides over a different honor, each one has its own character,
> And by turns they dominate while the time revolves.[6]

Comparisons of this sort are plausibly benign given how commonly they occur in both theoretical and ordinary contexts. But Aristotle argues that one cannot consistently conjoin the possibility of such comparisons with the previous pair of Empedoclean claims. If he is correct, Empedocles and

[4] DK B17.18: πῦρ καὶ ὕδωρ καὶ γαῖα καὶ ἠέρος ἄπλετον ὕψος. A complete list of elements/principles would include Love and Strife. But neither Love nor Strife is a kind of tangible body. So, "Empedocles makes the bodily elements four, but in all, with the sources of movement, his elements are six in number" (*GC* I 1, 314a16–17). Cf. DK A28, A30, A33, B6, and B71.

[5] At *GC* I 1, 314b7–8, Aristotle quotes DK B8.1 and B8.3 to support this attribution:
Something else I will tell you: of nothing is there birth, among all
Mortal things, nor is there an ending coming from baleful death,
But only mixture and separation of things mixed
Exist, and "birth" is a name given by mortal humans.
(DK B8.1–4; cf. B12, B17.34–35, and B21.13–14)

[6] DK B17.27–29:
ταῦτα γὰρ ἶσά τε πάντα καὶ ἥλικα γένναν ἔασι,
τιμῆς δ' ἄλλης ἄλλο μέδει, πάρα δ' ἦθος ἑκάστωι,
ἐν δὲ μέρει κρατέουσι περιπλομένοιο χρόνοιο.

his followers fail to recognize the consequences of their commitments and endorse each member of an inconsistent set of claims.

But what does it mean to say that the elements are comparable and why is comparability inconsistent with the conjunction of (E1) and (E2)? There are two plausible readings: (i) the elements are comparable *in quantity*; and (ii) the elements are comparable *in quality*. Aristotle argues that neither option is acceptable.

7.2.1 *Quantitative Comparisons*

On the most natural reading, Empedocles is comparing the quantity of the four elements. Aristotle considers this interpretation first and offers the following argument that it is incompatible with Empedocles' other commitments.

> If what is meant is that they are comparable in quantity, all the things which are comparable must possess some identical something by which they are measured; for example, if out of one cup of water come ten cups of air, both would have to be an identical something, since they are measured by the same thing. (*GC* II 6, 333a20–23)

According to Aristotle, quantitative comparisons entail that there is "some identical something" which both relata are (i.e., some one thing which belongs to or is predicated of them both). Whatever this common something is, it will serve the role of a material substratum. That is, the first body's material substratum will possess, in addition to those properties that make it the sort of material it is, whatever properties make it an instance of the first body's kind. And it is a substratum of the very same sort of material that possesses whatever additional properties make something an instance of the second body's kind. This guarantees that each of the bodies can change into the other. For each body's matter will be exactly the sort of thing that can, in principle, possess the other body's kind determining properties. If this is correct, one cannot quantitatively compare earth, water, air, and fire *and also* deny that they can change into one another.

This is a peculiar argument. It is far from obvious that quantitative comparisons entail the existence of a common matter. Presumably, there is no issue with comparing different quantities of a single element. Empedocles can say that the quantity of water in his bathtub is greater than the quantity of water in his drinking glass and even employ a single unit of volumetric measure to express this comparison more precisely without any problematic consequences.[7] There is a common matter in such cases, which is, water. But it is difficult to see how this material commonality could be a consequence of each being "measured by the same thing" in this sense. So why would the existence of a common material substratum follow from quantitative comparisons across bodily kinds? If I am baking a cake and the recipe calls for two teaspoons of salt and ¼ teaspoon of baking powder, does this recipe entail that these ingredients share a common material substratum? Does it entail that salt can turn into baking powder?[8]

Philoponus' interpretation of Aristotle's argument is helpful on this score.[9] When Empedocles says that the elements "are all equal," he is most likely referring to the total amounts of the four elements in the universe. But if this is correct, he cannot, says Philoponus, be comparing their volumes. First, the total volume of water is less than that of earth because we can observe that all of the universe's water is located in the earth's hollows. Second, given natural constraints on the positions from which we can make observations, we could never possess sufficient evidence to determine whether the total volumes of the universe's air and fire are equal. So Philoponus concludes that any claims of cosmic volumetric equality are either empirically false or empirically unwarranted.

Aristotle's concern, however, is not the truth of Empedocles' comparisons; his focus is what follows from their very intelligibility. And the key to understanding Aristotle's larger argument, Philoponus contends, lies in the following claim.

> [Ratio] For any two elements x and y, (i) there is a determinate ratio, r_1, that obtains between the total volume of x in the universe and the total volume of y in the universe, (ii) there is a determinate ratio, r_2, that obtains between the volume of any given portion of x and the volume of y into which that portion of x can be transformed, and (iii) $r_1=r_2$.

[7] I break from the majority and translate κοτύλη (the basic unit of dry measure) as cup rather than liter or pint. The estimated mean quantity of its various geographical and temporal uses is closer to the former (somewhere between one cup (U.S.) and one cup (Imperial)). Moreover, like cup, κοτύλη does not always refer to a unit of measurement; it often picks out a small vessel, drinking cup, or ladle and can even refer to the hollow of the hand. Ultimately, either choice will do since Aristotle's arguments require only that we employ the same unit of volumetric measure.

[8] Williams expresses a similar skepticism: "Nineteenth-century chemists could compare the volume of a given quantity of hydrogen with that of a given quantity of oxygen without supposing that one could turn into the other" (Williams 1982: 169).

[9] Philoponus, *In GC* 258.15–259.9.

This consilience of ratios is a position that Aristotle seems to accept and employ in *Meteor.* I 3. *Meteor.* I 3 is about the intervals between the heavenly bodies. Aristotle wants to know which of the simple bodies fills these intervals. Before he defends his answer (viz., *aithêr*), Aristotle argues that it cannot be any of the four terrestrial, simple bodies and focuses on the two most likely candidates, air and fire. Against air, Aristotle says that if air occupied the heavenly intervals, then "the air would far exceed the quantity required to maintain its proper proportion to the other elements" (*Meteor.* I 3, 340a4–5). For, when a given portion of water changes into air, we observe that the volume of the air that comes to be is greater than the water's volume by a constant ratio. And Aristotle insists that "the proportion between any given quantity however small of water and the air that is generated from it ought to hold good between the total amount of air and the total amount of water" (340a10–11). Let us say that the volumetric ratio that governs transformations of water into air is 1:10 (or even greater still). Whatever this ratio ends up being, it will be far less disproportionate than the ratio between the volume of the total amount of water in the universe and the volume of the heavenly intervals. For "the bulk of the earth (which includes the whole volume of water) is infinitesimal in comparison with the whole world that surrounds it."[10]

Why would the acceptance of [Ratio] entail that quantitatively compared bodies share a material substratum? Philoponus' explanation leads us to Aristotle's principal charge against Empedocles.

> If he [*sc.* Empedocles] says that they are comparable *in amount*, he would mean by this that they would change into each other if equal matter were underlying each of them; for example, if we were to say that the whole of water is equal to the whole of air because in each there is matter equal [in

[10] *Meteor.* I 9, 340a6–9. Cf. "the earth's mass is spherical in shape, and also, as compared with the stars, it is not of great size" (*DC* II 14, 298a18–20). My presentation of Aristotle's argument captures the most common interpretation. But Aristotle's considered position cannot be that a fixed ratio always obtains between the total volumes of the elements in the universe. If this were the case, then every elemental transformation would require a simultaneous and proportionate transformation in the opposite direction somewhere else in the universe. Aristotle gives no argument for any such claim. Moreover, it would involve a kind of action at a distance that is difficult to support on Aristotelian grounds. Perhaps he means that the ratio that governs local transformations also applies to global transformations. That is, if one cup of water yields ten cups of air, then, if all of the water in the universe were to turn into air, the total volume of air would be ten times that of the volume of water, were all of the air to be turned into water. This claim is not only something Aristotle would accept – why would the ratio that governs elemental transformations vary when different volumes are under consideration? – but also explains why air could not occupy the heavenly intervals. That is, there is no way enough air could ever come to be to fill the intervals without violating the ratios that govern the transformations of the other elements into air.

> amount] and the same [in kind], which can, if extended, produce air, and if contracted again, [produce] water. (Philoponus *In GC* 258.23–28)

If, for example, one cup of water always turns into ten cups of air, this is because there is a single kind of material substratum that is at one time water and at a later time air. A certain amount of this material substratum, when it is in a relatively dense condition, is one cup of water. The same amount of this material substratum, when it is appropriately rarefied, is ten cups or air. It is by the amount of this underlying matter, not by volume, that we measure the elements. We can say that one cup of water is equal in quantity to ten cups of air because the amount of matter that underlies each is identical. If Empedocles adopts this account, he would be able to say legitimately, though perhaps falsely, that the respective total amounts of the four elements are equal in quantity even though their respective volumes differ greatly. But what he would mean, then, is that the ratio that governs transformations between smaller volumes of the elements applies to the total volumes of the elements. And this, of course, would commit him to the view that the elements can change into each other.

Philoponus' analysis provides a plausible interpretation of Aristotle's position. But it fails to capture anything Empedocles would plausibly take himself to mean when he says that earth, water, air, and fire "are all equal." It fails for at least two reasons. First, rarefaction and condensation are not the mechanisms Empedocles employs in his analysis of the coming to be and ceasing to be of composites. It is Anaximenes, not Empedocles, who appeals to rarefaction and condensation. Empedocles speaks only of mixing and separation and no argument has yet been given that would force Empedocles to adopt these additional kinds of change. Second, though Empedocles could adopt this account for composite bodies – the four elements would serve the role of a common material substratum – he does not seem to be forced to apply it to the elemental bodies themselves. Neither Aristotle nor Philoponus make it obvious why he *must* do so. Without such an argument, Aristotle's criticism simply begs the question.

Can a plausible argument be made on Aristotle's behalf? What follows is my best attempt to do so. First, consider a broadly Anaximenean account. On such a view, the only difference between solid water, liquid water, and gaseous water is the density of the elemental body that constitutes each. On Anaximenes' view, this would be elemental air. But if we were to extend the general framework to a pluralist account, the most likely candidate would be elemental water. Solid water, liquid water, and gaseous water would be phases of elemental water. These phases are three ways *to be* elemental water. Moreover, there is no principled reason to privilege any of these phases over the others. That is, despite the elevated status liquid water has in fixing our conception of what elemental water is and is like, this status is unwarranted.[11] There is no level of density that is elemental water's natural condition. For how would we determine how much water is in a given receptacle? Measuring its volume will not do. For the volume changes whenever elemental water is condensed or rarefied. At best, we can say that a given volume of liquid water and a given volume of gaseous water comprise the same amount of elemental water. To say that one volume of liquid water and another volume of gaseous water comprise the same amount of elemental water is simply to say that all and only the elemental water present in one of the phases is preserved when it changes into the other phase. But there is no unit of volumetric measure that can apply to the preserved elemental water as such. There is no nonrelational way to measure the matter that undergoes rarefaction and condensation and, consequently, no nonrelational way to measure how much elemental water is present in a given receptacle.

Empedocles, despite his presumed protestations, must accept a parallel account. Let us assume that the processes of mixing and separation that constitute the coming to be and ceasing to be of composite bodies are closed systems.[12] That is, Love and Strife are principles of mixing and separation that are external to the bodies that are mixed and separated, but neither Love nor Strife add or subtract matter from the bodies they act upon. The ratio of ingredients present before being mixed will be the same within the mixture, and the amount of each of the ingredients will be the same before and after mixing as well. But the sum of the volumes of each of the ingredients need not be the same as the volume of the mixture. This can be established observationally, and it would be uncharitable to attribute its denial to Empedocles.

Empedocles need not appeal to rarefaction or condensation to explain this difference in volume. Expressions that typically refer to density and rarity occur throughout the extant fragments and testimonies (e.g.,

[11] This holds for us as well. Though we typically refer to water's liquid phase as water *simpliciter*, its solid and gaseous phases are equally H_2O. So while there may be a sense in which "water is wet" is true, there is also a sense in which it is false, since "H_2O is wet" is not always true.

[12] In thermodynamics, a closed system refers to any system in which energy can be exchanged with what is outside the system, but matter cannot. An isolated system refers to any system in which neither energy nor matter can be exchanged with what is outside the system. I am assuming that the activity of Love and Strife is external to the bodies that are mixed or separated, and that this activity involves something (loosely) analogous to energy exchange, making the former condition more appropriate.

πυκνός, μανός, πυκινός, ἀραιός, σφίγγω, πύκνωμα, and λεπτύνω) and are applied to the elements, various mixtures, and even the Sphere.[13] And Aristotle argues elsewhere that Empedocles must countenance rarefaction and condensation since his rejection of such phenomena renders him incapable of explaining these differences in volume.[14] But alternative interpretations of Empedocles' expressions are available that would allow him to avoid their typical meanings and it is at least possible that his analysis of mixing, once all of the details are spelled out, would suffice to explain changes in volume without an appeal to rarefaction or condensation.

Regardless, Empedocles must, at a minimum, accept that "the quantity of the ingredients is the same before and after mixing" is not synonymous with "the volume of the ingredients is the same before and after mixing." If so, Empedocles, like Anaximenes, will be unable to employ a single unit of volumetric measure to express an elemental body's quantity. Moreover, like Anaximenes, none of the conditions in which an element can occur is privileged over others as providing the natural state in which the quantity of the elemental body is manifest.[15]

So a given mixture and a given collection of unmixed bodies can differ in volume but nevertheless be equal in quantity. To say that they are quantitatively equal, then, is simply to say that all and only the elemental bodies present in one of them are preserved when it changes into the other. But there is no unit of volumetric measure that can apply to the preserved elemental bodies as such. There is no nonrelational way to measure the matter that undergoes mixing and separation and, consequently, no nonrelational way to measure how much of an elemental body is present in a given receptacle. Moreover, if Empedocles compares the quantity of the elements to each other, there will be no way to do so intelligibly without

[13] Cf. DK B75; B89; B92; B104; A49; A65; A78; A86.

[14] *DC* III 7, 305b9–19: "Again, when the mixed body is divided, they [sc., the followers of Empedocles] can show no reason why one of the constituents must by itself take up more room than the body did; but when water turns into air, the room occupied is increased. The fact is that the finer body takes up more room, as is obvious in any case of transformation. As the liquid is converted into vapor or air the vessel which contains it is often burst because it does not contain room enough. Now, if there is no void at all, and if, as those who take this view say, there is no expansion of bodies, the impossibility of this is manifest; and if there is void and expansion, there is no accounting for the fact that the body which results from division occupies of necessity a greater space."

[15] The condition in which all of the elements are uniformly mixed into the Sphere, the condition in which all of the elements are completely separated into nonoverlapping concentric rings, and the various conditions in which the elements are mixed and separated during intermediate cosmogenic periods, are all on *a par*. Cf. *GC* I 1, 315a19–25.

applying the same framework. If one cup of water is equal in quantity to ten cups of air, this means that there is something material that is present in both the water and the air which would be entirely and exclusively preserved whenever water is transformed into air in a 1:10 volumetric ratio (or whenever air is transformed into water in a 10:1 volumetric ratio). The matter preserved in such cases cannot be the elemental bodies themselves; it must be something that is not itself water or air but underlies each and can be, and therefore can come to be, either.[16]

7.2.2 Qualitative Comparisons

If Empedocles cannot consistently compare the elements quantitatively, perhaps he can still compare them qualitatively. Aristotle explores this possibility immediately after he dispenses with quantitative comparisons.

> If, however, they are not comparable in quantity (in the sense that from so much of one comes so much of the other), but in power (for instance, a cup of water and ten cups of air have the same cooling power), even then they are comparable in quantity, not *qua* quantity but *qua* having such a power. It would be possible to compare their power, not by a quantitative measure, but by an analogy: for example, "as *x* is hot so *y* is white." But "as" in quality signifies likeness and in quantity equality. Indeed, it is manifestly absurd that the bodies, though incapable of changing into one another, are comparable, not by analogy, but by a measure of their powers; namely, so much fire and many times that much air being equally or similarly hot. For the same thing, which is more, will, because it is of the same kind, have such a proportion. (*GC* II 6, 333a23–34)

Aristotle explores two ways to compare the qualities (i.e., powers/capacities [δυνάμεις]) of the elements: (i) comparing them directly by means of some quantitative measure and (ii) comparing them by means of an analogy (ἀνάλογον). I will discuss analogical comparisons first.

Analogies take the canonical form "*A* is to (or is in) *B* as *C* is to (or is in) *D*." For example, wing is to bird as fin is to fish. Analogies are proportional; they capture an "equality of ratios" between the pairs of relata.[17]

Some analogies are merely metaphorical. Though Aristotle praises metaphors that express a proportion as being especially vivid and rhetorically effective, they are unable to determine any causal or explanatory unity

[16] An interesting question that is beyond this essay's scope is this: Does the account of elemental transformation that Aristotle endorses in *GC* satisfy this general constraint and, if so, what is the underlying matter that the elements share?

[17] *NE* V 3, 1131a30–b4; cf. *Meta.* V 6, 1016b35.

among their relata that would be of use to a scientific investigation. And Aristotle chastises Empedocles for too frequently relying on metaphors in his explanations of natural phenomena. He says:

> It is equally absurd to suppose that anything has been explained by calling the sea "the sweat of the earth," like Empedocles. Metaphors are poetical and so that expression of his may satisfy the requirements of a poem, but as to knowledge of nature it is unsatisfactory. (*Meteor.* II 3, 357a24–29)

Despite these inadequacies, metaphorical analogies may still be didactically valuable insofar as they facilitate understanding. But to be properly explanatory, an analogy must express a more significant commonality.

Aristotle does not provide a detailed account of what commonalities suffice for an analogy to be suitably explanatory. The biological works offer numerous examples that involve different kinds of similarity: in relative bodily location, in order of biological development, in material character, etc. But the most important commonality concerns characteristic functions. For example, lungs and gills are analogous because they possess the same capacity, namely, cooling the body (*PA* I 5, 645b1–10).[18] The *Metaphysics* contains other analogies that reveal broader structural commonalities across genera. For example, white, black, and surface are to color as light, darkness, and air are to day and night. The members of each triad occupy the roles of the three fundamental principles of change: form, privation, and matter/substratum (*Meta.* XII 4, 1070b16–21).[19]

So would Empedocles be doing something wrong if he were to say "as *x* is hot so *y* is white"? Presumably not. This analogy is not merely metaphorical; it captures the metaphysical position that hot and white occupy with respect to the subjects that possess them – namely, both fall under the category of quality. And if Empedocles were to compare fire and water by saying "as fire is hot so water is cold" he would not commit himself to there being an underlying common matter and would not have to abandon the view that the elements do not change. For this analogy captures a similarity in the capacities, heating and cooling respectively, conveyed by such qualitative attributions to elemental bodies.

[18] Two excellent recent discussions of analogy in the biological works are Henry 2014: 145–169 and Leunissen 2014: 170–181. I discuss biological analogies as well as analogy in the *Metaphysics* in much more detail in Frey (forthcoming).

[19] Similarly, Aristotle claims that health, disease, the body, and the medical art are to medicine as the form of a house, the kind of disorder appropriate to houses, bricks, and the house building art are to house building. Again, the members of each series differ numerically and in kind but are nevertheless analogous because they occupy the roles of form, privation/disorder, matter, and efficient cause (*Meta.* XII 4, 1070b28–29).

If Empedocles went no further, he would avoid the problems with which Aristotle saddles him. But he does not stop with analogical comparisons. As Aristotle notes, Empedocles does not just want to say that the elemental powers are similar (ὁμοίως); he also wants to say that they are equal (ἴσος). And assessments of equality involve direct quantitative comparisons between the qualities of elemental bodies. If Empedocles says that the cooling power of one cup of water is equal to the cooling power of ten cups of air, he must employ the same quantitative measure to each power.

But quantitative comparisons between elemental powers entail quantitative comparisons between the elements themselves. Aristotle makes this point in the *Meteorology* as well.

> Nor does it make any difference if any one denies that the elements come to be from one another, but asserts that they are equal in power. For on this view it is certain amounts of each that are equal in power, just as would be the case if they actually came to be from one another. (*Meteor.* I 3, 340a13–16)

This is because quantitative comparisons, that is, any comparison in which one thing is said to be more, less, or equal to another thing, require that both relata belong to a single genus. Aristotle makes this point when he contrasts quantitative comparisons and analogical comparisons.

> For those animals that differ by degree and the more or less have been brought together under one kind, while those that are analogous have been kept apart. I mean, for example, that bird differs from bird by the more or by degree (for one has long feathers, another short feathers), while fish differs from bird by analogy (for what is feather in the one is scale in the other. (*PA* I 4, 644a16–22)[20]

According to Aristotle, analogous relata are discontinuous in a way that prevents any definition from applying to both. Analogically common features do not fall along a continuum ordered in terms of more or less and are thereby incommensurable. But if two qualities can be quantitatively compared, they occupy positions on a single continuum of change. Anything that possesses one will, in principle, be capable of possessing the other. And this requires that there be a single kind of underlying matter that is so capable. "For" as Aristotle says, "the same thing, which is more, will, because it is of the same kind, have such a proportion" (333a19–20). That is, if a body possesses a determinate quantity, x, of heat and another body possesses ten times x of heat, this is because the bodies share a

[20] Cf. *PA* IV 12, 692b3–9; *HA* I 1, 486b19–21; *Meta.* I 4, 1055a3–9.

common kind of underlying matter (i.e., matter that can be hot), and the amount of this underlying matter occurs in the same 1:10 ratio.

In sum, Empedocles is free to employ non-metaphorical analogies to compare different elemental bodies. But if he makes quantitative comparisons between either elemental bodies or the powers of elemental bodies, then he must accept that the bodies he compares share a single kind of underlying matter and, consequently, that instances of the different kinds of elemental body he compares can change into each other.

7.3 Nature and Chance in Generation

Aristotle's next criticism is quite brief and concerns growth (αὔξησις): Empedocles cannot countenance this kind of change at all.

> Moreover, growth is impossible according to Empedocles, except by addition. For fire grows by fire, "*earth grows its own bodily frame, and aithêr, aithêr,*" but these are additions. Things which grow do not, it is believed, grow in this way. (*GC* II 6, 333a35–b3; DK B37)

According to Aristotle, growth is not identical to substantial change, but it does require it.[21] For example, a living organism grows by altering and ultimately assimilating nutriment, that is, the nutriment comes to be one or more of the myriad living tissues that are present in the organism. What Empedocles describes involves neither alteration nor coming to be; it is simply the accretion of like elemental bodies. This can lead to an increase in size, but this increase is not growth. But given that Empedocles is already willing to deny that coming to be and ceasing to be, strictly speaking, occur, he would presumably deny that growth occurs without demur. As it stands, Aristotle's criticism marks a difference in their accounts, but he does not here give any reason to prefer one over the other.

The importance of Aristotle's next argument is far greater and strikes at Empedocles' second major thesis: (EMP2) Everything that comes to be by nature does so by chance. Aristotle distinguishes two ways something can come to be that fall within the purview of the study of nature and, in doing so, brings several interrelated questions into relief that any adequate account of coming to be must be able to answer.

> It is much more difficult to give an account of coming to be which occurs by nature. For things which come to be naturally all come to be in a particular way always or for the most part; those that depart from the always

[21] Aristotle provides his account of growth in *GC* I 5.

> or for the most part come to be spontaneously or by chance. What, then, is the cause of the human being always or for the most part coming to be from the human being, or wheat, rather than an olive, from wheat? And won't it be bone if things are put together in a particular way? For nothing comes to be when things come to be as chance has it, according to what he says, but only when they do so by a certain formula. What, then, is the cause of this? (*GC* II 6, 333b3–11)

If something comes to be, it will do so in one of two ways: (i) naturally or by nature (φύσει/κατὰ φύσιν) or (ii) spontaneously or by chance (ἀπὸ ταὐτομάτου καὶ ἀπὸ τύχης).[22] Aristotle appeals to only one feature to mark this distinction. Coming to be by nature proceeds as it does "always or for the most part" (ἢ ἀεὶ ἢ ὡς ἐπὶ τὸ πολύ) whereas coming to be by chance does not.[23]

According to Aristotle, something holds always if and only if it holds by necessity. "For 'of necessity' and 'always' are together," he says, because "what is of necessity is not such as to not be." The result is that "if it is of necessity, it is everlasting, and if everlasting, of necessity" (*GC* II 11, 337b35–338a2). But the features that Aristotle will argue are central to a proper characterization of coming to be by nature hold only for the most part and, consequently, will not hold in all cases (κατὰ παντός).[24]

Such exceptions have two main sources. First, they can arise when external bodies or environmental conditions act as impediments to the realization of those features. Second, they can arise because the "indeterminateness of matter" (τῆς ὕλης ἀοριστίαν) that underpins all embodied objects leads to inevitable variations (within a narrow range) in how, when, and even whether those features are realized.[25] Despite these possibilities, the features that are central to coming to be naturally will typically hold in most cases and, more importantly, their doing so is the default or normal condition insofar as the sources of deviation are neither internal to the accounts of the subjects of change nor internal to the accounts of the changes themselves.[26]

22 Aristotle provides detailed accounts of spontaneity and chance in *Phys.* II 4–6. But most of those details are not germane to the present argument and Aristotle uses the expressions interchangeably. Hereafter I will elide the first disjunct and use "by chance" alone.

23 Aristotle attributes this feature to coming to be by nature on numerous occasions (e.g., *PA* III 2, 663b28–29; *GA* I 19, 727b29–30; *GA* IV 8, 777a19–21). He explicitly notes its absence in coming to be by chance at *Phys.* II 8, 199b24–26; *DC* I 12, 283a33; *EE* VII 14, 1247a32.

24 Cf. *APo* I 4, 73a28–34.

25 Aristotle emphasizes the role that external circumstances play in impeding otherwise invariable processes at *Phys.* II 8, 199b14–26 and *GA* IV 8, 777a3–21. He emphasizes the role that material indeterminacy plays as a source of variation at *GA* IV 10, 778a4–9.

26 Though Aristotle often notes that exceptions to what occurs for the most part are improbable (*Rhet.* I 2, 1357a30–b1) or statistically infrequent (e.g., *Top.* II 6, 112b10–11; *GA* I 19, 727b29–30), there is nothing in his account which demands this.

What features belong for the most part to coming to be by nature? According to Aristotle, it is governed by at least three closely related regularities. Consider any putative case of something coming to be naturally, and let *A* refer to that from which it comes to be (viz., a body of some kind or a collection of different kinds of body). First, for the most part, one determinate kind of thing, and nothing of any other kind, comes to be naturally from *A*. For the most part, wheat, rather than an olive, comes to be from wheat.[27] Second, in many cases, the kind of thing that comes to be naturally from *A* is, for the most part, the same as the kind of thing that *A* is. For the most part, man comes to be from man.[28] And third, the changes involved when something comes to be naturally from *A* proceed, for the most part, in the same predictable and orderly manner and bring about, for the most part, something with the same orderly bodily configuration. For the most part, bone comes to be with a particular set of features, from a particular set of ingredients, by way of a particular change or series of changes.

Coming to be by chance does not realize any of these regularities. So, if Empedocles is going to countenance anything beyond coming to be by chance, which he must if he aims to explain the observable facts, his account must have the resources to explain why these three regularities govern coming to be.[29]

Aristotle argues that Empedocles lacks the resources to explain any of these three regularities and points to what is needed.

> For it is surely neither fire, at any rate, nor earth. But neither is it Love or Strife, for the former is a cause only of association, the latter of segregation. In fact, the cause is the substance of each thing, not merely "*mixing and separation of things mixed*," as he says. Chance "*is the name given to these*

[27] Elsewhere, Aristotle argues that anyone who thinks that "among the seeds anything comes to be at random" entirely "does away with nature and what exists by nature" (*Phys.* II 8, 199b15–16). For a thorough discussion of the Empedoclean zoogony, see Sedley 2016.

[28] Cf. *DA* II 4 and *GA* II 1, 731b24–732a3.

[29] "Let us admit that the soul knows and perceives the elements out of which each of these composites is made up; but by what means will it know or perceive the composite whole, e.g., what god, man, flesh, bone (or any other compound) is? For each *is*, not merely the elements of which it is composed, but those elements combined in a determinate mode or ratio, as Empedocles himself says of bone,

> The kindly earth in its broad-bosomed molds
> Won of clear water two parts out of eight
> And four of fire; and so white bones were formed.

Nothing, therefore, will be gained by the presence of the elements in the soul, unless there be also present there the ratios and the composition" (*DA* I 5, 409b30–410a8). The Empedoclean passage Aristotle quotes is DK B96. Cf. *DA* I 4, 408a18–23.

> *happenings*," not proportion; for things can be mixed by chance. The cause, then, of things which exist naturally is their being such and such, and this is the nature of each thing, about which he says nothing. So there is nothing "*About Nature*" in what he says. (*GC* II 6, 333b11–18)

The movements of the simple terrestrial bodies cannot, on their own, explain either the regularity of the coming to be of natural composites or the regularity of the characteristics of the natural composites that come to be. There is nothing about their movements that would lead to any one series of interactions being more likely or more regular than any other.[30] And adding Love and Strife as principles of movement will not overcome this gap. Love mixes bodies and Strife separates bodies and this is all that they do. Again, there is nothing about these processes themselves that would make one variety of mixing or separation more likely or more regular than any other.

But the problem with explanations limited to these resources is not simply that they make the frequency and regularity of observable changes and bodies a mystery. The deeper problem is that, even when the movements of elemental bodies acted upon by Love or Strife result in a composite with the characteristics of, say, bone, what results cannot properly be said to be bone at all. When discussing nutritive activity, Aristotle says that:

> Empedocles is wrong in adding that growth in plants is to be explained, the downward rooting by the natural tendency of earth to travel downward, and the upward branching by the similar natural tendency of fire to travel upward. For he misinterprets up and down; up and down are not for all things what they are for the whole world: if we are to distinguish and identify organs according to their functions, the roots of plants are analogous to the head in animals. (*DA* II 4, 415b28–416a6)[31]

What is functionally up in a plant grows cosmically down. This is not merely a verbal dispute. If an account of growth is blind to an organism's functional articulation and orientation, then it fails to pick out what it purports to explain. Unless one appeals to the plant's soul – the form for the sake of which a plant's roots come to be and by virtue of which they have their natural function – one has not understood what roots *are*.

[30] Cf. *GC* II 9, 335b31–33 and *Meta.* I 3, 984a21–25.

[31] "For, this distinction [*sc.* between up and down] is one of function, not merely of position relative to the earth and the sky. The superior is that from which flows in each kind the distribution of nutriment and the process of growth; the inferior is that to which the process flows and in which it ends" (*IA* 4, 705a30–b1; cf. *Juv.* 1, 467b32–468a12).

Empedocles cannot understand what roots are and so cannot explain either what it is for them to grow or why they grow in the manner and the order in which they do.

Similarly, even if Empedocles is able to explain why a composite body with the material characteristics of bone comes to be, he has not thereby explained why that body has the function it does within an organism. But this function is what makes bone what it is and so any explanation that cannot account for that function cannot explain the coming to be of bone. An animate tissue is not to be identified with a body with a given set of material characteristics. That is, an animate tissue's material capacities are determined by the ratio of the tangible contraries present in the ingredients from which it comes to be, but neither the natures of the ingredients nor anything which can be determined solely on this basis is the principle in virtue of which the animate tissue comes to be with those material capacities. Aristotle insists on numerous occasions that while, "we may allow that hardness and softness, stickiness and brittleness, and whatever other qualities are found in the parts that have life and soul, may be caused by mere heat and cold, yet, when we come to the principle in virtue of which flesh is flesh and bone is bone, that is no longer so."[32] So any account that, at best, explains the coming to be of bodies with a given set of material characteristics, cannot explain everything that comes to be by nature. And "chance '*is the name given to these happenings* [viz., mixing and separation that are due to the motions of elemental bodies and Love/Strife]' not proportion; for things can be mixed by chance."[33]

Empedocles problem is that, at best, he only countenances what Aristotle would call material and efficient causes. The first thing he is missing is the notion of a formal cause. Aristotle says that the cause of coming to be by nature in a particular and regular way, that is, by a "certain formula" (λόγῳ τινί), is "the substance (οὐσία) of each thing" and that the cause of things which exist naturally is, equivalently, "their being such and such" (τὸ οὕτως ἔχειν).[34] It is a thing's form/substance/essence

[32] *GA* II 1, 734b31–34. Aristotle makes similar points on numerous occasions, e.g., *Phys.* II 8; *GC* II 9, 335b30–336a12; *Meteor.* IV 12, 389b28–390a15; *DA* II 4, 416a9–18; *Meta.* I 3, 984b5–15.

[33] *GC* II 6, 333b14–16. This statement is an intentional parody of the last line of B8 (φύσις δ' ἐπὶ τοῖς ὀνομάζεται ἀνθρώποισιν); Aristotle replaces φύσις, which Empedocles uses to mean coming to be or birth, with τύχη. Cf. *GC* I 1, 314b6–8; *MXG* 2, 975b5–16.

[34] Elsewhere, Aristotle says, "For even Empedocles says bone exists by virtue of the ratio in it. Now this is the essence and the substance of the thing. But it is similarly necessary that the ratio should be the substance of flesh and of everything else, or of none; there it is on account of this that flesh and bone and everything else will exist, and not on account of the matter, which *he* names, – fire and earth and water and air" (*Meta.* I 10, 993a17–22; cf. *Meta.* I 4, 983b32–985a10; *DA* II 4, 416a13–15).

that determines the λόγος that governs both its coming to be and the characteristics it regularly possesses upon coming to be.

This view is also the culmination of *Meta.* VII (*Zeta*). In that book's final chapter, Aristotle finally recognizes that if one wants to know "what, and what sort of thing, substance is," one needs to know why individual substances are such (*Meta.* VII 17, 1041a6). We should ask questions like "Why are these bricks and stones, say, a house?" or "Why is this individual thing, or this body in this state, a man?" There is a way to answer such questions that reveals "the cause, which is (speaking abstractly) the essence (τὸ τί ἦν εἶναι)" (1041a27–28). That is, when we ask questions like this properly "we seek the cause, i.e., the form (τὸ εἶδος), by reason of which the matter is some definite thing; and this is the substance (οὐσία) of the thing" (1041b7–8). Aristotle concludes that "this [sc., form] is the substance of each thing; for this is the primary cause of its being" (1041b27–28). So Aristotle identifies form, essence, substance, and primary cause of being.

But Aristotle goes even further. He says that this form "is the nature of each thing" (*GC* II 6, 333b17). Aristotle makes the same identification in the *Metaphysics* when he says that "while some things are not substances, still such substances as there are, are composed (συνεστήκασι) in accordance with a nature and by way of a nature of their own, and their substance would seem to be this very nature (φανείη ἂν [καὶ] αὕτη ἡ φύσις οὐσία), which is not an element but a principle" (*Meta.* VII 17, 1041b28–31). In general, and *pace* Empedocles, Aristotle maintains that "the form indeed is nature more than [or rather than (μᾶλλον)] the matter."[35] And this appeal to nature, as Aristotle conceives it, explains the relevant regularities. For according to Aristotle, "those things are natural which, by a continuous movement originated from an internal principle, arrive at some end: the same end is not reached from every principle; nor any chance end, but always the tendency in each is towards the same end, if there is no impediment."[36]

[35] *Phys.* II 1, 193b6–7. Aristotle makes the posteriority of matter's status as nature explicit on numerous occasions. For example, he says, "it is plain that nature in the primary and strict sense is the substance of things which have in themselves, as such, a source of movement; the matter is called the nature because it is qualified to receive this" (*Meta.* V 4, 1015a13–15). Cf. *Phys.* II 1, 193a9–b12; *Phys.* II 8, 198b8–19; *Phys.* II 8, 199a30–32; *PA* I 1, 640b29; *GC* II 9, 335b34–35.

[36] *Phys.* II 8, 199b16–18. On generation's formal end determining its sequence, see *Phys.* II 8, 199a9; *PA* I 1, 639b16–21 and 640b1–3; *GA* V 1, 778b2–18; *GC* II 9, 335b30–336a12. Aristotle often notes the connection between nature and form when he discusses natural generation: "Universally, then, that from which they come to be is nature and that according to which they come to be is nature (for what comes to be has a nature, for example, a plant and an animal) and that by which

So, given that Empedocles does not appeal to forms *qua* natures in his account, Aristotle can jokingly say that, despite its title, there is nothing to be found about nature in Empedocles treatise "About Nature" (Περὶ φύσεως).

But Empedocles' purported failures do not end with the lack of formal causes. The second thing Empedocles is missing is the notion of a final cause.

> Moreover, it is this [*sc.* the form/substance/nature] that is the well-being and the good of each thing; but he commends only mixture. And yet it is not Strife but Love that separates the elements which are by nature prior to god, although they are also gods. (*GC* II 6, 333b19–21)

According to Aristotle, everything that occurs by nature occurs for the sake of some end.[37] More precisely, it is natures themselves that Aristotle says are for the sake of an end.[38] The formal and final causes are one because "the essence of a thing, i.e., the form for this (τὸ τί ἐστιν καὶ ἡ μορφή) is the end or that for the sake of which" (*Phys.* II 7, 198b2–3; cf. II 8, 199a30–33). And insofar as it is appropriate to think of ends as goods, it is reasonable to describe an explanation in terms of final causation as one that says something occurs "because it is better thus (διότι βέλτιον οὕτως) (not without qualification, but with reference to the substance in each case)" (*Phys.* II 7, 198b8–9; cf. *Top.* VI 8, 146b9–10). This notion of an end *qua* good in part explains the regularity of coming to be by nature; the condition that something of a given kind, for the most part, realizes is the condition that is best for things of that kind.[39]

they come to be is the nature, which is called according to the form, and it has the same form as that which comes to be (but the form exists in another; for man generates man)" (*Meta.* VII 7, 1032a22–25; cf. DK B37).

[37] Cf. *Phys.* II 5, 196b21; *Phys.* II 7, 198b4–5; *Phys.* II 8, 199a7, 11–12, 30–33; *DA* III 12, 434a31–32; *Somn.* 2, 455b16–17; *PA* I 1, 641b11–14 and 641b23–642a1. Aristotle often cites either a lack of interest in that for the sake of which an organ or vital activity is present or an inadequate grasp of this cause as the main reason why a purported explanation of a physiological phenomenon fails. For example, the reason why the explanations of respiration that Anaxagoras and Diogenes (*Resp.* 3, 471b23–29), Democritus (*Resp.* 4, 471b30–472a3), Plato (*Resp.* 5, 472b24–29), and Empedocles (*Resp.* 7, 473a15–17) provide all fail is the same: none of them understands that for the sake of which respiration occurs by nature.

[38] Cf. *Phys.* II 7, 198b4–5; *Phys.* II 8, 199a11–12; *Phys.* II 8, 199b13–26; *PA* I 1, 639b16–26, and 641b23–38.

[39] Empedocles, Aristotle notes, does speak of goodness in nature. But he applies this label only to Love and to mixing due to Love. This not only fails to explain the regularities that govern coming to be by nature, but also undermines Empedocles' own views. For as Aristotle interprets Empedocles, he is committed to the view that Strife can, under certain conditions, be a principle of mixing, and Love can, under certain conditions, be a principle of separation, warranting a reversal of the labels "good" and "bad" (cf. *Meta.* I 4, 984b32–985a28 and *DC* III 2, 301a15–10). Moreover, since the

So a nature determines (or is) the essence that exhausts what it is to be a natural substance of a given kind and is the primary cause of its being. The generation, being, and movement/activity of natural substances have a nature *qua* form as their principle, cause, and end. Empedocles cannot accept any of this. But without it, he cannot explain the regularities that govern coming to be by nature and cannot even properly grasp what it is that comes to be by nature.

7.4 The Principles of Movement

Aristotle's next target is Empedocles' third major thesis: (EMP3) Love and Strife are the principles of natural bodily movement. His first criticism concerns Empedocles' failure to present his account in a manner that reflects a sound scientific methodology.

> Again, his account of motion is too shallow. It is not enough to say that Love and Strife are the reason why things move, unless to be by Love means to move with this type of movement and to be by Strife to move with this other one. What was needed was to give definitions, or assumptions, or demonstrations, be they precise, relaxed, or in some other fashion. (*GC* II 6, 333b22–26)

One can hardly fault Empedocles for not fully anticipating the account of demonstrative science that Aristotle propounds in his *Posterior Analytics*. But there are ways to approximate the virtues of Aristotle's methodology that were readily available to Empedocles, and he largely eschews them. Both the method of inquiry and the manner of presentation Empedocles adopts blind him to the explanatory deficiencies of his account.

Empedocles does not seem to appreciate the difference between (a) a definition (ὁρισμός); (b) an assumption (ὑπόθεσις); and (c) a claim that one demonstrates.[40] At issue are a pair of Empedoclean assertions: (i) Love is the cause of a particular sort of movement, namely, mixing and (ii) Strife is the cause of a particular sort of movement, namely, separation. What is the status of these claims? On the one hand, they could be assumptions – being causes of these respective movements is simply part

coming to be of the Sphere involves the elements losing their separate status, there is a sense in which this constitutes the ceasing to be of the elements, and this would be due to Love, not Strife. The claim that the elements too are gods, stems from Empedocles' practice of identifying the four elements with the names of deities. Though the distribution is disputed (cf. DK A33), he attributes the names Zeus, Hera, Nestis, and Aidoneus to the four elements in B6 and identifies fire with Hephaestus in B96 and B98.

[40] Cf. *APo* I 3, 72a18–24 and *APo* I 10, 76a31–b39.

of what "Love" and "Strife" mean by definition and these claims stipulate that Love and Strife are as their definitions state. On the other hand, they could be claims in need of demonstration – they require a provision of reasons that adequately explain why Love must be such as to cause mixing and why Strife must be such as to cause separation. Empedocles' imprecision prevents us from knowing which of these he intends and perhaps reflects an ignorance that would render him incapable of definitively intending either.

The remaining complaints concern whether the two principles of movement Empedocles embraces suffice to explain the movements of natural bodies. The first is based on Aristotle's distinction between movement that is without constraint (ἄνω βίᾳ) and so in accordance with nature (κατά φύσιν/φύσει) and movement that is by constraint (βίᾳ) and so contrary to nature (παρὰ φύσιν).

> Again, since the bodies are seen to be moved both from constraint (namely contrary to nature) and in accordance with nature – for instance, fire moves upward without constraint, though by constraint downward – and that which is in accordance with nature is contrary to that which is by constraint, and there is such a thing as by constraint, so there is also such a thing as being moved in accordance with nature. (*GC* II 6, 333b26–30)

A movement is in accordance with nature if one must, in order to explain the movement properly, appeal to a principle that is internal (ἐντός) to that which moves. And a movement is due to an internal principle if and only if (a) the movement is of a determinate sort that is proper to bodies with the form or kind (εἶδος) that belongs to the entity to which the principle is internal; (b) the entity to which the principle is internal is the movement's proper subject; and (c) the movement will, when all preconditions for its exercise are met, occur immediately (εὐθύς) and necessarily if nothing external prevents it from doing so.[41] A movement contrary to nature occurs "only when something external (ἔξωθεν) moves a thing, or brings it to rest against its own internal tendency" (*EE* II 8, 1224b7–8).

For example, fire moves upward toward its proper (οἰκεῖος) location at the periphery of the sublunar realm without constraint and moves toward the cosmos' center only by constraint.[42] Though Empedocles need not

[41] Cf. *Phys.* VIII 4, 255b3–256a3; *Phys.* V 5–6; *GA* II 4, 740b21–24; *EE* II 8, 1224a15–22. A view of natural movement along these lines is defended in Kelsey 2003: 59–87; Frey 2015*a*: 88–126; Frey 2015*b*: 88–126.

[42] For Aristotle, the primary efficient cause of a terrestrial, simple body's natural motion is external, viz., the body from which it came to be (*Phys.* VIII 4, 255b32–256a3). Nevertheless, their characteristic natural movements are movements of the bodies themselves. The simple bodies

adopt the details of Aristotle's account of these two ways movement occurs, he must accept the distinction since, Aristotle maintains, it is something that we readily observe. But Empedocles doesn't seem to possess the resources to explain the distinction adequately.

> Is it this movement, then, which Love initiates, or not? No, for, on the contrary, earth's natural movement is downward and resembles separation, and Strife rather than Love is the cause of movement in accordance with nature; so, in general, Love would be contrary to, rather than in accordance with, nature. But unless Love or Strife moves them, there is no movement or rest of the bodies themselves at all. But this is absurd. Moreover, that they do move is seen. For, though Strife separated, *aithêr* was borne upwards not by Strife. But, he sometimes says it is as if by chance, "*for it was in this way that it chances to run, but often [it chanced to run] differently*," while at other times he says it is the nature of fire to be borne upward, but the *aithêr*, he says, "*sank down under the ground with long roots.*" (*GC* II 6, 333b30–334a5)

Love would seem to be the best available candidate for causing movement in accordance with nature. But Love causes only mixing and this activity, in fact, moves the elements away from the locations to which we observe them to move without constraint. That is, the ultimate outcome of Love's activity, the creation of the Sphere, involves moving all of the elements to a single location by force. It is Strife that separates the elements from the mixture that is the Sphere, and it is only upon being separated that the elements move by nature as we observe them to move.[43]

But Empedocles must still explain why the elements move as they do *once* they are separate. He cannot say that Love and Strife move them. The only things Love and Strife cause are mixing and separation. Moreover, Love and Strife are external to the elemental bodies and so could not be the only principles of their movement without, at best, making it movement by constraint. So unless Empedocles accepts something like Aristotle's account, he will find himself committed to a position that is both absurd (ἄτοπον) and at odds with what is observable, namely, that "there is no

possess their capacities for movement *per se* (*Phys.* II 1, 192b35–36), the exercise of these capacities "brings to actuality the proper activities that they potentially possess" (*Phys.* VIII 4, 255a29–30), and the movements that are the exercises of such capacities exhaust what it is for the simple bodies to be (formally) what they are most fully (*DC* IV 3, 310a33–34; cf. ἐντελεχείας at 311a3–6).

[43] Elsewhere, Aristotle notes that, "Even the man whom one might suppose to speak most consistently – that is, Empedocles – even he has made the same mistake; for he maintains that Strife is a principle that causes ceasing to be, but Strife would seem none the less to produce everything, except the One; for all things, with the exception of god, proceed from Strife" (*Meta.* II 4, 1000a24–29).

movement or rest of the bodies themselves at all." Given this, it is not surprising that Empedocles makes inconsistent claims about the movements of the elements: sometimes he says they move by their own natures, sometimes he says they are moved by Love and Strife, and sometimes he says they move as if by chance.[44]

Finally, Aristotle argues that Empedocles' cyclical cosmogony entails that his account of movement is incomplete.

> At the same time, he also says that the universe is in the same state now under the rule of Strife as it was formerly under the rule of Love. What, then, is the first mover and cause of movement? It is surely not Love and Strife. But of what movement are these causes, if that other is the principle? (*GC* II 6, 334a5–9)

The cosmos, under the rule of Love, proceeds from a state in which the four elements are separate and entirely stratified toward a state in which all of the elements are thoroughly mixed into the Sphere. The cosmos, under the rule of Strife, proceeds from the Sphere toward a state in which all of the elements are separate and entirely stratified. We are presently part way through the process Strife initiates, but how things now stand can be identical to the way things stand part way through the process Love initiates. If this is so, then neither Love nor Strife is necessary to explain why things are as they currently are. What Empedocles needs to provide is a cause of motion in general, a cause of both the motion that Love initiates and the motion that Strife initiates. This "first mover and cause of movement" (τὸ κινοῦν πρῶτον καὶ αἴτιον τῆς κινήσεως) would deserve the status of principle (ἀρχή). But Empedocles fails to include such a principle of movement in his account.

7.5 The Soul

Aristotle's final criticism of Empedocles' natural philosophy concerns his account of the soul: (EMP4) Soul is, or is constituted by, one or more of the four elemental bodies.

> An additional absurdity is that soul is constituted by the elements or is one of them. For how will the alterations of soul occur, such as being musical and once again unmusical, or remembering or forgetting? For clearly, if the soul is fire, it will come to possess only those properties that are appropriate to fire *qua* fire; if it is a mixture, only bodily properties. But none of these is bodily. (*GC* II 6, 334a10–14)

[44] Cf. DK B51; B53; B54; *Phys.* II 4, 196a20–23.

This argument is brief. This is largely because, as Aristotle goes on to say, "these matters are the work of another inquiry" (334a15).[45] Aristotle is clearly referring to *De anima* which includes (in *DA* I 4–5) a much more thorough discussion of several accounts of soul that prioritize the elements, including Empedocles'.

But the argument is nevertheless significant. Empedocles cannot, Aristotle contends, account for those alterations that are due to soul. For Empedocles would consider the examples Aristotle mentions – the transition from being musical to unmusical or from knowing something to no longer knowing it when one forgets – to be alterations of the soul itself. If, then, we identify the soul with the elements, the only alterations it can undergo will be thermal changes, hygric changes, positional changes, quantitative changes, or one of the several other varieties of change appropriate to bodies. And Aristotle flatly denies that alterations due to soul are, in this way, bodily alterations.

But there is more to Aristotle's argument than a simple denial. According to Aristotle's own account, the changes he mentions are not alterations of the soul itself but rather changes whose subject is the living organism that is ensouled.[46] More importantly, Aristotle identifies souls with natures and is consequently able to say that these changes have a nature as their principle, form, and end. Bringing these changes into the fold of his broader philosophy of nature allows Aristotle to maintain that these changes involve more than bodily movement. Changes due to soul typically do involve bodily movements, but there is a further aspect to such changes that does not consist in the replacement of some quality, quantity, or place. They are, instead, to be viewed as transitions from being something potentially or in capacity (δυνάμει) to being something actually or in activity (ἐνεργείᾳ/ ἐντελεχείᾳ). When understood in this way, a change is not a movement but is rather the preservation or salvation (σωτηρία) of what was present before, that is, a "development (ἐπίδοσις) of the thing into itself and its fulfillment (ἐντελέχειαν)." And "for this reason, it is not right to say that something which understands is altered when it understands any more than a builder when he builds."[47]

45 This sentence is naturally read as the first independent clause of a longer sentence whose second independent clause begins at 334a16. Joachim and Williams do so and place the entire sentence at the beginning of *GC* II 7. But Philoponus and Mugler 1966 split the sentence into two (the δέ clause and the μέν clause) and end *GC* II 6 with this first part as a complete sentence.

46 "It is doubtless better to avoid saying that the soul pities or learns or thinks, and rather to say that it is the man who does this with his soul" (*DA* I 4, 408b13–14).

47 *DA* II 5 417b2–9, cf. II 4, 416b1–3, Burnyeat 2002: 28–90, and Frey 2015*a*: 137–160.

This aspect of Aristotle's natural philosophy is notoriously enigmatic, and much more would need to be said to render it pellucid. I mention it only to underscore that the import of Aristotle's brief argument against Empedocles' account of soul is, like the arguments that precede it, not entirely negative. Aristotle's preferred account of soul is inextricably tied to his accounts of movement, change, and nature. This chapter's arguments don't only raise difficulties for Empedocles; each brings an aspect of Aristotle's preferred account to the fore as well. When viewed collectively, they make a strong case for its superiority over the worldview Empedocles offers us.

8

On Generation and Corruption *II 7*

Andreas Anagnostopoulos

8.1 The Place and Significance of *GC* II 7

GC II 7 might seem relatively unimportant, perhaps even dispensable. It is often treated simply as another source, alongside the more substantial discussion in *GC* I 10, of Aristotle's views about mixture.[1] One recent commentator describes both together as Aristotle's "attempt to say what mixture is," while another calls *GC* II 7 a "recap" of the earlier chapter.[2] In fact, *GC* II 7 has a quite specific and distinctive function in its immediate context. Earlier in *GC* II, Aristotle has identified, in rough agreement with Empedocles, earth, air, fire, and water as the elements, that is, the primary bodies, but posited that they possess a common matter that enables them to change into one another.[3] Presumably because of the superficial similarity of the two elemental theories, Aristotle turns in *GC* II 6 to a critique of Empedocles' natural philosophy. While that chapter is somehow digressive (334a15), *GC* II 7 shifts the critique to an issue germane to the treatise: the formation, from these somehow agreed upon elements, of more complex stuffs like bone and flesh – which I will call "elemental mixtures" without thereby taking a stance on the fraught issue of their precise relation to the elements. Aristotle argues that the Empedoclean conception of the four elements cannot accommodate the formation of

* I am grateful to Mary Krizan for sharing her then unpublished paper (Krizan 2018); to Peter Isépy, Christina Prapa, and Lara Trivellizzi for discussion of the Greek text; and to Timothy Clarke, Mareike Hauer, Sean Kelsey, Chris Noble, Christian Pfeiffer, and Christof Rapp for comments on earlier versions. I am especially grateful to Andrea Falcon for considerably improving a recent version and helping me bring the paper down to size, and to Laura M. Castelli for extensive comments and insightful discussion at every turn.

1 Two important exceptions are Frede 2004 and, especially, Krizan 2018.

2 Fine 1995: 305 and Cooper 2009: 157. References to Cooper 2009 in this chapter are to the second of its two "Notes," which did not appear in Mansfeld-de Haas 2004.

3 I follow Crowley 2013 in holding that by "element" (στοιχεῖον), Aristotle in the *GC* consistently means the primary *bodies*, rather than the basic contraries. See *DC* III 3, 302a14–19 and the chapter by Emily Katz in this volume.

elemental mixtures and explains how his own conception can. In this way, *GC* II 7 supports Aristotle's elemental theory over its closest competitor and highlights those distinctive features of the Aristotelian elements which enable them to mix: their common matter and the fact that they are characterized by contrarieties that admit of degrees.

To stop here, however, would be to underestimate the chapter's philosophical interest, which stems largely from its central role in Aristotle's project in the *GC*. This role, moreover, is bound up with the structure and unity of the *GC*'s investigation. Let us then briefly take up a broader perspective. The term "mixture" (μίξις) first appears in *GC* I 6, where Aristotle claims that more clarity about that concept (alongside those of acting on/being affected (ποιεῖν/πάσχειν) and contact) is needed in preparation for the upcoming treatment of "matter and the so-called elements: whether they exist or not, whether each is eternal or somehow comes to be, and if they do come to be, whether they all come to be from one another in the same way or some one among them is primary" (*GC* I 6, 322b1–6).[4] Aristotle is in particular motivating the need to delimit or define (διοριστέον: 322b25) these concepts, and after discussions of contact (*GC* I 6) and acting on/being affected (*GC* I 7–9), *GC* I 10 is the promised (327a30–31) discussion of mixture, concluding with a definition that explains what mixture is.

GC I 10 and *GC* II 7 are evidently doing quite different things. *GC* II 7 contains no definition of mixture and is anything but preparatory to Aristotle's treatment of the elements; instead, it presupposes significant familiarity with his elemental theory. By contrast, as we might expect from a preparatory discussion, *GC* I 10 does not specifically discuss elemental mixtures or even mention any particular ontology of the lowest levels of the sublunary universe.[5] While *GC* II 7 does not, therefore, simply complete or recap the earlier treatment of mixture, there is also good reason to doubt a diametrically opposed view of the relation between the two chapters. According to this view, *GC* I 10 provides an account only of mixtures of higher-level, non-elemental bodies, and so does not apply to

[4] An initial worry is that since it applies the term "mixture" to pluralist aggregation – surely it is such "mixture" that, as Aristotle claims, has been thought to effect generation (*GC* I 6, 322b6–9; cf. *GC* II 1, 329a3–8) – the passage does not tell us anything about genuine mixture. Looking ahead, I take this worry to be decisively defused by the fact the promised account of mixture, that of *GC* I 10, does not apply to aggregation, along with the fact that, for Aristotle, genuine mixture is in fact instrumental in effecting generation from the elements, while pluralist aggregation is not.

[5] For example, *GC* I 10 concludes that there is mixture on the rather unspecific grounds that "there are some such things, i.e., things affectable by one another and easily bounded and easily divided." (328b16–17)

the elemental mixtures of *GC* II 7.[6] This view makes it obscure why *GC* I 10's account of mixture should be a necessary preparation for the treatment of the elements. As I see it, *GC* I 10 offers a general account of the mixture of bodies, while *GC* II 7 involves a specific application of that account at the elemental level.[7] Though it is not the main focus of this essay, the point is important not only for the structure of Aristotle's *GC*, but also because it bears on the much debated issue of the metaphysical constitution of mixtures, and so the detailed analysis of *GC* II 7 to follow will highlight what I take to be decisive parallels with *GC* I 10.

If *GC* I 10 and *GC* II 7 are related as I suggest, we can understand why the former might be thought to be a necessary preparation for the treatment of the elements to come. *GC* I 10 spells out and legitimates a general concept of mixture whose applicability to the elemental level, as *GC* II 7 argues, requires, and thus supports, Aristotle's conception of the four elements as against the Empedoclean one. In fact, the significance of the possibility of elemental mixture in this respect, that is, as putting constraints on the nature of the elements and, in particular, supporting an Aristotelian elemental theory, is not confined to his dialectic with the Empedoclean, even if it is most effectively brought out in that context. A more general argument from the possibility of elemental mixture to Aristotle's elemental theory can be formulated using the resources of *GC* I alone, and its similarity to the argument of *GC* II 7 brings out conceptual parallels between *GC* I and *GC* II. Consider the more abstract (and so appropriate to *GC* I) possibility of elemental mixture, *whatever the elements*

[6] Krizan 2018 holds that while *GC* I 10 gives an account of mixtures of (i) *bodies*, in particular; (ii) only wet or liquid bodies; (iii) of higher complexity than the four elements; and (iv) which undergo mere alteration when mixed, the substance remaining intact, *GC* II 7 offers an account of elemental mixture, even of the dry primary bodies (as *GC* II 8 requires), but according to which it is strictly speaking the elemental *contraries* that are mixed. I register disagreement with different strands of her argument in the footnotes of this chapter. But let me say here, regarding issues (i) and (ii), that I do not see what would be at stake for Aristotle in claiming that elemental mixtures are really mixtures of contraries, rather than of earth, air, fire, and water, so as to preserve the alleged restriction to liquids in *GC* I 10; I also find it difficult to understand why Aristotle would so vehemently restrict mixture to separable bodies in *GC* I 10, if he were later to devote a chapter to mixtures of elemental contraries.

[7] I take *Sens.* 3, 440a31–440b4 to refer to *GC* I 10 as a general treatment of mixture of bodies, but not the "mixture" of qualities discussed in *Sens.* 3, which is however plausibly prior to that of bodies (see Fine 1995: 306–340). My understanding of the relation between *GC* I 10 and *GC* II 7 follows the observation in Burnyeat 2004, that *GC* I is largely devoted to general accounts of basic processes, including generation/destruction, mixing, and affection, even if their application to more complex sublunary entities requires "suitable adaptations and refinements" (11).

turn out to be. This possibility, I suggest, can be shown to require Aristotle's conception of the elements, understood, again somewhat abstractly, as a middle way between traditional monism and pluralism: that there is a plurality of elements possessing a single, common matter, which however is not itself an element or separable body.[8] Aristotle claims that (i) genuine mixture involves mutual affection among the ingredients (*GC* I 10, 328a31–33 and 328b14–22), and in particular, I take this to be an affection in the ingredients' natures (*GC* I 10, 328a30).[9] But (ii) mutual affection requires a common matter (*GC* I 7, 324a34–b9 and *GC* I 10, 328a19–21). While mutual affection between two things in respect of their accidental qualities might not immediately require a commonality in their substantial matter, that is, that which underlies their substantial natures, mutual affection in respect of (parts of) those natures plausibly would; for I take it that the implied common matter must lie "further down", as a metaphysical subject for those natures.[10] But (iii) this common matter cannot be a separable body, for this would make Aristotle a traditional monist, and the mixtures in question would turn out not to be *elemental* mixtures. Therefore, (iv) elemental mixture requires a common matter of the elements that is not a separable body. Perhaps one ought, more cautiously, to limit this conclusion to pairs or sets of elements that can mix with one another, and perhaps one should conclude only that such pairs or sets of elements share a common aspect to their matter; whether the commonality extends to the matter of *all* the elements, and whether it *exhausts* the matter of the relevant elements, will plausibly depend, respectively, on which elements can mix with one another, and whether the implied mutual affection involves features that exhaust their natures. It is clear at least that neither traditional pluralism nor traditional monism can explain elemental mixture, because neither allows for affection in an element's nature.

Whether understood in relation to the four standard elements or more abstractly, then, elemental mixture showcases crucial features of Aristotle's elemental theory as against competing ones. But even this does not fully

[8] This middle way is implicit in *GC* II 7's presentation of the Aristotelian position: to say that "the elements" (334b16) have a common matter implies that this matter does not threaten their status as elements.

[9] On the affection reaching the ingredients' natures, see Anagnostopoulos 2021: 8–10. As I point out below, this also holds of elemental mixtures as described in *GC* II 7.

[10] Those who hold that one contrary within an element serves as the matter, with no additional underlying subject for the contraries, will disagree. See Furth 1988: 221–227, and Gill 1989: 41–82 and 243–252. See also note 51 below.

capture mixture's relevance for elemental theory. In fact, Aristotle motivates the initial investigation of mixture in a different way. In *GC* I 6 and *GC* II 1, Aristotle points to the apparent or alleged role of various processes – mixture, aggregation, alteration, acting on/being acted on, and "another change" that remains unnamed – in generation, either generation of the so-called elements or, more relevantly for the case of mixture, generation of more complex substances from the elements.[11] Most importantly, just before embarking on the analysis of the elements and their constitution, Aristotle writes:

> So let us agree that it is an excellent thing to say that the primary are principles and elements – the things from which, by their changing either by combination and separation or by some other kind of change, generation and perishing result. (*GC* II 1, 329a5–8)[12]

If mixture is such a change, that is, one by which generation from primary things occurs, then, to that extent, showing that the four Aristotelian elements undergo mixture is a way of showing that they really do function as elements. Moreover, understanding mixture, in this role, will be a crucial part of understanding the entirety of natural generation – what the opening lines of the *GC* announce as the treatise's main undertaking (*GC* I 1, 314a1–2).[13] Now, for Aristotle, living substances are constituted by uniform stuffs like bone and flesh, and these are in some sense mixtures of the elements.[14] Elemental mixture is thus in some sense a condition of the generation of natural substances above the level of the elements. With regard to elemental theory, mixture is at issue in the *GC* not just because it is a phenomenon that Aristotle's elemental theory can uniquely accommodate (as if any other such phenomenon would do); rather, insofar as it is instrumental to the generation of more complex bodies from the elements, elemental mixture exhibits the elements *as* elements, in particular, the elements *of bodies* – the topic announced at *GC* II 1, 328a31–32.

The importance of elemental mixture in this role, that is, as instrumental to the generation of higher-level substances, may explain one feature of *GC* II 7 that my description of the chapter as concerned with elemental

[11] *GC* I 6, 322b6–26; *GC* II 1, 329a2–8. Thanks to Timothy Clarke for stressing the importance of these passages for the role on mixture in the *GC*.

[12] See also *DC* III 1, 298b6–11, with Andrea Falcon's Introduction to this volume.

[13] On the intimate relation between generation and elemental theory as topics of the *GC*, see the introduction to this volume.

[14] A major complication is that genuine flesh, bone and the like are characterized functionally by reference to their roles in living organisms and are not, it seems, formed immediately through mixing. See Anagnostopoulos 2021: 174–175 for a brief discussion and further references.

mixture might obscure. *GC* II 7 aims, in the first instance, to resolve a difficulty about flesh, bone, and the like coming to be from the elements, rather than about the possibility of the "mixture" of the elements as such. From the perspective outlined here, it is more important that the building blocks of higher-level substances can come to be from the Aristotelian elements than that they are appropriately called "mixtures."

8.2 Mixture in *GC* I 10

Given that *GC* II 7 takes up many of the issues raised in *GC* I 10, it will be useful to summarize the latter's main points. The first part of the chapter (up to 327b31) addresses an *aporia* challenging the possibility of mixture, on the grounds that the ingredients can neither continue to exist, nor can one or both of them be destroyed; the argument seems to demand an explanation of how mixture differs from generation and destruction (327b6–8). After specifying that only separable bodies are subject to mixture, he answers the initial *aporia* by taking a middle way: "what has come to be from them [the mixture] is something else in *energeia*, but is still each of them in capacity, [i.e.,] those things that were, before having been mixed, and [which] have not been destroyed" (*GC* I 10, 327b22–26). Aristotle later explains that the ingredients can be recovered from the mixture (327b28–29), and points to the preservation of the capacity of or for them (ἡ δύναμις αὐτῶν: 327b31) as evidence for their not having been destroyed, once mixed.[15]

A second *aporia* asks whether mixture is relative to perception, by involving the ingredients being placed next to one another in imperceptibly small bits, or rather involves the minimal parts of the ingredients being placed beside one another. Against the second view, Aristotle cites the impossibility of smallest parts, while both views conflate mixture with mere composition (σύνθεσις): unlike composition, a mixture is *homoeomerous*, and each part has the same account (λόγος, 328a9) as the whole, just as any portion of water is water.

Faced once again with the threat that there is no mixture, Aristotle proposes that mixture involves the mutual affection of the ingredients, so as to reach an intermediate qualitative condition. Mutual affection requires

[15] On the translation of the sentence quoted above (327b22–26), and on the meaning of ἡ δύναμις αὐτῶν (327b31), see Anagnostopoulos 2021: 150–153.

that the ingredients (i) have the same matter (328a18–23); (ii) exhibit contrariety (328a32–33); (iii) are present in roughly comparable quantities, that is, commensurate in their powers (328a28). In addition, though not constitutive of mixture, division into small parts facilitates the requisite affection, and so mixable things are also (iv) easily divided (328a24), which is a result of being easily bounded (328a33–b4), and liquids are especially so.[16] Things which meet the above conditions qualify as mixable, so that Aristotle can close *GC* I 10 with a definition of mixture as "the unification (ἕνωσις) of the mixable things, once they have been altered" (328a21–22).

8.3 Aristotle and the Empedoclean: Elemental Transformation and Common Matter

Aristotle opens *GC* II 7 by contrasting his own and the Empedoclean conceptions of the elements.

> But regarding[17] the elements out of which the bodies are composed, for those who hold that there is something common, or that the elements change into one another, if one of these is the case, then the other follows necessarily. (*GC* II 7, 334a15–18)

He characterizes the two positions according to where they stand on two questions: (i) whether there is something common underlying the

[16] Mary Krizan argues forcefully that *GC* I 10, 328b14–22 restricts mixtures to liquid ingredients, in contrast to *GC* II 8's insistence that most or all sublunary elemental mixtures involve all four elements (Krizan 2018: 200–204). It is true that the conditions of divisibility, mutual affectability, and being easily bounded seem to be "*definitive* of the bodies that are able to mix" (Krizan 2018: 201), and that *GC* II 2, 329b30–2 characterizes the wet/liquid and the dry as "easily bounded" (εὐόριστον) and "difficult to bound" (δυσόριστον) respectively. However, that passage is drawing on *GC* I 10, 328a33–328b5, which claims only that what are "especially" (μᾶλλον) or "most" (μάλιστα) subject to mixture are bodies that undergo mutual affection easily (ῥᾷον) and quickly (θᾶττον), and that these are the "most" (μάλιστα) easily bounded: liquids (see Frede 2004: 296n23). I prefer to posit a slight discrepancy with *GC* II 2 rather than one within the continuous argument at *GC* I 10, 328a18–b22. In fact, there is a broader unclarity in Aristotle's use of being easily bounded in the *GC*. Whereas *GC* II 2 associates being easily bounded with taking on external boundaries (and it is not clear why *all* the ingredients of a mixture need to be like this), in *GC* I 10, it is invoked as a condition of divisibility. To complicate matters further, *GC* II 8 says that without any moisture, earth would crumble or fall apart (334a1–3; cf. Andrea Falcon on *GC* II 8 in this volume) and *Meteor.* IV 4, to similar effect, characterizes the wet and the dry as "glue" for one another (381b29–382a3).

[17] Here I read περί (with Joachim) rather than ἐκ (with Rashed), which, despite the attempts of Rashed 2005: 69 ("*Revenons maintenant aux éléments*") and Buchheim 2010: 69 ("*Aufgrund der Elemente*"), seems too difficult a reading. The two readings are of equal stemmatic value according to Rashed 2005: CCXXXVII.

elements and (ii) whether the elements change into one another. The possible answers to these questions specify just two possible positions, since, as Aristotle claims, an affirmative answer to the one question implies an affirmative answer to the other.

Why does intertransformability imply a common matter (as also suggested, albeit in qualified ways, at *GC* I 3, 319a34–b1 and *DC* IV 5, 312a30–33)? Although this question cannot adequately be settled here, let me simply point out that the justification for this thesis earlier in the *GC* is not as clear as one might like. Aristotle does not explicitly appeal to elemental matter where we might expect it most, that is, in the account of elemental transformation (*GC* II 4), though he has discussed it in *GC* II 1.[18] Aristotle might intend that elemental transformation be viewed as requiring the elements' ability to affect one another, which he has claimed requires a common matter in *GC* I 7, but he has not, up to now, indicated specifically that elemental transformation can only be brought about by another element. One might instead take Aristotle to be appealing to the idea that all generation requires a pre-existent underlying thing that persists into the product, perhaps from *Phys.* I, but some have suggested that the persistence of matter is rejected or qualified in the *GC*.[19] Another possibility is that Aristotle is assuming that the four elements' formal natures are exhaustively characterized by shared qualitative contrariety (either based on the very supposition of their intertransformability, or drawing on the analysis of their natures in *GC* II 2), so that the requisite matter will be no more and no less than what is receptive of those contrarieties; if so, one cannot, on these grounds, posit a common matter for the elements independently of positing that their natures are exhaustively characterized by contrariety.

Does common matter *entail* elemental transformation? I take it that a common matter is essentially receptive to the substantial natures of all four elements, so there is in principle no obstacle to their intertransformation. On the supposition that these natures are constituted by the poles of the basic contrarieties specified in *GC* II 2, very basic empirical assumptions entail that elemental transformation will occur.

[18] See Crowley 2019: 1–26 for complications related to the idea that *GC* II 1 concerns the matter of the elements, as well as Timothy Clarke, this volume.

[19] See especially *Phys.* I 7 and *GC* I 4, along with, recently, Broadie 2004: 123–150, O'Connor 2015: 45–75, Charles 2018: 178–205, and Henry 2019: 40–76.

8.4 The Empedoclean Case

Aristotle next points out that both conceptions of the elements face a difficulty; I will depart from the order of the text and discuss his full treatment of the Empedoclean opponent before looking at how the difficulty arises for his own conception of the elements.

> Those who, by contrast, do not make coming to be [of the elements] from one another, not even as from each (μὴ ποιοῦσιν ἐξ ἀλλήλων γένεσιν μηδ' ὡς ἐξ ἑκάστου), except in the way that bricks come to be from a wall – [for them] it is baffling (ἄτοπον) how, from those [elements] there will be bone and flesh and any of the others. (*GC* II 7, 334a18–21)

The denial that the elements come to be from one another (ἐξ ἀλλήλων) is familiar from earlier discussions as something Empedocles claimed (*GC* I 1, 315a4–5; *GC* II 6, 333a16–18). But Aristotle adds that the Empedoclean denies that the elements come to be "as from each" (ὡς ἐξ ἑκάστου), with the qualification "except in the way that bricks come from a wall" (334a19–20). A passage in *DC* III 5 is more explicit, saying that for Empedocles and Democritus alike, earth, air, fire, and water come to be from one another (ἐξ ἀλλήλων) only in the sense that what is already present in some mass is separated out (ἐκκρίνεσθαι) (305b1–5). This suggests that the qualification "except in the way that bricks come from a wall" is meant to apply to both coming to be "as from each" and coming to be "from one another."

But what do these two phrases mean? Williams claims that "from one another" refers to elemental transformation while "as from each" refers to what he sees as the contrasting case of extraction from a compound; he supports the latter claim by observing that the brick wall example is a "generation of simple from compound."[20] That is clearly a fallacious inference and, as we saw, even elemental "transformation" for the Empedoclean will turn out to involve extraction from a larger complex. Furthermore, Aristotle will go on to *argue* that the Empedoclean is committed to the claim that elements come from the compounds flesh, bone, and the like only as bricks from a wall *as a consequence of* (συμβαίνει) the Empedoclean conception of how the elements combine to form those compounds in the first place (334a31–33); that claim should not already be contained in the characterization of the Empedoclean view. The obvious alternative is that ἑκάστου means each *element*, and this is more

[20] Williams 1982: 174

natural insofar as the elements are mentioned in the opening line (334b16), and this mention is picked up just before and after the line in question.[21]

But how is ὡς ἐξ ἑκάστου then different from ἐξ ἀλλήλων, and why does Aristotle mention both? One possibility is to understand the ἐξ ἀλλήλων claim as "for any pair of elements such that one comes to be from the other, the latter comes to be from the former as well," which is consistent with there being no transformation between certain elements (e.g., there might be transformation between dry elements and between wet ones, but not between wet and dry). But this makes the elements coming to be ἐξ ἀλλήλων a logical consequence of their coming to be ἐξ ἑκάστου, while the μή [. . .] μηδέ structure suggests that by denying coming to be ἐξ ἀλλήλων, one is not necessarily already denying coming to be ὡς ἐξ ἑκάστου (in the strong sense, not the "as bricks from a wall" sense). At the same time, the fact that Aristotle writes <u>ὡς</u> ἐξ ἑκάστου ("*as* from each") suggests that coming to be "from each" can also be described as coming to be "from one another." I suggest that, on the one hand, ἐξ ἀλλήλων could be understood to indicate that whenever one element is transformed into another, that very product can be changed back into the first element. On such an understanding, one could deny that elemental transformation is ἐξ ἀλλήλων without yet denying that it is ἐξ ἑκάστου, for there might be cases in which, for example, air comes from fire, and cases in which fire comes from air, but not cases in which air comes from fire that originally came from air.[22] On the other hand, coming to be ἐξ ἑκάστου can certainly be described as coming to be ἐξ ἀλλήλων in a weaker sense: in the sense that fire generally comes from air and air from fire. Aristotle's thought, then, is that the Empedoclean denies that the elements come to be "from one another," even if that is understood in the weaker sense, that is, "as from each."

After claiming that the challenge arises for his own conception of the elements as well, Aristotle returns to the challenge for the Empedoclean:

> For the former, who speak like Empedocles, what way will it be [*sc.* in which flesh, bone and the like come to be from the elements]? For it must be composition, just like a wall from bricks and stones. And this mixture, then, will be from elements that are preserved, but lying next to one another

[21] Namely, it supplies the subject of μεταβάλλειν (17) and the reference of ἄλληλα (334a17), ἀλλήλων (334a19), ἐκείνων (334a20), ἀλλήλων (334a22), αὐτῶν (334a22–23), and αὐτά (334a23).

[22] In *Phys.* I 4, Aristotle ascribes to Anaxagoras the view that the extraction of one stuff from another is always potentially reciprocal in this stronger sense of ἐξ ἀλλήλων – it is always possible to extract some of the initial stuff from the product.

> in small bits. In this way, then, flesh and each of the others will come to be. It follows, then, that fire and water do not come to be from any part of flesh whatsoever, in the way that, from wax, a sphere might come to be from this part, a pyramid from some other one, but either could also have come to be from either part. But this indeed occurs in this way: both from any part of the flesh whatsoever. For those who speak in the former way, however, this is not possible, rather, like stone and brick from a wall, each comes to be from a different place and part. (*GC* II 7, 334a26–b2)

Aristotle first asks how the Empedoclean might explain flesh and marrow coming to be from the elements. Their only option is composition (σύνθεσις), the description of which clearly recalls that at *GC* I 10, 237b31–328a17. One difference, which may prove important, is that *GC* II 7 does not broach the question of how small the neighboring bits of the ingredients are. Note that while Aristotle here uses the term σύνθεσις for the process (τρόπον), he later uses it, as he did in *GC* I 10, to characterize the product: a certain sort of combined entity. Aristotle does not say why the Empedoclean is limited to composition as the process by which flesh comes to be from the elements. The fact that Aristotle understands composition as preserving the ingredient-elements suggests that he is not drawing on the Empedoclean denial of elemental transformation (which does not automatically rule out other forms of destruction), but rather on the denial of a common matter, together with his earlier determination that mutual affectability requires a common matter (*GC* I 7, 324a34–b9). And if flesh does preserve the elements in minute portions within it, then, according to the Empedoclean, each ingredient can only emerge from the portions of flesh occupied by that very ingredient – just as from a part of a wall that consists of a brick one cannot extract a stone.

Flesh, bone, and the like, Aristotle claims, are not in this respect like a wall, but are instead like wax, each part of which can be made into various solid figures. That is, Aristotle posits:

> *Everywhere Recoverability*: Each ingredient is recoverable from any portion of flesh, bone, and the like.[23]

But what exactly does *Everywhere Recoverability* claim, what is its precise role in the argument, and is it plausible as an unsupported premise? These issues are less clear than they may seem. Philoponus seems to take *Everywhere Recoverability* to provide a direct empirical refutation of – a

[23] *Everywhere Recoverability* is in one way stronger than *GC* I 10's claim that the ingredients of a mixture "can be separated again" (327b28–9), which does not specify from which individual portions of the mixture each ingredient is recoverable.

sort of counterexample to – the Empedoclean denial of elemental transformation.[24] And it is surely sufficient for that purpose to observe that masses of flesh can turn entirely (or nearly enough) into one element, for example, water, as, Philoponus seems to do (*In GC* 269.22–24). At the same time, Philoponus (cited approvingly by Williams) constructs an objection to this argument on behalf of the Empedoclean, that if the individual elements are distributed throughout some mass of flesh in imperceptibly small bits, we cannot observe whether an element can be recovered from every part of that flesh.[25] This kind of objection seems best suited for the case of a mixture being resolved into the original ratio of ingredients, such that empirically confirming *Everywhere Recoverability* would require distinguishing *from which* bit each element-ingredient is extracted. However, if the function of Aristotle's argument is as Philoponus initially suggests, that is, to directly refute the denial of elemental transformation, so that it can be grounded in the observation that masses of flesh can be turned entirely (or nearly enough) into water, then the objection seems beside the point.[26] At the same time, if that is the observational basis of Aristotle's argument, then the argument does not tell us anything about whether flesh is a composition, since, if there is elemental transformation, then even a composition of elements can be turned entirely into any one of its ingredients. The notion of composition turns out to be dispensable in the argument, and for all that has been said, Aristotle himself could hold that flesh is a composition of elements.[27]

What we have been considering, in effect, is a reading of *Everywhere Recoverability* as:

> *Weak Everywhere Recoverability*: For each ingredient, that ingredient can come to be from every portion of flesh (by some process or other).

What if, by contrast, *Everywhere Recoverability* is meant directly to contradict the view that flesh, bone, and the like are compositions, and only on that basis, the Empedoclean denial of elemental transformation? If so, I suggest, it is most plausibly understood as:

[24] See Philoponus, *In GC* 269.24–25, whose ὑπόθεσις is that posited at 269.6–7 and mentioned at 12–13.

[25] Philoponus, *In GC* 269.25–35; Williams 1982: 174.

[26] The observation that a mass of flesh can turn entirely (or nearly enough) into water is also problematic, but it does not beg the question against the Empedoclean opponent.

[27] A more flexible notion of composition, whose limits are fixed by the denial of elemental transformation, is not threatened by the above concern: this notion of composition allows any affection of the element-ingredients that does not provide a path for transformation. This is not, however, the notion of composition that Aristotle employs (334a29–30).

> *Strong Everywhere Recoverability*: For each ingredient, that ingredient can come to be from every portion of flesh *by the same process.*

Strong Everywhere Recoverability seems to imply that the flesh is uniform with respect to the features responsible for the recoverability of each ingredient. The recovery of an ingredient from some portion of flesh is plausibly taken to be a manifestation of that ingredient's presence in that portion of the mixture – according to Aristotle, a kind of potential presence. If so, *Strong Everywhere Recoverability* implies that each ingredient is present uniformly throughout the flesh, which contradicts the composition model he attributes to the Empedoclean. As a point against the view that elemental mixture is composition, *Strong Everywhere Recoverability* plays roughly the same role here as homeomery at *GC* I 10, 328a10–12.

While *Weak Everywhere Recoverability* can be given a reasonable observational basis, *Strong Everywhere Recoverability* is subject to a worry close to, but distinct from, that of Philoponus' Empedoclean objector: a defender of the composition model could object that if flesh is a composition of sufficiently small bits, we cannot empirically rule out that when it becomes water, different bits of it are undergoing different processes – for example, bits of earth are being moistened, bits of air being cooled – while some bits remain unaffected. To this extent, *Strong Everywhere Recoverability* is questionable as an ungrounded premise in Aristotle's argument, and we might thus favor the weaker reading, along with the idea that the argument targets the Empedoclean denial of elemental transformation alone.

But Aristotle does seem to be invoking the stronger version. The course of his argument suggests that his immediate target is the composition view of mixture. For he makes a point of bringing out the Empedoclean commitment to viewing flesh, bone, and the like as compositions, and only then draws the consequence that their ingredients could only come to be as bricks from a wall. Relatedly, Aristotle emphasizes the "way" (τρόπος) in which bone and flesh arise from the elements throughout the discussion. Note also that, a few lines below, Aristotle will help himself to the claim that flesh, bone, and the like are not compositions. Though that claim might be arrived at by other means (see the next Section), the most natural reading is that he is relying on the anti-Empedoclean argument in the first half of *GC* II 7. Finally, the wax example, like other wax examples, seems to involve wax that is uniform and not already pre-shaped, at least not in such a way that one would require different processes to turn different portions of it into a sphere (this bit of wax can be made directly into a sphere, this bit needs to be smoothed out first, this one melted down, etc.); there is no mention of such processes.

Aristotle's argument, then, is most plausibly taken to invoke *Strong Everywhere Recoverability*, despite its questionable empirical basis. It is nevertheless reassuring that even *Weak Everywhere Recoverability* is sufficient to refute, not the claim that flesh, bone and the like are compositions – which Aristotle has other reasons for rejecting anyway – but the Empedoclean conception of the elements. And doing that is the main goal of this part of the chapter.

8.5 The Aristotelian Case: The *Aporia*

The upshot of the discussion of the Empedoclean position is that if flesh, bone, and the like are to come to be from the elements, the elements must be intertransformable by virtue of a common matter. But near the beginning of the chapter, Aristotle points out that the challenge arises even on that Aristotelian conception of the elements:

> But what was said involves difficulty (ἀπορία) even for those who generate [the elements] from one another: in what way does something else, over and above them, come to be from them? I mean, e.g., water can come to be from fire and fire from it, for there is something common underlying; but also flesh comes to be from these, and marrow too. How then do these come to be? (*GC* II 7, 334a21–26)

After his discussion of the Empedoclean position, Aristotle develops the challenge for his own conception of the elements as follows:

> How there will be something from both of them, e.g., from hot and cold or fire and earth, similarly involves difficulty even for those who posit a single matter for these. For if flesh is from both and is neither of them, nor yet a composition of things preserved, what remains except that the matter is what is from them? For the destruction of the one produces either the other or the matter. (*GC* II 7, 334b2–7)

Aristotle does not mean that positing a common matter, and thereby the intertransformability of elements, *creates* an obstacle, real or apparent, to accounting for the formation of flesh, bone, and the like from them. Rather, positing a common matter does not by itself obviously furnish such an account, since it does not obviously add a plausible option for what might come to be from two element-ingredients to those options already on the table.

Aristotle seems to be treating two options as already ruled out. First, flesh cannot be (a) one of the element-ingredients (the other having been transformed into it). In *GC* II 7, Aristotle takes for granted that flesh, bone, and the like, will each be something "different from, over and above" (334a23) the ingredients, as stated in the first passage above, or "neither of them [the ingredients]" (334b4) as he puts it in the second. Second, flesh cannot be (b) a composition of element-ingredients (both being preserved). This option, I suggested, has been ruled out by *Everywhere Recoverability* in the first part of the chapter.[28] The Aristotelian conception of the four elements presented in the chapter's opening lines provides a third option: that flesh is (c) the matter common to the elements. Aristotle does not say why he rejects this third option: perhaps its falsity is evident; or perhaps he has in mind that elemental matter would become a separate perceptible body and usurp the elemental status of earth, air, fire and water, or that we would have to identify all of flesh, bone and the like with the common matter, and so with one another.[29]

Aristotle's statement of the *aporia* might seem innocent of the resolution of *GC* I 10's initial *aporia*, in particular, of the idea that the ingredients are potentially present in the mixture; indeed, the three options mentioned here seem to mirror the three options considered in that earlier *aporia* – that both ingredients remain, that one remains while the other is destroyed, and that both are destroyed (*GC* I 10 327a34–b6).[30] And so this part of *GC* II 7 might seem to take several steps backward, replaying the initial moves of *GC* I 10. Now, since the common matter of fire and earth – option (c) – is potentially each, at least some kind of potential presence of the ingredients is under consideration and evidently insufficient to resolve the *aporia*. But also, while *GC* I 10's initial *aporia* concerned the fate of the ingredients in a mixture, here the initial question is in terms of what might result *from* a combination of element-ingredients. Finally, the *aporia* here presupposes Aristotle's identification of earth, air, fire, and water as elements and their intertransformability through a common matter, both of which would be out of place in *GC* I 10.

8.6 The Aristotelian Case: The Solution

Thus, a crucial part of Aristotle's answer is that *his* elements can combine so as to give rise to something other than the three options mentioned so

[28] Aristotle is not, I think, relying on *GC* I 10's claim that composition is not mixture, since flesh, bone, and the like have not yet been characterized as Aristotelian mixtures.

[29] I owe the last suggestion to Sean Kelsey.

[30] See Krizan 2018: 193–195 on these parallels.

far. But first recall that I will flag indications that Aristotle is attempting to show that *mixture* – in the sense of *GC* I 10 – can be instantiated at the elemental level, though that is not the chapter's explicit aim. I will also argue in passing that certain passages that focus on contrary properties do not, as is sometimes suggested, show that these, rather than the four elemental bodies, are the genuine ingredients of elemental mixtures.

In the first part of Aristotle's solution, the claim that elemental contrariety admits of intermediacy makes room for a product of elemental combination distinct from the three options considered earlier.

> Is it, then, since something can be more and less hot and cold, that whenever (ὅταν) something should be the one without qualification in actuality, it will be the other in capacity, but whenever it is not completely hot or cold, but rather in the one case hot-cold, so to speak, and in the other cold-hot, because things being mixed destroy the excesses of one another (διὰ τὸ μιγνύμενα φθείρειν τὰς ὑπεροχὰς ἀλλήλων), then (τότε) it will be neither the matter nor either of the contraries in actuality without qualification, but in between, and according to its being in capacity more hot than cold or *vice versa* – according to this ratio – twice as hot in capacity as cold, or three times, or in some other such way? (*GC* II 7, 334b7–16)

Aristotle contrasts the qualitative condition, both potential and actual, of individual elements with that of elemental mixtures. First, whenever something is "without qualification" hot in actuality, that is, "completely" (παντελῶς) hot, like fire, it is "the other," that is, cold, potentially (δυνάμει). But during the process of mixture, the element-ingredients destroy each other's "excesses" – in this case hot and cold without qualification.[31] The result is neither extreme in actuality but rather something intermediate. These claims recall the claims in *GC* I 10 that the ingredients of mixture affect one another (328a31–33; 328b14–22), and that such mutual affection displaces the ingredients out of their own natures, leading to something intermediate (328a29–31). The phrases "hot–cold" and "cold–hot" indicate different positions within a range of intermediate positions between the extremes. In the next lines, Aristotle claims that the mixture is intermediate also with respect to potentiality: it is somehow

[31] I have supplied a subject for forms of εἶναι throughout this passage, so that it concerns whether some body is actually or potentially hot or cold or in between, just as it must concern whether some body is the matter; similarly, I take μιγνύμενα ("things being mixed") to refer to the ingredient-elements. On both points I follow Williams 1982 and Buchheim 2010. Rashed 2005 and Krizan 2018: 207–209 (and possibly Joachim 1922: 242) read διὰ τὸ μιγνύμενα φθείρειν τὰς ὑπεροχὰς ἀλλήλων (334b11–12) as claiming that the contrary properties mix and destroy one another's excesses, which would contrast with *GC* I 10's claim that only bodies mix. I find the notion of contrary qualities *having* excesses difficult: those qualities just *are* the relevant excesses.

potentially both hot and cold, and in a particular ratio, for example, potentially twice as hot as cold. Aristotle must mean to link the actual character of the mixture with what it is potentially – the kind of link Aristotle makes in the case of what is completely hot and thereby potentially (only) cold. That is, the description of the mixtures' intermediate potentialities should be seen as filling out the consequent of the ὅταν … τότε ("whenever […] then") construction (334b9–11). As with their actual conditions, Aristotle here implies a range of possible intermediate conditions with respect to potentiality.

What does it mean to be potentially hot and cold in a particular ratio? Two different answers could accommodate the idea that what a mixture is potentially corresponds to its intermediate actual character. It is natural to think that since fire, being "altogether hot," is potentially just cold, the degree to which a mixture is potentially hot will increase according to its increasing unlikeness to fire – according to its increasing coldness and its having less fire as an ingredient. Thus, a mixture that is actually more hot than cold will be potentially more cold than hot. On this first reading, a mixture's being potentially hot would seem to reflect the extent to which it is hot *merely potentially*. One might say that the presence of heat is *more potential* in a relatively cold mixture than in a relatively hot mixture. However, Aristotle's text suggests not varying degrees to which a feature's presence is potential, but that, according to potentiality (which is not clearly a matter of degree), a mixture is, for example, more hot than cold. On a second reading, the degree to which a mixture is potentially hot will increase according to its *nearness to* absolutely hot fire – according to its increasing hotness and its having more fire as an ingredient. But somewhat paradoxically, increasingly hot, and so increasingly potentially hot, mixtures, approach a limit case of something that is actually hot but (as here described) potentially only cold. There is no real tension here, because Aristotle has in mind two different processes. Absolutely hot fire is potentially cold because it can become qualitatively different than it is, as a whole, while the mixture is potentially twice as cold as hot because it can give rise to (absolutely) cold and hot things in the appropriate ratio (see below), where this process does not involve a net qualitative addition or subtraction. The mixture's potentiality is, in a sense, directly reflective of its actual condition, which is in turn a function of the ratio of its original ingredient elements.[32]

[32] Contrast both these readings with that in Bogen 1996, which ties Aristotle's use of δυνάμει to the presence of the elements' characteristic capacities: "the presence by ability [δυνάμει] of an element

An advantage of the second reading is that it accommodates, and indeed suggests, that Aristotle has in mind not just the mixture's being potentially, for example, hot and cold in some ratio, but in particular, its being potentially the *element-ingredients* in the original ratio, again, by reference to a process that does not involve a net qualitative gain or loss; that is, what is at issue here is the mixture's ability to be *resolved* into its ingredients. This, as we will see, fits well with the discussion of a mixture's potentiality and of the extraction of the ingredient-elements to come.

If this second reading is correct, and the elemental mixture's intermediate potentiality is that reflected in its resolution into its ingredients, then it is quite natural to take the contrast case, the potentiality of what is "without qualification" or "completely" hot, to be the potentiality of one element for another; indeed, the context suggests as much. Moreover, this contrast case will cover all three types of elemental transformation specified in *GC* II 4, since they all involve in a net qualitative addition.[33] Thus, on the reading proposed, *GC* II 7 ascribes to elemental mixtures a distinctive sort of potentiality for the elements, which will be further elucidated below.[34]

Why then does Aristotle here speak of being potentially hot and cold, rather than being potentially the element-ingredients? In fact, this is the only passage that speaks unambiguously and exclusively of contrary properties where we might expect the elements or contrary things. The reason, I suggest, is that Aristotle wants to compare different numerical ratios of recoverable ingredients, and speaking of the ratio of qualities, that is, of elemental capacities, provides a convenient way of doing so; it also mirrors his reference to commensurate capacities when describing the requirements of mixture at *GC* I 10, 328a28–29. By contrast, when a quantitative specification of that ratio is not needed, for example, when specifying the recoverability (or potential presence) of the ingredients as such (no matter in what ratio), or their recoverability in the *original* ratio, Aristotle speaks instead of the potential presence or recoverability of the ingredient bodies themselves (see below).

Aristotle asserts not only that what is (actually) intermediate in temperature is neither completely hot nor completely cold – that is, neither of the

in a compound consists of (nothing more than) the possession by the compound of the relevant non-maximal abilities" (192).

33 On the types of elemental transformation in *GC* II 4, see Krizan 2013: 195–226 and Margaret Scharle in this volume.

34 See especially note 38 below, on the possible significance of this result for a troublesome passage in *GC* I 10.

ingredients – but also that it is not the matter, thereby recalling the third option canvassed in the *aporia*. The need to distinguish his answer from matter arises because both are in a sense between the extreme elemental contraries, and because elemental matter is also somehow potentially all of the elements. Thus, in the following lines Aristotle emphasizes that flesh, bone, and the like are potentially the elements in a distinctive way.

> The other things [*sc.* flesh, bone, and the like], then, will be from the contraries or the elements, these having been mixed, and the elements will be from those [*sc.* flesh, bone, and the like], which are in capacity [the elements] in a way; not however in that way, namely as the matter [is in capacity the elements], but in the way stated. And what is coming to be [something] is in this way mixture, but in the former way matter. (ἔσται [17] δὴ μιχθέντων τἆλλ᾽ ἐκ τῶν ἐναντίων ἢ τῶν στοιχείων, καὶ [18] τὰ στοιχεῖα ἐξ ἐκείνων δυνάμει πως ὄντων, οὐχ οὕτω δὲ ὡς [19] ἡ ὕλη, ἀλλὰ τὸν εἰρημένον τρόπον· καὶ ἔστιν οὕτω μὲν μίξις, [20] ἐκείνως δὲ ὕλη τὸ γινόμενον.) (*GC* II 7, 334b16–20)

I take "the other things" (τἆλλα: 17) to be flesh, bone, and the like, and treat the participle μιχθέντων ("having been mixed") at line 17 as governed by τῶν ἐναντίων ἢ τῶν στοιχείων ("the contraries or the elements": 17). Further, I read "contraries" here in a loose sense that can pick out the four elements (cf. *GC* II 8, 335a3–9); indeed, Aristotle specifies that he means to refer to them with the phrase ἢ τῶν στοιχείων, which might be paraphrased as "or, more precisely, the elements."[35] This means that flesh, bone, and the like are mixtures of the four elements. In the other direction, the elements are from "those" (ἐκείνων: 334b18), that is, flesh, bone, and the like, on account of the latter being potentially, and specifically, "in the way stated" (334b19). This reference can only, I think, be to the intermediate potentiality of a mixture described above at 334b7–16, which not only confirms that we should take "those" to be flesh, bone, and the like, but also suggests that these stuffs are being said to be *the elements* potentially; we should read ὄντων (18) transitively, with an understood complement, τῶν στοιχείων. The fact that Aristotle specifies an elemental mixture's potentially being the elements (334b17–18) by reference to an

[35] For the move between looser and stricter senses of "contrary" see *Phys.* I 5–7 and *Phaedo* 103 B1–C2. Krizan 2018: 187–226 reads "contraries" here more strictly and takes "the contraries or the elements" to indicate that the strict contraries are the genuine elements. Against that, in the very next clause τὰ στοιχεῖα refers exclusively to the material elements. On the broader question whether Aristotle characterizes the elemental contraries as "elements" in *GC*, see Crowley 2013. For a reading of *GC* II 7's account of elemental mixtures without prime matter, and on which the strict contraries are the real "elements," see Frede 2004: 303–308.

earlier discussion of its being potentially *hot* and *cold* in various ratios (334b7–16) suggests once again that we should not make too much of his use of contrary properties in the former passage. Aristotle's main contrast in 334b16–19, then, is between two ways of being potentially the elements, the one characteristic of an elemental mixture and the other characteristic of elemental matter. If so, the following distinction between matter and elemental mixture as things that are "coming to be" (γινόμενον: 334b20), which seems to draw on the immediately preceding contrast ("in this way ... in the former way": 334b19–20), is best taken to concern matter and elemental mixture not as the products of generation, but rather as the starting points that undergo a process of generation, namely, generation of the elements, and according to the transitive use of γίγνεσθαι.[36]

We have, then, three different sorts of potentiality for the elements: (i) the *mixture's intermediate potentiality for the ingredient-elements* is what ties together 334b7–20. This potentiality is contrasted first with (ii) *one element's potentiality for another*, and then with (iii) *matter's indeterminate potentiality for the elements.* Earlier I suggested that (i) the mixture's intermediate potentiality for the ingredient-elements is to be understood by reference to the recovery of the ingredients in the original ratio, with no net qualitative addition to the whole. Note also that from the Aristotelian-Empedoclean identification of earth, air, fire, and water as the elements, it follows that flesh, bone, and the like, themselves perceptible bodies, are in some sense "out of," and of higher complexity than, the elements, so that, importantly, flesh, bone, and the like will be potentially their ingredient-

[36] My understanding of ἐκείνων (334b18) as referring to flesh, bone, and the like, and of ὄντων (334b18) and γινόμενον (334b20) as transitive, follows Williams 1982, de Haas 1999: 27–28, and Buchheim 2010. I believe it captures the tight relations between the two contrasts in this passage better than the readings of Joachim 1922: 243, Rashed 2005, and Krizan 2018: 210–214, who take the passage roughly as follows: "The other things [*sc.* flesh, bone, and the like], then, will be from the contraries or rather the elements, these having been mixed, and the elements will be from those [*sc. the contraries*], which exist potentially in a way, not however in that way, namely as the matter [exists potentially], but in the way stated. And what comes to be [i.e., the product] is in this way mixture, but in the former way matter." Joachim and Krizan take Aristotle's alleged claim that the elements are "from" the contraries to capture the dissolution of a mixture, but the claim I ascribe to Aristotle, that elements are from the mixtures, does so more straightforwardly. It is also unclear in what sense "matter" is to be understood as a product of generation "in that [former] way" (ἐκείνως: 20). Joachim 1922: 243 takes ἐκείνως to refer back to the statement of the *aporia* at 334b6–7, where matter appears as an option for the result of the combination of elements, but that is, first of all, too far back in my view, and second, that passage does nothing to legitimate thinking of matter as "generated" from the elements. Krizan preserves a connection with 334b16–19 by taking the "matter" here and at 334b18 to be the elements, which would depart from the usage of "matter" elsewhere in this chapter.

elements in connection with a *downward* process of resolution. That is, the (i) distinctive potentiality of mixture is exercised in a transformation from something of greater complexity to simpler element-ingredients. The first contrast case, (ii) one element's potentiality for another, concerns a lateral change to something of the same level of complexity, and a net qualitative change that is one contrary's replacement of another. Though Aristotle does not explain how, (iii) elemental matter's potentiality seems to differ from both cases and in both respects. First, matter seems to be potentially the elements in an upward direction, to the extent that the elements are of greater complexity than matter. Second, matter's potentiality is in a certain sense realized through an addition of contrariety, but not in the same sense as an element's potentiality for another element: matter is, in itself, not qualified by contrariety at all (even though it only exists with contrariety), so that *qua* potentiality of *matter*, the potentiality is not essentially realized in the replacement of an intrinsic contrary (i.e., one intrinsic to matter); in addition, matter's potentiality can be realized in a qualitative change that is not between contrary poles.

Of course, the hesitant formulations above reflect the fact that though Aristotle speaks of matter as "coming to be" (334b20) the elements, he cannot have in mind a case of bare matter taking on contrariety; matter's potentiality is present in something with a higher and comparatively more formal nature, and its realization as a taking on of some feature requires that it be characterized initially by some opposed feature. Nevertheless, in a case of fire turning into earth, it is the matter that is fundamentally potentially earth and receptive of coldness, so that even if (iii) matter's potentiality is not understood in terms of exchanging contraries, it is the fundamental passive principle involved in such a change. The exercise of (ii) an element's potentiality for another element fundamentally involves the exercise of (iii) matter's potentiality.[37] I would like to suggest that the same can be said – at least nothing in the above passage speaks against it – of (i) an elemental mixture's potentiality for its ingredients in the original ratio. For the dissolution of an elemental mixture can be understood to consist in small portions of matter, which currently possess an intermediate character, exchanging it for a contrary property along the same qualitative dimension, that is, bits of matter realizing (iii) their characteristic potentiality for the elements (mixing will involve the opposite process).

[37] A small selection of relevant passages: on matter's passivity (*Phys.* I 6, 189a21–27; *GC* I 7, 324a32–b10); on matter's potentiality (*Phys.* I 9; *GC* I 4, 320a2–5; *GC* II 1, 329a33), and on matter's relation to contrariety (*Phys.* I 6, 189b2–8; *GC* II 1, 329a24–32).

We might even want to go one step further and say that the *only* passive principle involved in both elemental transformation and the dissolution of mixtures (as well as their formation, on which, see below) is (iii) elemental matter's potentiality for each element. At any rate, the (i) distinctive sort of potentiality Aristotle ascribes to an elemental mixture, for its ingredients in the original ratio (like (ii) that of one element for another), can be thought of as specifying particular ways that (iii) the basic potentiality of elemental matter can be exercised because of its current configuration. To this extent, it does not seem to require that a given ingredient-element enjoy a stronger or otherwise special grade of potential existence in certain portions of a mixture than it does in another element.[38]

The final passage of the chapter is puzzling because Aristotle seems largely to repeat himself: he explains, once again, how a mixture with intermediate character arises through the mutual affection of elements, given that they are in some sense contrary to one another. I would like to suggest that one main function of the passage is to situate the formation of elemental mixtures within the broader framework of his elemental theory, and in particular, to view it as a consequence of the elements' ability to be affected in their natures, an ability that is exercised also in elemental transformation.[39] As such, it is the most basic ability that explains *all* generation from the elements and, according to the suggestion at *GC* II 1, 329a5–8, thus enables them to serve *as* elements. As I have emphasized, this ability is also unique to the elements as *Aristotle* conceives them.[40]

> Since the contraries are also affected according to the earlier account (for what is in actuality hot is in capacity cold, and what is in actuality cold is in capacity hot, so that if they are not commensurate, they will change into one another, and similarly in the case of the other contraries), the elements

[38] The contrast between an elemental mixture's being potentially its ingredients in the original ratio (without a net qualitative change) and one element's being potentially another provides one way of fulfilling a common interpretive desideratum regarding Aristotle's solution to first *aporia* of *GC* I 10 (327b24–25; cf. 328b17–19). There, the potential presence of the ingredients in a mixture is meant to provide an alternative to their "destruction" and interpreters have often viewed elemental transformation as the relevant case of destruction (a claim I have questioned in Anagnostopoulos 2021). In particular, Aristotle is often thought to require an especially strong potential presence of the ingredients in mixture (see Section 8.7), in order to distinguish elemental mixing from elemental transformation (each element is potentially the others, after all). I have argued that *GC* II 7 ascribes to mixtures a distinctive potentiality for their element-ingredients, one which does not require such a strong presence.

[39] Note that, in *GC* I 6, Aristotle introduces mixture as one of the phenomena appealed to by "both those who generate the elements and those who [generate] the things that are from the elements" (322b6–7).

[40] These are surely good reasons for including a substantial treatment of being affected (*GC* I 7–9) in preparation for the treatment of the elements.

> also change primarily in this way (καὶ πρῶτον οὕτω τὰ στοιχεῖα μεταβάλλει), and from them flesh and bone and such things [come to be], the hot becoming cold, the cold hot, when they go toward the intermediate; for neither [hot nor cold] is there [in the intermediate], though the intermediate is wide and not indivisible. And in the same way, dry and wet and such things produce flesh and bone and the others according to intermediacy. (*GC* II 7, 334b20–30)

Aristotle first appeals to the idea from *GC* I 7–9 ("the earlier account") that contrary things can affect one another (*GC* I 7, 323b29–324a8); there, this condition is subsequently spelled out in terms of a common matter (*GC* I 7, 324a34–324b9) and being potentially such and such (*GC* I 9, 326b29–327a1), the second of which is close to the illustration that at *GC* II 7, 334b21–23: that what is actually hot is potentially cold, and *vice versa*. Thus, as in *GC* I 7, the "contraries" he has in mind here are not contrary qualities, but what we might call contrary things, and to this extent, the matter underlying the relevant contrary qualities, though unmentioned in this passage, is still in view. Now, I take Aristotle to be reminding us of a *general* claim, that contrary things affect one another. This general claim is then illustrated in a parenthetical remark, which involves the particular case of the hot and the cold: what is hot is potentially cold and vice versa, so that when one of these is preponderant, the other will turn into it. Even this parenthetical remark need not be taken specifically to concern elemental transformation. The following claim, that "the elements change primarily in this way" (πρῶτον οὕτω τὰ στοιχεῖα μεταβάλλει: 334b24–25) is non-parenthetical; it is the apodosis of "since the contraries are affected according to the earlier account" (334b20–21), on the assumption that the elements are contraries. I take the claim to be simply that the elements are affected by one another ("change . . . in this way"); for the elements, such change will be more specifically affection in their natures. It is perhaps because of the significance of this kind of affection for the status of the elements *as* elements, noted above, or because it is the most basic kind of change they undergo, that Aristotle describes it as the "primary" way in which they change.[41] Given that the elements are subject to such affection by one another, they can combine to form flesh, bone, and the like, provided that they are

[41] Williams 1982: 178, by contrast, takes Aristotle to indicate that the "first" way the elements change is by becoming other elements, with mixing being a second way, though Aristotle does not explicitly speak of a "second" way. See also Joachim 1922: 244 ("two different consequences"). Marwan Rashed and Thomas Buchheim understand the text as saying that it is primarily the elements that change in this way, but the word order favors my reading.

roughly commensurate in their initial quantities. For in that case, and unlike the case mentioned earlier in the passage (334b23–24), the elements will affect one another reciprocally, so that each is brought part of the way toward the other, to one among a wide range of intermediate positions. In appealing to mutual affectability (once again) and its conditions, as well as to the need for roughly commensurate quantities to avoid mere assimilation to the preponderant ingredient, the text closely parallels *GC* I 10.[42]

Aristotle's explanation of the formation of elemental mixtures – his response to the second version of *GC* II 7's *aporia* – does not merely point out that the elements are subject to affection by one another. It also claims that in certain circumstances, they affect one another reciprocally, and in doing so, reach an intermediate qualitative condition. And this goes beyond not only positing a common matter for the four elements, but also the entire elemental theory as presented earlier in *GC* II, which nowhere mentions that the two basic contrarieties admit of intermediacy at the elemental level. For example, *GC* II 4 is consistent with the view that elemental transformation is a matter of portions of elements switching instantaneously between contrary poles, without passing through qualitative intermediates. *GC* II 7, then, not only supports, but also adds to the elemental theory presented earlier in *GC* II, by raising and responding to a genuine, even if not particularly troublesome, challenge for that theory.[43]

8.7 The Nature of Elemental Mixtures: Uniformity and Stratification

According to a prominent approach to mixture in Aristotle from late antiquity to the present, the ingredients enjoy an especially strong grade of potential presence – stronger than what is ordinarily meant by potential presence and stronger than the sense in which one element exists potentially in another. Most famous is Philoponus' comparison of the presence of the ingredients in a mixture to the "tempered" (κεκολασμένως) presence of geometrical knowledge in a drunk geometer (*In GC* 188.16–26). This strong presence is sometimes understood in terms of a kind of *stratification*, such that the ingredients are present as subjects of a sort below the level of appearance. Thus, Theodore Scaltsas compares the ingredient-natures in a

42 See *GC* I 10, 328a18–33 and 328b14–22.

43 In particular, it is a genuine challenge even if a common matter cannot be motivated independently of the idea that the elemental natures are characterized by shared contrariety.

mixture to "two springs pressing against each other" once they have reached equilibrium.[44] Like the drunk geometer's knowledge, the ingredient-natures, according to Scaltsas, are being exercised in the mixture; they are thus not absent, but rather present within the mixture at a lower level than the mixture's intermediate character. But unlike the drunken geometer's knowledge, the ingredient-natures are natures of substances or at least bodies, so that their presence below the level of appearance suggests that the ingredients remain as (something approximating) metaphysical subjects for the supervening intermediate character. This subject-like presence makes it especially difficult to see how two different ingredients or their natures could both be present uniformly throughout the mixture. Thus, Scaltsas also claims that mixtures are *non-uniform* with respect to the strong potential presence of the ingredients: different ingredients are present (potentially) in different portions of the mixture.[45] Now, it is unclear whether Aristotle's elemental theory allows for the presence of *any* element-ingredient or its nature "below" the mixture's qualitative character, given that they are, respectively, intermediate and extreme with respect to the same qualitative contrarieties.[46] That issue aside, I will here focus on the text of *GC* II 7 and argue, based on the foregoing analysis, that there is no hint of, and much to tell against, both the stratification and the non-uniformity of mixtures. *GC* II 7 suggests that elemental mixtures are fully uniform, even in their being potentially each ingredient, and that they are unstratified in the sense that the mixture's intermediate qualitative character is predicated directly of the common elemental matter.[47]

44 Scaltsas 2009: 252.

45 Note that Fine 1995: 266–369 and Sharvy 1983: 439–457 attempt to account for a strong (for Sharvy, actual) presence of the ingredients without non-uniformity. I address these issues in relation to *GC* I 10 in Anagnostopoulos 2021.

46 For a negative verdict, see Cooper 2009: 148–160. It should be noted that Aristotle seems to allow that an element can have one contrary property "intrinsically" while possessing a different quality along the same range of properties "extrinsically." See *GC* II 2, 330a13–24 and *PA* II 2–3 (along with Fine 1995: 348–349). Aside from the distinction between a level of outward appearance and a lower level at which the elements or their natures somehow persist, another key distinction found in late ancient and medieval commentators is between the elemental "form" and the characteristic qualities of the elements. See de Haas 1999: 21–46 and Weisberg-Wood 2004: 681–706 on the late ancient and medieval interpretive tradition.

47 Mixtures must nevertheless enjoy a greater complexity than the elements in one sense, but this is not a predicative complexity. According to *Meta.* X 7, intermediate qualities are "from" or "out of" (ἐκ), and in a certain sense "compounds" (σύνθετα) of, the corresponding contrary poles (See Castelli 2018: 170–198 as well as *Sens.* 3). The view that the mixture's intermediate character is predicated directly of elemental matter is what Kit Fine calls "leveling" (Fine 1995). I owe much to his defense of this view.

To start, *GC* II 7's elemental mixtures have a uniform and intermediate *actual* character, while the ingredient-elements are said to exist only potentially in them. This is after all what distinguishes Aristotelian mixtures from Empedoclean compositions. Moreover, Aristotle clearly says that *each ingredient* is changed, its "excesses" destroyed, so as to arrive at an intermediate temperature. What about the characteristic *powers* of the ingredient-elements, such as fire's capacity to heat? What Aristotle here describes as heat and coldness in actuality just *are* such powers, so that insofar as the mixture is uniform in actual temperature, there does not seem to be room for it to possess – at least not in actuality – the characteristic powers of the ingredients.[48] And whatever we make of Aristotle's theory of elemental matter, he stresses here that it is something common to the elements, so that its presence in the mixture will not introduce non-uniformity.

Of course, the view I am opposing posits non-uniformity concerning the (strong) *potential* presence of the ingredients, understood in terms of the presence of their natures at a lower level. However, it is not easy to see how *GC* II 7 might leave room for such a difference between the potential and actual character of a mixture. As we saw, Aristotle links the actual condition of an element or mixture with what it is potentially: the mixture's intermediate actual temperature, for example, translates into being potentially hot and cold in a certain ratio, which I have suggested is just a way of characterizing the mixture's being potentially (in a distinctive sense) the ingredients. If the ingredients were non-uniformly present, the intermediate potentiality would have to be understood as a property of the mixture as a whole, but not of (many of) its parts: an intermediate potentiality of the whole would correspond to, or consist in, some portions of the mixture being potentially just hot and others being potentially just cold. These portions would not conform to Aristotle's correspondences between intermediate actual temperature and intermediate potentiality, as well as between absolute actual temperature and absolute potentiality (i.e., for just hot or just cold). For while intermediate in actual temperature, some of them will be potentially just hot, others potentially just cold, albeit differently from the way in which fire and water are potentially just hot and cold. Positing non-uniformity in the element-ingredients' potential presence thus limits and departs from the scheme of actual and potential qualitative conditions that Aristotle offers us here.

[48] Again, see Cooper 2009: 148–160.

A second relevant passage is the wax analogy, which tells us that each of the ingredients can be extracted from each part of the mixture. I have argued that this claim should be understood as sufficient to counter the composition view of mixtures, that is, as *Strong Everywhere Recoverability*, which, as I noted above, seems to imply that the mixture is uniform with respect to the potential presence of each ingredient.[49]

Finally, concerning the alleged subject-like presence of the ingredients and persistence of their natures below the surface, note that Aristotle is especially concerned to understand mixture as arising through mutual affection, and in particular, a mutual affection that is explained by the elements' mutual contrariety and by the fact that, for example, "what is in *energeia* hot is in capacity cold, and what is in *energeia* cold is in capacity hot" (*GC* II 7, 334b21–22). Aristotle refers to the general phenomenon that contraries affect one another, so that they are "driven out of their own natures" (*GC* I 7, 323b28–29; cf. *GC* I 10, 328a29–30). In the case of elemental mixture, the ingredients destroy each other's "excesses" (334b11–12). As I have claimed elsewhere, this explanation of mixing does not sit well with the idea that the ingredients or their natures remain present at a level below the new intermediate character.[50] It suggests instead that contrary features – in this case, the elemental natures – *give way to* intermediate ones.[51]

GC II 7 seems to present, then, a picture of flesh, bone, and the like as fully uniform, without the element-ingredients present in a subject-like way, mediating the predicative relation between a mixture's intermediate qualitative character and elemental matter. We have also seen that the chapter provides plentiful indications that flesh, bone, and the like meet various conditions on mixture specified in *GC* I 10; moreover, attention to the role of mixture in the *GC* suggested that *GC* I 10's account must be intended to apply to the elemental mixtures of *GC* II 7. Thus, my analysis of *GC* II 7's elemental mixtures, if correct, gives us good reason to reconsider interpretations of *GC* I 10 on which it requires that mixtures are non-uniform or stratified.

[49] Scaltsas 2009: 256–257 considers the two options for the wax analogy but opts, cautiously, for the weaker reading.

[50] Anagnostopoulos 2021: 158.

[51] If the elemental contraries do not persist below the surface, the formation and dissolution of elemental mixtures presents a challenge for those who have appealed to the need for a persistent contrary property in order to buttress the "stepwise" account of elemental transformation (on which, see Krizan 2013: 201–203 and Scharle in this volume), as well as for those who have grounded non-traditional accounts of prime matter in the availability of a persistent contrary in every change of the elements.

9

On Generation and Corruption *II 8*

Andrea Falcon

9.1 Introduction

Janus Bifrons (*Ianus Geminus*), the god of ends and new beginnings, and more generally of transitions and passages, is often represented as having two faces looking in opposite directions – a graphic way to illustrate his capacity to look simultaneously at the past and the future. Like the Roman deity, *GC* II 8 looks backward and forward at the same time. Consider its opening:

> All mixed bodies – as many as exist in the region around the center – are composed of all simple bodies. (*GC* II 8, 334b30–31)[1]

The reference to "mixed bodies" suggests continuity with the discussion of elemental mixtures offered in *GC* II 7. But the elaboration "as many as exist in the region around the center" points to a progression of thought beyond the case of elemental mixtures such as flesh, bone, and marrow mentioned in *GC* II 7. Moreover, this qualification introduces a cosmological dimension that becomes even more prominent in the subsequent discussion of the efficient causes of generation and corruption in *GC* II 9–10 (with *GC* II 11 as an attempt to deal with the further question of whether there are things that come to be *of necessity* in the unbroken chain of generation and corruption). It is telling that the opening sentence of *GC* II 9 reminds us that generation and corruption take place in the region around the center.[2]

Admittedly, the circumlocution "the region around the center" is far from crystal clear. In *GC* II 3, Aristotle has already described the upward

* A draft of this chapter was presented at the MUSAPh/LMU research workshop on the *GC* held in Munich on January 7 and 8, 2020. I would like to thank Andreas Anagnostopoulos, Laura Castelli, Joe Karbowski, Timothy Clarke, and Leah Rotsia-Dimou (my commentator) for their comments and remarks on my presentation. Special thanks to Andreas Anagnostopoulos, Sean Kelsey, and Emily Katz for their written feedback on an earlier version of this piece.

[1] Aristotle repeats this point in the concluding sentence of the chapter.

[2] *GC* II 9, 335a24–26.

motion of air and fire as motion toward the boundary, and the downward motion of earth and water as motion toward the center.[3] If one understands the boundary and the center as two regions of the sublunary world, one may conclude that Aristotle is now considering those mixed bodies that inhabit a particular region of the sublunary world. Upon reflection, however, it is more likely that Aristotle is referring to the whole sublunary world as the region around the center. Consider how Aristotle paints, in a few strokes, his conception of the natural world in *Meteor.* I 2. He reminds us that there is one celestial simple body out of which all the heavenly bodies moving in a circle are composed. In addition, he tells us that there are four sublunary simple bodies arranged in the following way: fire and earth occupy the highest and the lowest place, whereas air is nearest to fire and water nearest to earth. Aristotle concludes from these considerations that the whole world around the earth is composed of these bodies.[4] The world in question is the sublunary one, which is contrasted *as a whole* to the celestial region. While the latter is inhabited by a celestial simple body moving in a circle around the center, the sublunary world consists of four simple bodies arranged in a definite way. Moreover, all the sublunary bodies inhabiting this world are a mixture of all four simple bodies.[5]

At the outset of *GC* II 8, Aristotle is moving away from the discussion of the generation of elemental mixtures offered in *GC* II 7. He focuses on the generation of all mixed bodies that inhabit the region around the earth. He explicitly mentions plants in the second part of the chapter.[6] We can think of plants as mixed bodies insofar as they are composed of homogeneous

[3] *GC* II 3, 330b32–33. Aristotle does not stop to elaborate on this claim. Rather, he expects familiarity with his theory of elemental motion from the *DC*. Aristotle has a more economical (and more precise) formulation of this theorem: if unimpeded, a simple body moves either toward the center (πρὸς τὸν μέσον) or away from the center (ἀπὸ τοῦ μέσου). See, for instance, *DC* I 2, 268b21–22 and *Meteor.* I 2, 339a14–15. The advantage of this second formulation is that the center can be unambiguously identified with the center of the universe occupied by the earth. By contrast, the reference to the periphery is ambiguous. When Aristotle says that fire and water move toward the periphery (πρὸς τὸν ὅρον), he cannot have in mind the outermost part of the universe. For the outermost part of the universe is occupied by the celestial simple body that naturally moves in a circle around the center (περὶ τὸ μέσον). Hence, Aristotle must mean that fire and water move toward the outermost part of the *sublunary* world.

[4] *Meteor.* I 2, 339a19–20.

[5] One might object that the mixed bodies in the so-called kapnosphere, a remote region of Aristotle's sublunary world, are a highly combustible mixture of only air and fire giving rise to phenomena such as the shooting stars, torches, goats, and comets (*Meteor.* I 4–8). In fact, Aristotle's position is more nuanced. While the above phenomena are predominately made of a dry exhalation, there can be no dry exhalation without its opposite, the moist exhalation (*Meteor.* II 4, 359b32–34). This claim, far from isolated, is an essential piece of Aristotle's general theory of contrariety and opposition: whenever one contrary is present, the other too must be present.

[6] *GC* II 8, 335a12.

parts – those whose generation is discussed in *GC* II 7. Aristotle considers those parts elemental mixtures and explains their generation out of the mixture of the four sublunary elements. For Aristotle, accommodating their generation out of earth, water, air, and fire is a litmus test for adjudicating between two competing elemental theories: his own theory of the elements and the one that can be traced back to Empedocles. Clearly, the elements remain Aristotle's main concern in *GC* II 7, which are now regarded as the ingredients of elemental mixtures such as blood, bone, and marrow. They continue to be his focus in *GC* II 8. Here Aristotle offers arguments for the claim that all of them must be present in all the composite bodies that exist in the region around the earth. Considering what Aristotle says in the rest of the chapter, the emphasis in its opening sentence should be placed on the second instance of "all":

> All mixed bodies – as many as exist in the region around the center – are composed of *all* simple bodies. (*GC* II 8, 334b30–31)

Indirect evidence that this is the intended emphasis is what Aristotle says at the end of *GC* II 8. There, Aristotle recalls the physical theorem that he has proven and expects us to take away from this chapter: all bodies are composed of *all the simple bodies.*[7] Aristotle no longer speaks of "mixed bodies" even though it is abundantly clear that the composite bodies at the end of the chapter are the mixed bodies mentioned at the outset. At this point of his argument, Aristotle probably does not feel the need to make contact with the discussion of elemental mixtures offered in *GC* II 7.

A close reading of *GC* II 8 suggests that the chapter is a coda to the search for the principles of perceptible bodies – a search that is already under way at the outset of *GC* II 2. By the end of *GC* II 6, Aristotle has clearly fulfilled the promise made in the key programmatic passage at the outset of *GC* I 6, where he has announced a study of matter and the so-called elements with a focus on three questions: (i) whether there are elements; (ii) whether they are everlasting or rather they come into being in some way; and (iii) if they come into being, whether they all come into being from one another or if one of them is prior.[8] If this reading is correct, the treatment of the *aporia* concerning the generation of elemental mixtures in *GC* II 7 and the treatment of mixed bodies in *GC* II 8 form an appendix to the main discussion on the physical foundations of the sublunary world.

[7] *GC* II 8, 335a22–23. [8] *GC* I 6, 322b1–4.

A final clarification on the language adopted in the opening sentence of *GC* II 8 is in order. Aristotle establishes a contrast between *mixed* and *simple* bodies. The simple bodies in question are the four elements that have been his main focus in *GC* II 1–7. But the expressions "elements" and "simple bodies" convey different ideas. By speaking of earth, water, air, and fire as elements, Aristotle emphasizes their combinability. Recall what Plato says in the *Timaeus* about those who claim that earth, water, air, and fire are the elements and provide an account starting from them as from the principles of everything (*Tim.* 48 B–C). He says that, in reality, earth, water, air, and fire should not even be compared to syllables, that is, to the first level of complexity one obtains when one combines the letters of the alphabet (στοιχεῖα).[9] We do not need to engage in a full discussion of Plato's possible critical target(s). What matters, for our present discussion, is that the elements (στοιχεῖα), like the letters of the alphabet, are not only primary but also by definition combinable.[10] In the introduction to this volume, I argued that Aristotle may have Plato in mind as a critical target when he speaks of the so-called elements.[11] While Plato rejects the use of the word "στοιχεῖον" for earth, water, air, and fire, Aristotle endorses it. Certainly, he has no qualms in *DC* III, where he adopts this term and argues that the elements are the primary bodies into which other bodies are analyzed.[12] By calling earth, water, air, and fire the elements, Aristotle indicates that they are the primary constituents of all sublunary bodies. Admittedly, he seems to be less interested in their combinability, which was surely important in the earlier use of the term.[13] In light of the claim defended in *GC* II 8, we way wonder whether the idea that the elements can be *mixed* is compatible with the idea that they are *combinable*. If it is not, our chapter as a whole would constitute further evidence that Aristotle has moved away from the original use of the term "στοιχεῖον."

Instead, when Aristotle refers to earth, water, air, and fire as the simple bodies, he insists on their simplicity (334a32). At this stage of the argument, the use of the expression "simple bodies," and hence the emphasis on their simplicity, is far from obvious. In *GC* II 3, Aristotle shows that

9 According to Eudemus of Rhodes, Plato was the first to refer to the principles of natural and generated things as elements (Simplicius, *In Phys.* 7.10–17; Eudemus fr. 31 Wehrli).

10 For more on this point, see Burkert 1959: 167–197. *Contra* Crowley 2005: 367–394.

11 See Introduction.

12 *DC* III 3, 302a14–19.

13 The connotation of combinability is surely lost when Aristotle extends the use of the term "στοιχεῖον" to the celestial simple body. See, for instance, *DC* III 1, 298b6: περὶ μὲν οὖν τοῦ πρώτου τῶν στοιχείων εἴρηται. Since the celestial simple body is not subject to generation and corruption, this body cannot be combined with the sublunary elements.

these bodies display the following combination of primary qualities: hot and dry (fire), wet and hot (air), cold and wet (water), and dry and cold (earth). Aristotle's insistence – in *GC* II 3 as well as in *GC* II 8 – on using the expression "simple body" cannot be a mere coincidence.[14] By Aristotle's lights, the fact that earth, water, air, and fire are susceptible to physical analysis in terms of four primary qualities is no threat to their simplicity. More directly, these bodies remain simple even if they display some internal structure. For instance, the dry and the cold in the case of earth, and the cold and the wet in the case of water, are not regarded as the constituents of the elemental bodies. In fact, we may even think that the account of the four elements as the material conditions for generation and corruption advanced in *GC* II is an indirect vindication of the use of the expression "simple bodies" for them. This account shows that earth, water, air, and fire cannot be analyzed into simpler or more basic constituents even though a full account of their mutual transformation reveals that they must possess a combination of active and passive qualities and also a common matter.[15]

9.2 Every Simple Body Will Be Present in Every Composite Body

The heading I have chosen for this section is a *verbatim* quotation from *GC* II 8, 335a9. This sentence brings the first part of the chapter to a natural conclusion. The task Aristotle sets for himself is to show that all four simple bodies must be present in all mixed (or composite) bodies. His argument has two parts: first, Aristotle shows that earth and water must be present in any composite body; then, he argues that the generation of any composite body also requires the other two simple bodies insofar as they are contrary to earth and water.

We have already seen that Aristotle holds that earth and water move toward the center of the universe, while air and fire toward the periphery. Given this theoretical commitment, it is quite natural for Aristotle to start his argument from earth and water. Both simple bodies are naturally present around the center of the universe. However, he offers separate reasons for their presence in any composite body. Earth must be present because each simple body is present – above all (μάλιστα) and mostly

[14] In *GC* II 3 Aristotle also speaks of "observable simple elements." For a discussion of this expression, see Emily Katz in this volume.

[15] See Crowley 2013: 161–182 for a study of the relation between the four elements and the four basic qualities. For more on prime matter, see the coda to the Introduction to this volume.

(πλεῖστον) – in its own natural place.[16] Since the center of the universe is the natural place of earth, earth must be the primary constituent of any composite body that inhabits this region. In addition to earth, water too must be present for two separate reasons. First, among the simple bodies, water alone can be easily bounded (εὐόριστον). Moreover, water *qua* wet (ὑγρόν) holds earth together since, if moisture could be completely extracted from earth, the latter would fall apart.[17]

Let us engage in a close examination of this stretch of text starting from the last claim. The earth Aristotle has in mind cannot be one of the elemental bodies. As a simple body, earth does not contain any moisture. In *GC* II 3, however, Aristotle distinguishes the simple bodies from the *apparently* simple bodies. The latter are what we might call "everyday" earth, air, fire, and water – namely, the earth in a garden, atmospheric air, the fire in a fireplace, and the water in a lake.[18] The earth Aristotle refers to in our passage must be an *observable* (and as such *apparent*) simple body. Since the latter is a mixed body, there is no problem in assuming that earth *qua* observable simple body contains some moisture. If we could successfully and completely take this moisture out of the observable earth, we would see this earth to disintegrate into dust.

The distinction between simple bodies and *apparently* simple bodies may turn out to be useful as we try to make sense of the other claims made in this stretch of text. The reasons given for the presence of water in any composite body have puzzled the interpreters of *GC* because they are difficult to reconcile with the elemental theory advanced in *GC* II 2–4. Recall that Aristotle has reduced all tangible properties to two pairs of primary qualities: hot–cold, and wet–dry. While hot and cold are contraries that have the capacity to act on something else (active properties), wet and dry are contraries that have the capacity to be acted upon by something else (passive properties). To understand how these primary properties are distributed over the four elements, we must keep in mind

[16] It is not difficult to see why a simple body is present mostly (πλεῖστον) in its natural place. Dislocating this simple body from its own natural place would require a force acting against nature. By contrast, the intended meaning of μάλιστα is not immediately obvious. Perhaps the idea is that the earth in the center of the sublunary world is the earth *par excellence*. As such, it exhibits the nature of the earth in the clearest and most perfect way.

[17] *GC* II 8, 335a2–3: εἰ γὰρ ἐξαιρεθείη τελέως ἐξ αὐτῆς [*sc.* γῆς] τὸ ὑγρόν, διαπίπτοι ἄν. The most natural reading of this sentence is as follows: "If what is wet were completely extracted from earth, *earth* would fall apart." In other words, "earth" (γῆ) is the implicit subject of διαπίπτοι. I don't think that it is open to us to complete the sentence in the following, more creative, way: "if what is wet were completely extracted from earth, *the compound of earth and water* would fall apart."

[18] *GC* II 3, 330b1–7. For a full discussion of this passage, see Emily Katz in this volume. Cf. Crowley 2013: 170–173.

that each element must possess both an active and a passive property. The presence of both kinds of properties renders the elements such as to act and be acted upon by one another, which is physically required for their ability to change into one another. To water, Aristotle assigns two primary qualities: coldness (an active property) and wetness (a passive property).[19] Moreover, according to *GC* II 3, even if each simple body participates in an active and a passive property, it is most properly characterized by only one of the two. Surprisingly enough, Aristotle singles out coldness rather than wetness as the primary quality of water.[20] Unlike wetness, which is a passive property and is defined as being not limited by any boundary but being easily bounded,[21] coldness is an active property and is defined as the capacity to unite homogeneous and heterogeneous things alike.[22]

Considering these theoretical pronouncements, we would expect that water *qua* wet (a patient) explains why any mixture of earth and water can be easily bounded, whereas water *qua* cold (an agent) explains why any mixture of earth and water is held together by water. Instead, Aristotle says that water is the only simple body that can be easily bounded.[23] We can fix the apparent dissonance between *GC* II 2 and *GC* II 8 by positing that in *GC* II 8 Aristotle means that water *qua* wet can be easily bounded. I note, in passing, that the property of being easily bounded is a physical requirement of what is mixable.[24] So it makes sense that this property is recalled when Aristotle discusses the simple bodies as the ingredients of mixed bodies. More perplexing is what Aristotle says next. He says that water *qua* wet is that which holds the mixture of earth and water together.[25] At least at first sight, this statement is baffling because Aristotle regards wetness as a *passive* property. It is not obvious – at least based on the definition of wetness given in *GC* II 2 – how wetness can unite things and hold them together.

The statements that Aristotle makes in *GC* II 8 can be corroborated by simple empirical observations. Think of some clay and how water

[19] *GC* II 3, 330b5. For a discussion of the *empirical* reasons for this assignment, see Emily Katz in this volume.

[20] *GC* II 3, 331a5. From Philoponus we learn that some ancient interpreters objected to the characterization of water as something cold rather than something wet (Philoponus, *In GC* 230.23–25, no name is named).

[21] *GC* II 2, 329b29–30: ὑγρὸν δὲ τὸ ἀόριστον οἰκείῳ ὅρῳ εὐόριστον ὄν.

[22] *GC* II 2, 329b28–29: ψυχρὸν δὲ τὸ συνάγον καὶ συγκρῖνον ὁμοίως τά τε συγγενῆ καὶ τὰ μὴ ὁμόφυλα.

[23] *GC* II 8, 334b34–335a1: "water alone of the simple bodies is easily bounded."

[24] Full discussion of this requirement in Krizan 2018: 200–204.

[25] *GC* II 8, 335a1–2: "earth cannot remain together without what is wet, but this is what holds it together."

(*qua* wet) is not only what makes the clay easy to model but also what is ultimately responsible for holding it together. The problem is how to make empirical observations of this kind cohere with the theoretical pronouncements of *GC* II 2–4. In his commentary on *GC* II 8, Williams suggests a radical solution: in *GC* II 8, Aristotle – against his own theoretical pronouncements made in *GC* II 2–4 – relies on a pre-theoretical conception of water as something primarily and eminently wet. As a result, any active or passive property of water should be traced back to the distinctive wetness of water.[26]

Although tempting, this suggestion cannot be accepted – not only because it would be, to say the least, odd for Aristotle to be forgetful of what he has said earlier in the book, but also because Aristotle appears to be mindful of the results achieved in *GC* II 2–4 as he turns to the second part of his argument. More on this front in a moment. First, I would like to suggest a solution to the perceived inconsistency between *GC* II 8 and *GC* II 2–4. Recall that Aristotle distinguishes between *apparently* simple bodies and *genuine* simple bodies in *GC* II 3. The whole section on earth and water in *GC* II 8 contains an argument about the *genuine* simple bodies based on observations concerning the *apparently* simple bodies. The genuine simple bodies are open to empirical observations insofar as they are the principles of the empirical simple bodies. To this extent, the divide between genuine and apparently simple bodies is not as strong as one might initially think. One may object that recalling the distinction between *apparently* simple bodies and *genuine* simple bodies (that is, elements) is not different from invoking a pre-theoretical conception of water. In fact, recalling this distinction is equivalent to reminding ourselves of a corollary of the elemental theory presented in *GC* II 2–4. Williams is happy to conclude that Aristotle is forgetful of his own elemental theory. By his lights, the whole passage is innocent with respect to the elemental theory offered in *GC* II 2–4. My approach reads *GC* II 8 as an attempt to apply the theory in order to evaluate its overall explanatory power.

By my lights, the arguments for the presence of earth and water in any mixed body inhabiting the sublunary region are empirical arguments, based on simple observations, and thus concern the *apparently* simple bodies. The arguments for the presence of air and fire in any mixed body are of a different nature. Recall that air and fire move upward toward the periphery of the sublunary world. As a result, it is far from obvious that air

[26] Williams 1982: "the present argument can be seen as relying on common sense and views expressed by Aristotle elsewhere and betrays a forgetfulness of what he has said earlier in this book." (179)

and especially fire are present in all the simple mixed bodies around the center of the universe. How does Aristotle secure this result? He does so by means of an argument based on the language of opposition and contrariety. First, Aristotle states that air is contrary to earth, and fire to water;[27] then, he argues that the generation of any composite body is from contraries. According to Aristotle, of any given pair of contraries, if one of the two contraries is present, the other one has to be present as well.[28] Since the argument is based on the language of contrariety, the considerations offered in this stretch of text no longer apply to the *apparently* simple bodies; rather, they apply to simple bodies as outlined in *GC* II 2–4.

To fully understand what Aristotle is claiming in this stretch of text, we need a clear and precise answer to the following two questions: (1) What are the contraries Aristotle has in mind? (2) How does Aristotle apply the language of contrariety and opposition to the simple bodies? Let us look at the second question first. The language of contrariety applies to simple bodies in a derivative way. Recall that the simple bodies are substances and that, according to the official doctrine of the categories, substances do not have, strictly speaking, contraries.[29] Without explicitly referring to the *Categories*, Aristotle recognizes that he is going beyond the letter of the theory of the categories. Tellingly, he qualifies his claim: earth is contrary to air, whereas water is contrary to fire, *as it is possible for a substance to be contrary to another substance.*[30] Still, he is willing to apply the language of contrariety and opposition to his four elements because, as per *GC* II 3, they exhibit contrary affections (παθήματα).[31] Air and earth are contraries because earth is cold and dry, whereas air is hot and wet. Fire and water are contraries because water is cold and wet, whereas fire is hot and dry. In other words, the primary contraries singled out in *GC* II 2, organized in two pairs and distributed over the simple bodies in *GC* II 3, license the use of the language of contrariety and opposition in the case of the four simple bodies.

Let us now turn to the first question, which is concerned with the contraries Aristotle has in mind when he says that the generation of any

[27] *GC* II 8, 335a5–6: "air and fire are present because they are contraries to earth and water. Earth is contrary to air, whereas water is contrary to fire (in the way that it is possible for a substance to be contrary to another substance)."

[28] *GC* II 8, 335a6–8: "since instances of coming to be are from contraries, and one or the other extreme of the contraries is present, then the other must be present as well."

[29] *Cat.* 5, 3b24–27. [30] *GC* II 8, 335a6. [31] *GC* II 3, 331a2–3.

composite body is from contraries.[32] The contraries in question are the four basic qualities organized in two pairs of primary contrarieties – namely, hot–cold and wet–dry. If a member of one pair of contraries is present (say, cold) the other must be present as well (hot). One way to understand is to think that one of the two contraries is actually present, while the other is present only potentially. This idea fits well with what Aristotle says elsewhere (most notably, in *Phys.* I). Unfortunately, it fits less well with *GC* II 8. If we apply this idea to our argument, we obtain that the relevant mixed bodies are composed of earth and water in actuality and are potentially fire and air in the sense that they could *be turned into* fire and air. But in *GC* II 8 Aristotle wants to reach a stronger conclusion: the mixed bodies are *composed* of all four elements. Hence, his argument from contraries should not be understood in terms of being actually one contrary and potentially the other.

A more promising strategy consists in assuming that the generation out of contraries mentioned in our passage is restricted to the generation of mixtures. In *GC* II 8, Aristotle has argued that earth and water must be present in any mixed body that inhabits the region around the center of the universe. In *GC* II 3, he has also argued that each of these two bodies is associated with a specific primary quality: while earth is dry, water is cold. Now, the presence of dry (earth) implies the presence of wet (air), and the presence of cold (water) implies the presence of hot (fire). Therefore, all four simple bodies must be present in any mixed body.[33] This does not make the argument compelling because it remains unclear why there needs to be mixture along *both* dimensions of elemental contrariety. To rescue the argument, one might focus on the idea that each element is characterized by one contrariety. It turns out that the primary contrary of earth (dry) and that of water (cold) cannot by themselves facilitate mixture; hence, the remaining two simple bodies are needed to generate a mixture, namely air (wet) and fire (hot).[34]

There is one final aspect of Aristotle's argument that deserves a few words of comment. How are we to understand these generations (γενέσεις) out of contraries? The text is elliptical, but we do have a model for

[32] We need to supply "of composite body" or "of a mixed body" after generations (γενέσεις) in line 335a7.

[33] Alternatively, one could argue that the presence of cold and dry (earth) implies the presence of hot and wet (air), while the presence of cold and wet (water) implies the present of hot and dry (fire). This route is less economical. It also generates some redundancies. Hence, my preference is for the argument I have offered in the main text.

[34] I owe this suggestion to Andreas Anagnostopoulos.

understanding how these generations occur. This model is the generation of elemental mixtures discussed in *GC* II 7.[35] Recall that the generation of elemental mixtures poses a serious problem (*aporia*) because we must understand how, by mixing simple bodies, we obtain homogeneous bodies that are ontologically distinct from simple bodies. Presumably, we are expected to extend the solution defended in *GC* II 7 to all mixed bodies that occupy the region around the center of the universe.

If we adopt this reading, we see that the conceptual link between *GC* II 7 and *GC* II 8 is closer than we may have initially thought. Emphasizing the conceptual proximity, if not outright continuity, between *GC* II 7 and *GC* II 8 helps us explain why Aristotle begins the chapter by focusing on the notion of mixture. He opens his discussion by speaking of *mixed* bodies and means by this *mixtures* of the four simple bodies. As the argument proceeds, his language changes. He now speaks of "composite bodies." The overall impression is that he considers the two expressions "mixed bodies" and "composite bodies" interchangeable. My contention is that, at least in the context of *GC* II 8, these expressions are interchangeable because Aristotle understands composition in terms of mixture and understands mixture in light of the account of elemental mixtures offered in *GC* II 7.

9.3 Nourishment as Additional Evidence

I argued that Aristotle places the first part of *GC* II 8 in continuity not only with *GC* II 7 but also with *GC* II 2–4. Aristotle builds on the results achieved in his search for the material principles of perceptible bodies. Moreover, he uses his own elemental theory to show how the mixed bodies inhabiting the region around the earth must be made of all four sublunary simple bodies.

In the second part of the chapter, Aristotle offers additional evidence in support of this conclusion. He argues that all the composite bodies that inhabit the region around the earth are nourished by the same things out of which they are composed. Plants are explicitly mentioned because they are a possible counterexample to his general claim. Someone might object that plants take in only water. Aristotle replies that plants are nourished by a mixture of water and earth. This response can be unpacked in two different ways. We can take the water and the earth that the farmers mix

[35] For a full discussion of the generation of elemental mixtures, see Andreas Anagnostopoulos in this volume.

when they irrigate their plants as two *apparently* simple bodies. On this reading, water and earth are *mixed* rather than simple bodies. If so, we can conclude that plants are nourished by all of the four genuine simple bodies out of which they are composed.

As an alternative, we can take the nourishing earth and the water to be the simple bodies that are present in all the mixed bodies that occupy the region around the earth. Recall that Aristotle has already offered an argument for the view that the presence of water (cold) and earth (dry) entails the presence of their contraries – namely, fire (hot) and air (wet). While the first reading applies to the distinction between simple bodies and *apparently* simple bodies introduced in *GC* II 3, the second implicitly invokes the argument given in *GC* II 8 for the two simple bodies, the presence of which in any mixed body is less obvious. Either reading takes us to the desired conclusion, namely that plants take in nourishment that contains all four simple bodies. According to both, Aristotle does not give us a complete argument for the thesis that nourishment contains all of the four simple bodies. Rather, he responds to a potential objection to his general thesis based on the case of plants, and his response contains an argument leading to the conclusion that all four elemental bodies must be present in the nourishment not only of plants but of all composite bodies.

Joachim reads our passage slightly differently.[36] By his lights, nourishment is additional evidence that the composite bodies are made of all of the four simple bodies because nourishment must be wet and dry. This piece of information is not offered in our chapter but can be extracted from what Aristotle says elsewhere.[37] If nourishment is wet and dry, it must contain at least water (cold and wet) and earth (cold and dry). But if nourishment contains earth and water, it must also contain fire (hot and dry) and air (hot and wet). It is only at this point that plants would enter into the argument as a potential counterexample. Aristotle would reply that plants would be no exception to the rule because they too are nourished by a mixture of earth and water. We have seen that this mixture can be understood either as a mixture of apparently simple bodies or as a mixture of simple bodies. Either way, we come to the conclusion that the four simple bodies must be present in all composite bodies, including plants. In response to this reading, I note that there is absolutely no evidence in our text that Aristotle is relying on claims made outside the *GC*. More directly, there is no evidence that Aristotle is expecting his

[36] Joachim 1922: 246 (endorsed in Rashed 2005: 169n1).

[37] *PA* II 3, 650a3–4.

reader to have familiarity with his biological writings even though the biological relevance of what Aristotle says is indisputable.[38]

We should also be careful not to overinterpret this passage. For instance, I do not think that this passage is at loggerheads with Aristotle's official account of nourishment (τροφή).[39] Aristotle surely has the conceptual resources to recompose the perceived discrepancy between the general account of nutrition offered in the *De anima* and the claim that every composite body is nourished by the same thing out of which it is made. In the *De anima*, Aristotle tells us that it matters whether by "nourishment" we mean the first or the last thing that is added to the living body. While the first thing added is unconcocted (undigested) food, the last thing added is concocted (digested) food.[40] When Aristotle says that all things are nourished by the same things out of which they are made he must mean concocted food.[41]

GC II 8 ends with a final reflection that combines the topic of the elemental bodies with that of nourishment. This reflection is prompted by the ancient view (οἱ πρότεροι λέγουσιν) that fire is the only simple body that is nourished (or rather nourishes itself).[42] This view is also recalled in the *De anima:*

> To some, the nature of fire seems to be the cause of nutrition and growth. The reason is that fire alone among the bodies, or rather the elements, appear to nourish itself and to grow. (*DA* II 4, 416a9–11)

In *GC* II 8, Aristotle seems to think that he is somehow able to explain this ancient belief while holding onto one of the main physical doctrines established in *GC* II, namely that there is no barrier to the phenomenon of elemental generation and all the simple bodies are generated by one another.[43]

His explanation can be restated as follows:

(1) What nourishes itself is not the matter but rather the enmattered form and the shape.

[38] The biological orientation of the *GC* is a major aspect of the interpretation developed in Rashed 2005: CXL–CLXXXVI.

[39] Williams 1982: 179. [40] *DA* II 4, 416b2–4.

[41] When he gives his general theory, Aristotle does not distinguish between the different kinds of living beings. By Aristotle's lights, plants take in only *concocted* nourishment. This means that either the water and the earth the farmers mix are already concocted nourishment or they are nourishment that will be concocted in the soil before being taking in by the plants via their roots.

[42] The Greek τρέφεσθαι can be understood as a middle (rather than passive) voice.

[43] This is one of the major theorems proven in *GC* II 4. See the essay by Scharle in this volume.

For this claim, Aristotle relies on his general account of growth in *GC* I 5, where he has argued that while it is the living body that grows, the growth of the living being is to be understood as the growth of its form, not of its matter.[44]

> (2) Fire alone (μόνον) nourishes itself among the simple bodies because fire alone (μόνον) is akin to the form and the shape, or rather fire is the simple body that above all (μάλιστα) is akin to the form and the shape.

The qualification "above all" (μάλιστα) is introduced because the considerations that Aristotle is going to offer in support of his claim apply to air as well. As a result, one might conclude that air too is akin to the form and the shape. By stating that these considerations are *best* exemplified in the case of fire and saying that fire is *above all* akin to the form and the shape, Aristotle is still entitled to reach the desired conclusion. I note, in passing, that the claim that fire and air are akin to the form and the shape is not a *unicum* in the Aristotelian corpus. Aristotle elaborates on this claim in *DC* IV 4, where he claims that what contains (τὸ περιέχον) is akin to the form, while what is contained (τὸ δὲ περιεχόμενον) is akin to the matter.[45] There Aristotle goes on to say that the distinction between form and matter applies to spatial distinctions: while the above region (τὸ ἄνω) is what is akin to form insofar as it contains, the lower region (τὸ κάτω) is akin to matter insofar as it is contained.[46] We can take our claim (2) to be an application of the general theorem defended in *DC* IV 4.

What grounds the claim that fire and air are akin to the form is the following thesis:

> (3) Fire and air are the simple bodies that are borne toward the boundary.

We have already seen that this physical theorem is discussed in the context of the theory of natural motion offered in *DC* IV.[47] While both air and fire are borne toward the boundary, it is only fire that, according to the theory of natural motion, *rises to the surface*.[48] Put differently, if unimpeded fire alone rises above all the other bodies that are moving up. This licenses the further claim that:

> (4) Fire alone of the four elements is akin to the form and the shape.

[44] *GC* I 5, 321a24–28. [45] *DC* IV 4, 312a12–13. [46] *DC* IV 4, 312a13–16.

[47] We have seen that it is recalled also at the outset of the *Meteorology* (*Meteor.* I 3).

[48] The Greek ἐπιπολάζειν is *terminus technicus*. For a discussion of its significance, see Bodnár 1997: 81–117.

10

On Generation and Corruption *II 9*

Panos Dimas

10.1 Introduction

The central aim of *GC* is to account for the phenomena of generation and corruption (*GC* I 1, 314a1–3). Moreover, as Aristotle makes plain in the opening lines of *GC* I 2, it is vital to this end that he arrives at a firm view on whether these phenomena exist, and if they do exist, how they come to pass (*GC* I 2, 315a25–28). He considers these questions to be pressing for two reasons. Firstly, though most of his predecessors recognized that generation and corruption are different from alteration, none of them treated these phenomena so as to bring out that difference (*GC* I 2, 315b15–20). Democritus was the only previous thinker to devote serious attention to the problem, enough to honor it by providing distinct accounts of each of these phenomena. But the Democritean theory is regrettably untenable, for it is premised on a thesis that has been proved false, namely that there are indivisible bodies (*GC* I 2, 315b6–9, 317a12–22). This brings us to the second, most important, reason why these questions are pressing for Aristotle: to establish properly that generation, corruption, and alteration are genuinely distinct processes, one must account for the way these phenomena, and in particular generation and corruption, come to pass. Therefore, we are told in *GC* I 2, unless we reach such an account we will have to concede "either that generation [and corruption] does not exist or that it is just alteration" (*GC* I 2, 315b20–24).[1]

* I am very grateful to Andrea Falcon, Sean Kelsey, Spyridon Rangos, and Stasinos Stavrianeas for their helpful comments. Thanks are also due to the anonymous readers for their help in improving this chapter and the volume.

[1] What I take Aristotle to be saying in this passage (315b20–24) is that since Democritus is the only thinker who accounts for generation and corruption as being different from alteration, and his account is untenable, we risk having to admit either that there is no generation or else that it is simply alteration.

Hence, Aristotle sets himself a twofold task in the *GC*. Firstly, he must account for generation and corruption so as to reveal that they are distinct from one another and from alteration. He must, therefore, pinpoint causes responsible for these two phenomena. Secondly, he must show how these causes work, both individually and jointly, so as to bring about generation and corruption. *GC* I 3 offers a specification of the reference of the terms "generation" and "corruption": generation is the coming to be of a thing that did not exist before, while corruption is the ceasing to be of a thing that did exist before. Both are instances of change, but the kinds of change that these terms refer to are (a) the coming to be and (b) the perishing of *substances*. *GC* I 3 offers a general account of how these changes ought to be understood and distinguishes them from alteration, which is discussed in *GC* I 4. What *GC* I 3 does not do is identify the causes of these changes and explain exactly how those causes produce those changes.

GC II 9 begins this task by fleshing out, however cryptically, the role played by the material cause, and by arguing for the need of an efficient cause. *GC* II 10 completes this phase of the task by discussing the efficient cause, while *GC* II 11 completes the task with an account of how the causes of generation and corruption produce their results (namely, *substances*) forever without end.

GC II 9 makes progress toward providing such an account, first by pointing to matter as one of the causes any investigation must focus on if it is properly to account for these three kinds of change (i.e., generation, corruption, and alteration). The second main concern of the chapter is to bring out the importance of the efficient cause and to chide previous thinkers for ignoring it. The chapter thus divides into two main thematic sections, the second of which is also divided in two. The first section provides a characterization of the kind of matter that is a cause of generation and corruption (335a32–b5). This section, though it leaves no doubt about the importance of matter, is scarce in detail. It indicates matter's role with a cryptic statement and leaves it to the reader to do the work needed to arrive at an account. The remaining, and by far the larger, part of the chapter is dedicated to Aristotle's censure of his predecessors (335b8–336a14). First, Aristotle criticizes Plato for neglecting the all-important efficient cause (the account of which Aristotle reserves for the next chapter, *GC* II 10) (335b9–24). Second, he criticizes certain other predecessors – the materialists, as I will call them for brevity – for venturing to account for generation and corruption by pointing to matter and ignoring form altogether (335b24–336a14). In addition, Aristotle prefaces both sets of criticisms by saying that none of his predecessors stated the efficient cause, though he does say that they did dream of it vaguely (335b7–8). We will have to understand whether this is a fair statement. All three thematic sections are discussed in what follows, but the first one is explored in some detail.

10.2 The Scope of Generation and Corruption

10.2.1 The Place of GC II 9 in Aristotle's Account of the Causes of Generation and Corruption[2]

Before I begin my critical examination of *GC* II 9, I would like to draw attention to a potentially important but largely neglected fact regarding both the *GC* as a whole and the contribution of *GC* II 9 to Aristotle's account of the material cause of generation and corruption. As already mentioned, *GC* II 9 says remarkably little about the workings of the material cause in the coming to pass of these phenomena. Though *GC* II 1–8 offers an important outline of matter's contribution in generation and corruption, *GC* II 9 attempts to capture matter as a cause of these phenomena; it does so in a very concise statement, the import of which will have to be unpacked. The chapter begins with the statement that "some things" (ἔνια) are subject to generation and corruption, and that these things are found "in the region around the middle" (ἐν τῷ περὶ τὸ μέσον τόπῳ) (335a24–25). The preposition περί signifies, as it standardly does with accusative, the borders of the spatial region in which these things are found; here it identifies the sublunary region as distinct from the superlunary one. One might have thought that the fact that the things subject to generation and corruption are found in *this* region is a point too obvious to bear mentioning. And yet, just two lines further down, Aristotle picks up the point, remarking that the principles of generation "are equal in number to and the same in kind as those of the eternal and first things" (*GC* II 9, 335a28–29). This reference to the superlunary region in a work, indeed in a chapter, dedicated to the explanation of the *sublunary* phenomena of generation and corruption, cries for an explanation.

The point of this reference to the heavens in general, and to the heavenly bodies in particular, becomes apparent once it is seen in light of the fact that the *GC* is likely to be a sequel to the *DC*. Though one may rightly be suspicious of grand claims regarding the chronological order of Aristotle's works, we should at the same time pay heed to the rich and

[2] See also the Introduction to this volume.

complex system of cross-references found in the *DC*, *GC*, and *Meteorology*. These references certainly place these works in that order thematically. So much at least is explicitly confirmed by the opening lines of the *Meteorology*, which refer to these works as preceding it in exactly this order, and the *Physics* as preceding all three of them (*Meteor.* I 1, 338a20–25).[3] Such thematic ordering would explain the apparently cavalier reference to the heavenly bodies as "first" (πρῶτα) in our text. "First" is the term used several times in the *DC* to refer to the celestial simple body.[4] As significantly, this is the body that comes first in the order of exposition (and indeed explanation) in this sequence of works on nature, which also confirms that at least the first two books of the *DC* (*DC* I–II) ought to be read before the *GC*. But the fact that *DC* I–II comes first in thematical exposition does not, alone by itself, explain the mention of the heavenly bodies here in *GC* II 9 – not unless the message Aristotle wants to drive home is that understanding how and why the first bodies are "eternal" (ἀίδια), explained in detail in the *DC* I–II, is a prerequisite for understanding how and why the sublunary substances are subject to generation and corruption (and, no less importantly, that the same kind of causes can have such radically different effects in these two very different regions of the natural world). The suggestion lies near that Aristotle expects his reader to keep in mind, and indeed reflect on, the relevant lesson from the *DC* I–II.

10.2.2 Opening Remarks and Puzzles

GC II 9 opens with the statement that "we must investigate the number and nature of the principles of *all* generation alike" (335a25–26).[5] We are not told exactly what kind of things the "some things" (ἔνια, 335a24) that are subject to generation and corruption are. But Aristotle is clear that the ensuing account applies to *all* of them (335a25). So they must include not only the substances that *exist on account of nature* but also artifacts.

After stating the aim of the chapter – to account for *all* generation and corruption – Aristotle mentions side by side those things that *are* subject to

[3] On the research program outlined in *Meteor.* I 1, see Burnyeat 2004: 17–24 and Rashed 2005: CXLIV–CLII.

[4] See, for instance, *DC* I 3, 270b1–4; *DC* I 9, 278b9 and 279a32.

[5] Though there is no mention here of corruption, corruption is implicitly assumed in that generation presupposes it. In several places, Aristotle explicitly mentions only generation but *GC* II 9 is plainly about both generation and corruption, as its first line makes clear (γενητὰ καὶ φθαρτά, 335a24; see also 335b4 and 16).

generation and corruption (namely, those found "in the region around the middle") and the heavenly bodies (literally, "the eternal and primary things") which are *not* subject to generation and corruption (335a28–29). Both are substances. Now, in addition to the sublunary substances, the so-called elements (namely earth, water, air, and fire) are also subject to generation and corruption.[6] Any one of the elements can, in Aristotle's account, stop being the element that it is and become any other.[7] That raises two possibilities. The first is that we have an intentional juxtaposition of what is generable and corruptible with the superlunary bodies (all of which are substances); which, together with the previous statement that the aim is to account for *all* generation, might seem to imply that the elements too are substances. Alternatively, it might rather be that, although the account aims to apply to all generation and corruption, there is a reason for juxtaposing the sublunary and the superlunary substance that would *not* imply that (in Aristotle's view) the elements are not substances. This raises a large issue beyond the scope of this essay.[8] In what follows then I will take it that when Aristotle speaks of *all* generation and corruption, he has in mind *also* the generation and corruption of the elements (whether or not the elements are properly speaking substances).

10.3 Aristotle's Opening Statement and the Need for an Efficient (aka Moving) Cause

GC II 9 begins with a clear indication that it aims to throw light on the causes of all generation (and corruption). Then we are offered the seemingly peculiar remark that the target is these causes as universals (περὶ τῶν καθόλου, 335a27–28). In fact, the reference to universals in this context does no more than indicate the nature of the task at hand, which is to grasp the relevant causes and their explanatory function at the most general level, namely the level of being. Put another way, we are seeking the causes that make the generable and perishable things *be* generable and perishable. That is why Aristotle refers to these causes, not with *aition* or *aitia*, each

[6] "So-called elements" as opposed to the "elements unqualified" (στοιχεῖα ἁπλᾶ) that Aristotle attributes to Empedocles at *GC* I 1, 314a26–30, presumably because, unlike the latter, Aristotle conceives of earth, water, air, and fire as subject to change – indeed, as subject to generation and corruption. See Crowley 2008: 223–242. I will henceforth be referring to them as simply elements.

[7] See Scharle on *GC* II 4 in this volume.

[8] For an entry into the literature on this topic see Cohen 1996.

one of which means "cause," but rather with *archê*. While *archê* may designate a cause, what it means, strictly speaking, is a beginning in the sense of a principle; when it is used to designate a cause, at least by Aristotle's lights, this cause is not any chance cause but rather a first cause in the sense that there is no other cause beyond it.[9] In *GC* II 9, Aristotle is after the first causative principles responsible for the coming to be or perishing of *any* particular thing. Such principles are not temporally first; rather, they are first at the level of being, so that the thing that is explained with reference to them, whatever it is, would not have come to be or perish were it not, ultimately, for the workings of these principles.[10]

After the announcement of the goal of *GC* II 9, we have seventeen Bekker lines, amounting to virtually one third of the remaining of the chapter, where the generable and perishable things of the sublunary region are contrasted with the bodies of the superlunary one (335a27–b7). We are given no such contrast in *GC* I 3, which is the other chapter that discusses generation and corruption. If contrasting in this way these two sets of substances were not striking in itself, this stretch of text offers us all that the chapter has to say on the topic of matter as a cause of generation and corruption virtually in one single phrase. Aristotle intends to position his account of the principles of generation and corruption in relation to his account of the principles of the eternal bodies. This intention is further confirmed by the fact that, as we will see, matter is the cause responsible for the difference between sublunary and superlunary bodies in respect of their being subject (or not subject) to generation and corruption. The implication is clear: understanding the grounds for the categorical distinction between the two sets of bodies making up the universe is part of understanding the way in which generation and corruption come to pass.[11] To stress this, Aristotle remarks that the principles of both kinds of things are the same – a remarkable claim given how radically different the things which these causes are said to be causing are in their mode of being. So we need to understand the way in which these causes bring about diametrically different results.

[9] See, e.g., *NE* I 7 where Aristotle quite clearly distinguishes between αἴτιον and ἀρχή, calling the latter "first," thereby implying that though the former too is a cause, it need not be the *first* cause (*NE* I 7, 1098b2–3: τὸ δ' ὅτι πρῶτον καὶ ἀρχή).

[10] Thus, making Joachim's rendering of ἀρχή as "originative cause" thoroughly apt.

[11] Commenting on the relationship between *DC* and *GC*, Jaeger remarks: "Here [in the particular field of scientific thought which is physics], far more even than elsewhere [in particular philosophy], he [*sc.* Aristotle] must be understood through his development, if we are really to grasp him in his individuality" (Jaeger 1948: 293).

Before that, however, a question arises as to the number and nature of the principles of generable and perishable things – *how many* (πόσαι) and *which ones* (τίνες) they are (335a25–26). Aristotle's answer is that the principles are equal in number and the same in genus (γένος) as those of the eternal and first substances (335a28–29). The point about number is straightforward enough: the principles, in the sense explained above, of eternal substances are as many as the principles of the generable and perishable ones. By contrast, the point about "genus" is less straightforward. It cannot be that the discussion takes aim at organizing the principles in a genus-and-species structure in the usual sense. Still, there is a clear message here: though the *number* of principles in these two domains is the same, there may be a sense in which the *natures* of the principles in these two domains are not the same. What the reader is being alerted to, I suggest, is that some of the principles, specifically the material principle, manifests itself differently in these two sets of cases.

That said, one of Aristotle's remarks about principles in these opening lines has puzzled some scholars. One of the principles, Aristotle says, is a principle "as matter," while another is "as form" (335a29–30). But to these, he goes on to say, there must be added a third one, for these two are not enough to generate things that are generable, just as they are not enough in the case of primary substances either (335a30–32). The principle he implies is missing, though he does call it this explicitly, is the efficient or moving cause. A first worry is that Aristotle here speaks of three causes, while, given what he says *Phys.* II 3–7, we would expect him to say that there are four.[12] A further worry is that in the previous sentence Aristotle tells us that the number of the principles of the generables and corruptibles is the same as that of the eternal ones, which might be taken to suggest that the principles of the eternal substances too are three. The possibility that *GC* is posterior to *Phys.* II seems compatible with the possibility that there may have been a change of heart with respect to the number of causes. But taking these lines to indicate anything of the sort would be to overreach. Aristotle in fact says as much in the first line of the next chapter, where he remarks that he has already offered his general account of the causes in an earlier work (*GC* II 10, 336a13). This is a clear statement to the effect that the account presented in *Phys.* II still stands. This is also implied by the opening lines of the *Meteorology*, where the *Physics* account of "the first principles of nature" is recalled (338b20),

[12] For example, Williams 1982: 183 seems willing to consider the possibility that this chapter is evidence that the doctrine of four causes is replaced by a doctrine of three.

without any hint that the *GC*, which it mentions in the same sentence, has done anything to alter it.[13]

Moreover, Aristotle does mention the fourth cause, namely the final cause, later on in *GC* II 9. He refers to it "as that for the sake of which is found in the shape and the form" (*GC* II 9, 335b6). This repeats the point from *Phys.* II 7 that the final cause is inseparable from the formal cause. Indeed, regarding the natural substances, which are the primary focus of the discussion in *GC* II 9, *Phys.* II 7 states that the efficient, the formal, and the final cause "often come to be one" and that "the formal and final causes are one" (*Phys.* II 7, 198a24–26).

The clue to understanding why Aristotle speaks about the causes and their number in the way he does at the outset of *GC* II 9 lies first in the facts about form as they are laid out in *Phys.* II, and second in what we are told in *GC* II 9. To begin with, the mention of matter and form at the beginning of *GC* II 9 (335a29–30) includes all four causes responsible for generation and corruption; for (as we learn in *Phys.* II) all four causes are found in these two (as Aristotle conceives of them). Moreover, the two causes not mentioned at the outset of *GC* II 9 are referred to later on (335a30–32, b6–7). But more significant still is the way these two causes are eventually referred to. The efficient cause, never mentioned by name, is referred to by pointing to the need for the work which it does, while the final cause is only mentioned in passing. We need to be sensitive to the possible significance of the configuration in which the four causes appear in our chapter.[14]

Second, form and matter, both mentioned by name at the beginning of *GC* II 9, are each proposed as being one of the causes of generation and

[13] It may, therefore, seem odd that Aristotle here appears to be claiming that matter and form cannot be sufficient to account for generation and corruption, when *Phys.* II 7 makes it clear that the form of a substance is the seat of *both* the end that is in the nature of substance to fulfill *and* the moving cause that directs it toward that end. According to Devin Henry, "*GC* II 9 must mean to treat the form and the mover as two distinct principles of generation" (Henry 2019: 82–83). If so, Aristotle departs from the *Phys.* II 7 view, in a way that may raise the question whether the final cause too is now thought of as a distinct fourth principle, and we have seen reason to think that such a radical turn of events is unlikely. Alternatively, as I propose, while adhering to the conceptual framework outlined in *Phys.* II, Aristotle picks out the efficient cause for special mention for the sole purpose of driving it home that his predecessors fail to identify and account for this cause, even though it is indispensable for generation, which is what they claim to be explaining. Here the target may be Plato, precisely because he introduces the Forms but fails, according to Aristotle, to identify the efficient cause of which the form is the seat.

[14] So, I agree with Henry that Aristotle "is interested exclusively in the role that form plays in substantial generation" (Henry 2019: 82), though, as explained previously (in footnote 13), I am skeptical that, compared to *Phys.* I, *GC* II 9, Aristotle "extends the role of form in generation" (Henry 2019: 80).

corruption, not only by Aristotle himself, but also by the thinkers he criticizes later in *GC* II 9. While the form is singled out by Plato, matter is proposed by the materialists. But the focus of this chapter is on the moving or efficient cause, namely the cause that sets off and completes the motions that result in a thing's being generated as a particular kind of thing. The need to account for generation and corruption is fully recognized by Aristotle's predecessors and motivates their theorizing. But they failed to identify the efficient cause. Aristotle focuses on the efficient cause in this chapter and the next one precisely because, although the importance of this cause is universally recognized, it remains as yet unidentified (*GC* II 9, 335b8). Much talk about the final cause in a context such as this is out of place. As Aristotle has already noted in *GC* I 7, "the that-for-the-sake-of-which is not active (ποιητικόν)" (*GC* I 7, 324b 14); in other words, it is not the cause that is responsible for imparting motion.

10.4 Classes of Things

According to Aristotle, the Forms as Plato conceives of them are "not enough" to account for generation, "just as they are not enough for the case of the primary things" (*GC* II 9, 335a32). Appearances notwithstanding, the point here is not that the Forms can account for generation neither in the sublunary nor in the superlunary region. As we are going to see, Aristotle holds that that there is no generation and corruption in the superlunary region. The point is rather that the Forms as conceived by Plato fail to account either for the generation and corruption of substances in the sublunary region or for the substances in the superlunary region (where there is neither generation nor corruption). Since they do not have a moving cause to point to, they also fail to account for their motion.

The purpose of lines 335a32–b6 is to identify the nature of the principle-as-matter of generable things, and to do so while *contrasting* it with the nature of that same principle "in the eternal and primary things" (ἐν τοῖς ἀιδίοις τε καὶ πρώτοις). Aristotle thereby makes good on the point, implicit in the statement that these principles are the same "in kind" (335a29), that they are not the same in *every* respect. So he begins by pointing out that the principle-as-matter of generable (and corruptible) things is what is *capable of existing and not existing* (335a2–3). He concludes by repeating the point: "so this is the principle-as-matter of the generables," where "this" (τοῦτο, 335b5) refers to "what is capable of existing and not existing" (τὸ δυνατὸν εἶναι καὶ μὴ εἶναι, 335b5–6).

Aristotle now distinguishes three classes of things (335a33–b3). Two of them contain things that share in being, though in radically different ways,[15] whereas the third class contains things that do not share in being at all. First are the things that belong to one of the two classes of things that *do* share in being, namely the ones that are said to exist by necessity (335a33–34). They are the celestial bodies, which are eternal and first. They exist by necessity in the sense that they were never generated and will never perish. Second are the things that Aristotle says necessarily do not exist (which by Aristotle's lights is stronger than saying merely that they never have existed and never will exist). Last are the things that are generable and perishable, about which Aristotle says that it is possible for them to be and not to be.

What kinds of things exist (including the way they exist [or do not exist]) is, as we should expect, a result of their respective principles and causes. Aristotle's earlier claim, that the principles he is seeking are equal in number to and the same in kind as the principles found "in the eternal and primary things" (335a28–30), leads us to expect that he will say what he has to say about the principles of generation by comparing and indeed contrasting them with the principles "in the things that are eternal and first." Moreover, at least as regards the material principle of generation, that is exactly what he does: unlike the material principle of the things that are eternal and primary, the material principle of *generable* things is what is capable of *existing and not existing* (335b5–6).

Little is said about what does not exist by necessity. The class of things diametrically opposed to this is that of things that exist necessarily; it is "impossible [for them] not to exist" (ἀδύνατον μὴ εἶναι, 335a35). These are the celestial bodies. To be eternal in this sense does not simply mean to exist for an infinite time, as this alone does not make it true that it is impossible for them not to be. It does not guarantee that they never came into being *and* will never cease to exist. For they could have existed for an infinite time if they existed forever in the past but perish at some time in the future or come into being and carry on existing forever after. It would be true in both cases both that they have an infinite life span and that it is possible for them not to be. Aristotle's claim is that the celestial bodies have existed forever in the past and will exist forever in the future. In fact, Aristotle's claim is stronger: it is impossible for them ever to not have existed in the past and not always to exist in the future (see *DC* I 12, 281b20ff.). For Aristotle, it is an empirical fact that the celestial bodies are

[15] Aristotle refers to these two classes as the class of sensible substances in *Meta.* XII 1, 1069a30–32.

eternal in the sense that they were never generated and will never perish.[16] This is only one of the premises for the inference that it is necessary for them to so exist. The other premise, which is central to the concern of *GC* II 9, is based on the notions of capacity and actuality, not mentioned here but fully developed elsewhere (most notably, in *Meta.* IX). Capacity and actuality are properties of matter and form respectively, the only two causes named at the beginning of the chapter. Central to the account of the principle-as-matter for generable things in *GC* II 9 is the notion of capacity, and the impossibility of their non-existing in question here is grounded in that notion. As argued in *Meta.* IX, things are endowed with capacities, and among the things so endowed are substances, and that precisely because one of their constituents is matter.[17] These capacities may become actualized in them and make them thus be something that they were not before. So, a thing *X* may possess the capacity the actualization of which would make it be *F*, and *X*'s possessing that capacity makes it possible for *X* to be *F*. Also, on this view, to be impossible for *X* to be *F* is for *X* to lack the capacity the actualization of which would make *X* be *F*; if so, it is impossible for *X* to be *F* and (therefore) *X* will never be *F*.

A capacity, Aristotle states in *DC* I 12, is also a capacity for its opposite: for instance, having the capacity to sit, as humans do, implies having the capacity to not sit (281b15–17). Now, this is an example of a capacity with respect to a category other than substance; having this capacity, a substance (a human being, say) can place or displace itself in a particular way. But capacities may also be possessed with respect to the category of substance, and such capacities must be possessed by things that are not substances. Aristotle is concerned with this second type of capacity in *GC* II 9 – namely, the one possessed by the matter of generable beings, which Aristotle identifies as what is capable both of being and of not being (335a32–33, b5–6), and which must, therefore, be a capacity of this sort.

[16] And in fact he says as much in *DC* I 11, 281b4ff. Aristotle points to evidence from the senses for the truth of the claim that the celestial bodies are eternal, or immortal as he says at *DC* I 3, 270b6–17. No empirical proof can substantiate the claim that there exist such eternal bodies. Arguably, Aristotle's conviction is not the result of empirical evidence alone but of that plus a conviction that some version of the principle of plenitude is at work. This would ensure that in addition to the things that are impossible to exist and to those that are possible to exist, there are also things that cannot possibly not exist.

[17] By introducing the account of actuality by analogy in this way in *Meta.* IX 6, 1048b6–8, Aristotle says that actuality is related to capacity as substance is to matter (ὡς οὐσία πρός τινα ὕλην), thus placing capacity solidly in matter. Cf. *Meta.* IX 7, 1049a18–26 where Aristotle makes the point that if something is made of something else, casket, say, from wood, or wood from earth, and "if there is something first which is no further called after something else nor said to be of it, this is first matter" (εἰ δέ τί ἐστι πρῶτον ὃ μηκέτι κατ' ἄλλο λέγεται ἐκείνινον, τοῦτο πρώτη ὕλη; 1049b24–26).

This is a capacity for being, period, that is, for being a substance, the actualization of which makes its possessor to be (period), that is, it makes it be a substance.[18] Indeed, one way in which we may have a particular substance come into existence, indeed *the* way all generable substances do come into existence, is via the actualization of something's capacity for existing as a substance. But, as is the case with all capacities, the capacity of being this or that substance implies also its opposite, namely that of not being this or that substance.

10.5 Celestial and Sublunary Substances Contrasted

10.5.1 Celestial Substances

According to the account offered in the *DC*, then, what owes its existence to the actualization of a capacity for existing also possesses a capacity for not existing. But if something exists and does *not* owe its existence to an actualization of a capacity for existing, it does not possess a capacity for its opposite, *not* existing.[19] In that case it is not possible for it ever not to exist. This implies that the first class of things mentioned above (335a33–34) – namely, the celestial bodies that exist of necessity – are eternal *because* their existence is not due to the actualization of a capacity for existence. If it were, since every capacity is also a capacity for its opposite, there would be no warrant for the claim that they are eternal, the sense of which claim is that there is no time in which these things possibly do not exist. That is Aristotle's reason for saying that they exist of necessity.

The constitutive principles of any substance are ontologically prior to the substance they constitute. The celestial substances owe their being to what causes them to be the things they are. These causes are the same in number and genus as the causes of the generable and corruptible things. Now, *DC* I 3 makes the obvious, virtually analytic, point that immortal substances have immortal constituents (*DC* I 3, 270b8–9). Ontologically speaking, however, what makes the celestial bodies immortal is that their

[18] Cf. *Meta.* IX 8, 1050b16: ἁπλῶς δὲ τὸ κατ' οὐσίαν. As I understand this, to say that a substance exists by having had a capacity for existing which *it* possesses turned to actuality is nonsense. True, a capacity for existing as some substance turned to actuality lies at the ground of the existing of something as a substance; but this is a capacity that belongs to something other than the substance in question, as for instance the capacity had by an embryo to become a man belongs to something other than the substance that is the man.

[19] This is not to say that they possess no capacity whatsoever. They do, as I point out further down, possess the capacity for motion, but also, of course, the capacity to be seen. See also Frey 2015*b* for a discussion of this issue.

constituents are immortal. Their constituents are, as are the constituents of all substances, matter and form, and to imply that the form is immortal is straightforward. The interesting implication is that the matter of the celestial bodies is immortal, and the key feature on account of which it is immortal is that it is not subject to alteration.[20]

Lacking the capacity that would make the celestial matter subject to alteration entails that whatever characteristic/property this matter has it has always had it. Therefore, the celestial bodies have always been the way they are now. It also entails that they will never lose the characteristic/property they have, and they never acquire new ones. They will therefore never be different than they are now. More importantly, at least for certain purposes, this same celestial matter's lack of the capacity for alteration grounds Aristotle's claim that the celestial bodies were never generated and will never die. For he is committed to the view that generation and corruption are conditioned on something's sustaining alteration.

Somewhat unexpectedly, perhaps, their matter proves to be the reason why the heavenly bodies are immortal, a point implicitly also made in *Meta.* XII with the remark that the matter of the eternal things is unsuited for generation and corruption (*Meta.* XII 2, 1069b24–26). But, as we also read in the same lines, the matter of the celestial bodies is suited for locomotion from one place to another (1069b26).[21] What makes it so suited is that though it has no capacity for alteration it has a capacity for locomotion. The fact that the matter of the celestial bodies has this capacity is an essential part of the explanation of why the heavenly bodies move from place to place – though not of the explanation that their motion is circular.

This alerts the reader to a subtler detail, easy to overlook, regarding Aristotle's views on the causation of motion. For Aristotle, the celestial matter's capacity to move from place to place is a necessary condition for the circular motion of the celestial substances. At the same time, the fact that celestial matter is incapable of being altered makes it impossible for celestial bodies to be in any other way than the way they are, and so to be

[20] *DC* I 3, 270b1–3 cites several properties that he claims the first body has, namely that it is "eternal and has no growth or diminution but is ageless and unalterable and impassive" without ranking them in terms of causal priority. Still, if the first body is unalterable and ungenerated, it follows that it has the other properties, including that it is eternal (cf. 270a12–18). But Aristotle cannot derive the eternality of the first body from its unalterability, because he arrives at its unalterability on the supposition that it is eternal (hence the reservation "if one believes our assumptions," εἴ τις τοῖς ὑποκειμένοις πιστεύει, 270b3).

[21] For further discussion of Aristotle's conception of the matter of the celestial bodies, see Charles 2000: 88–97 and Falcon 2005: 101–109.

subject to generation or corruption. These facts shed light on an unappreciated feature of matter: namely that, in addition to being what is in motion, matter partly causes the motion in which it is involved. Though form is *a* cause of why the celestial bodies move the way they do, another essential part of the "because" of the fact that they move is that their matter has the capacity for locomotion.[22]

10.5.2 Sublunary Substances, Matter, and Motion

The capacity of superlunary matter for motion from one place to another is a necessary though only partial cause of the fact that the first bodies can engage in such motion. But the fact that this matter is incapable of any other kind of change is the cause of the first bodies' being eternal. We now have our clue to understanding how Aristotle thinks matter is one of the causes of the generation and corruption that we observe in the sublunary region (*GC* II 9, 335a24–30). Matter in the sublunary region is endowed with a richer set of capacities, which includes a capacity for alteration in addition to a capacity for locomotion.[23] *GC* II 9 signals this fact with the statement that the "cause as matter for the things that are generable is that which is capable of being and not being" (335a32–33).[24] The grounds for this being so are offered in *Meta.* XI and are encapsulated in the statement that matter "does not exist in actuality but only in capacity" (*Meta.* XI 2, 1060a20–21).

Matter's involvement in generation and corruption is well-attested empirically. It is what *suffers* generation and corruption. Not so well-attested is the part matter plays in *causing* generation and corruption, which is due to its capacity for alteration. Recall that all change presupposes that a substrate survives to underwrite it; so the fact that matter is "capable of being and not being" cannot mean that in generation and corruption matter itself perishes and comes into being again. Matter survives the change while suffering it. Just as substances have capacities with respect to categories other than substance and can thus be or not be qualified in this or that way, so does matter have a capacity with respect to the category of substance and can thus be this or that substance. So, all that the remark that matter is capable of being and not being can mean is that

[22] The issues are admittedly difficult and complex. See Judson: 1994: 157–161.

[23] I will be referring to the matter found in the sublunary region simply as "matter," and to that found in the superlunary region as "first body."

[24] "Generable" here renders the Greek term "γενητά." The Greek could also be rendered "generated." The theoretical point in question remains the same on either rendering.

matter is capable of being some particular thing (some particular substance) and of not being that thing.

Here then is how the capacity of the matter of generable substances, namely its capacity for being and not being, makes this matter a *cause* of generation and corruption of substances. Matter causes corruption on account of its capacity *not* to be one of the constituents that, together with the other, the form, makes up a particular substance. Matter causes corruption when this capacity is actualized, with the result that the substance of which that matter was a constituent no longer exists. For this capacity to be actualized is for the matter that it is a capacity of no longer to be informed by the form that, together with that matter, was the substance of which the matter in question was a constituent. Similarly, matter is a cause of generation in that it has the capacity to *be* something it was not; it causes generation when this capacity is actualized.[25] Actualizing this capacity amounts to receiving a form together with which the matter in question constitutes some particular substance that did not exist previously. But this should not be taken to mean that matter is endowed with a capacity for being and not being only with respect to the category of substance. For it is not only substances that are generated and pass away. Matter in the sublunary region is the seat of every capacity that may be had by beings in that region, and this is also the sense of the statement in *Meta.* XI 2, that matter "does not exist in actuality but only in capacity" (1060a20–21). The nature of matter is to make itself available for the actualizing power of form. So, the matter of any substance in the sublunary region, considered in and of itself, just as matter and apart from the form together with which it constitutes substance, is thus endowed. The matter constituting a particular substance too is thus endowed. But while it is informed by a particular form to constitute a substance, the range of its capacities that may be actualized, in tandem with the actualization of its capacity to be that particular substance, is determined by the actualizing powers of the form together with which it constitutes this substance. Herein lies the reason why matter does have, though substances do not

[25] Henry 2019: 78–80 is right that Aristotle's remark in *GC* II 9 that matter is capable of being and not being means being or not being *a substance*. Aristotle certainly holds that. It needs to be pointed out, however, that this is not the only thing that the remark can mean, and the fact that Aristotle says "existing and not existing" (εἶναι καὶ μὴ εἶναι) without further qualification suggests that he has in mind matter as the seat of capacity quite generally. Thus, the matter of generable beings is what is capable of existing and not existing, not only as substance but also as other things (e.g., as something of such and such quality or quantity or in such and such place, and also – if, as one my reasonably hold, the elements are not substances, or not substances in an important sense – this or that element).

have, capacities with respect to the category of substance. Still, any change that takes place in some numerically distinct something, be it a change with respect to the category of substance or with respect to some category other than substance, is permitted by the fact that matter has the capacity to be subject to that change.

10.6 The Form as an Agent of Motion and Plato's *Phaedo*

After the short, cryptic statement on matter's role in generation and corruption, Aristotle offers another, less cryptic, but still unclear, statement to introduce form as a principle for the explanation of generation and corruption. We expect form to inform matter so that substances come to be and to leave matter so that substances perish. And *GC* II 9 does state that in addition to matter, a cause of the generation of substances is "that for the sake of which (οὗ ἕνεκεν), the shape and form; and this is the account of the essence of each substance" (335b6–7). Now, this description gives us the form in two of the three causal aspects we have from *Phys.* II – namely, the formal and final cause – the second of which is not mentioned at the outset of the chapter (335a30). The phrase "that for the sake of which," the final cause, identifies the end to which substances of the sort specified by the form in question aim, that is, the precise kind of "state" (ἕξις) that is appropriate for a substance of this kind to be and maintain itself in. But the description Aristotle provides in this passage also gives us the formal cause, namely the definition or account (λόγος) of the kind to which this particular substance belongs. It is substances of this kind that appropriately are and maintain themselves in states such as the ones specified by the final cause.[26]

[26] Henry is right that "forms play a different causal role depending on whether we are talking about *being* or *becoming* a substance" (Henry 2019: 81). But the way he understands this, when Aristotle says at 335b6–7 that the form is a principle of generation as (1) "that for the sake of which" (οὗ ἕνεκεν), *and* (2) "the account of the essence of each substance," he does *not* mean to include (2) in the "the role that forms play in the context of generation" (Henry 2019: 81). His reason is that (2) "picks out the form in its role as a formal cause of *being* a substance," a state which the thing generated has not yet attained but is maturing towards; by contrast, Henry argues, (1) picks out "the form in its role as a final cause of its *becoming*" a substance (which is just as you would expect in a discussion of the principles of generation). But this flies right in face of the fact that what Aristotle says in these lines is, first, that what is capable of being and not being is a cause *for the generables* as matter, and second, that the cause *as* that for the sake of which is their shape and form. (Here is the Greek: διὸ καὶ ὡς μὲν ὕλη τοῦτ' ἐστὶν αἴτιον τοῖς γενητοῖς, ὡς δὲ τὸ οὗ ἕνεκεν ἡ μορφὴ καὶ τὸ εἶδος· τοῦτο δ' ἐστὶν ὁ λόγος τῆς ἑκάστου οὐσίας.) Aristotle says both these things in the same, very short sentence, evidently referring to both (1) "that for the sake of which" *and* (2) "the account of the essence of each substance" as causes of generation; moreover, in so doing, he says implicitly, but also clearly, that the seat of the final cause is *the form*. True, as Henry points out, "that for the sake of

We now have three of the four causes, though we still lack the efficient cause. The form as a formal cause tells us the kind to which the particular substance generated belongs throughout its maturation and at its fully mature state, while in its aspect as final cause it specifies the right state at which it aims and is right for this substance to be in. In case one is inclined to think that the latter supplies also the efficient cause, *GC* I 7 makes it plain that it does not: the "that for the sake of which is not a productive [sc. active] agent (ποηтικόν)" (*GC* I 7, 324b14). The final cause specifies the end, but it does not set off and maintain the movement that brings the substance to its proper end; for that task an active or productive agent is needed, the efficient cause. That is what makes the subject *become* something (γίγνεταί τι), and it is only when the subject reaches the state prescribed by its end, and toward which the active agent was working to bring it about, that the subject stops *becoming* something, having as it were finally become the something that it was to be all along. The example Aristotle uses in *GC* I 7 is that of the ailing body brought back to the state of health. Closer to the argumentative concerns of *GC* II 9 would be an example that focuses on the active agent (τὸ ποιητικόν) that brings a newly born living being through the stages of growth that such a being must go through in order to reach the state that it is appropriate for it to be in, and brings it to that state while at the same time securing that its development happens within the bounds appropriate for beings of the particular kind specified by the form.

Already at the beginning of *GC* II 9, where Aristotle says that matter and form are not sufficient to account even for the first bodies, he hints that this chapter will be concerned with the efficient cause. To demonstrate its importance, he will now move on to point to weaknesses in the accounts of Plato and the materialists that are due to their failure to identify it properly. He begins by targeting a passage in the *Phaedo* where Plato has Socrates propose that the Forms are *the* cause of generation and corruption. At *Phd* 99A–102D Socrates complains that previous accounts fail to answer seemingly easy questions such as, for instance, what is it that

which" is crucial in generation. It directs what is generated toward the mature being it is meant to be. But the form, one would have thought, is also crucial in generation, inasmuch as it governs and oversees the maturation process of the thing generated at each and every step. It does so by providing the specific framework within which form (in its role as moving cause) is to work, and by seeing to it that the necessary "workings" are aligned with the goal aimed at. Form ensures that the maturation process stays on the path that it, the form, lays down, so that what is generated succeeds in reaching the state of being a full-blown substance. Hence, the infant or child or adolescent, though on its way to becoming a mature adult, *is*, at every stage of this process, a *human* infant or child or adolescent.

makes one thing and some other thing to be *two* things? It cannot be, Socrates says, that they are brought close together, for each is one even when they are close to one other. Moreover, they can still be said to be two even when they are pulled apart. On the hypothesis of the Forms proposed in the *Phaedo*, what makes them to be two is that they partake of the Form of Two. And they cease to be two when they no longer partake of the Form of Two. As Aristotle reports this view, there are (1) the Forms (εἴδη) and (2) the participants in the Forms,[27] so that any particular thing is whatever it is in virtue of participating in the relevant Form. It comes to be the thing it is by *coming* to participate in the Form; it ceases to be that thing by *ceasing* to participate in it.

A look at the relevant passages in the *Phaedo* shows that Aristotle reports the view accurately and that his objections are fair. The first objection is that if all there were to a particular substance's being generated is that something participates in the relevant Form, and the Form is always there and participants are always available, why "do the forms not always generate continuously, but they do sometimes and others not" (*GC* II 9, 335b18–9)? As I read it, the objection is that if Plato's hypothesis is right, we should expect to have humans being generated nonstop and everywhere, which is clearly not the case; what we have is humans being born in intervals, in fact orderly such intervals, an orderliness which the hypothesis of the forms as laid out in the *Phaedo* comes nowhere near explaining. Plato could perhaps counter that a continuous generation of humans or other substances is prevented by the fact that the world is full of substances, so that there is not enough stuff to participate in the Form of Human Being and for human beings to be generated nonstop. However, as I take it, the point of Aristotle's complaint is that Plato's view does not do justice to the orderliness in which substances of any particular kind beget substances of that same kind. Consequently, on Plato's view, there is nothing to preclude the possibility that substances of any sort just pop up randomly and haphazardly, while they disappear in the same way and do all this non-stop. But substances, Aristotle points out, are not generated continuously; rather, they sometimes are and sometimes are not (335b19: ποτὲ μὲν ποτὲ δ᾽ οὔ). When Aristotle says that substances are generated at

[27] Plato does not say what kind of thing these participants are, according to Aristotle, but it is reasonable to think that he understands them to be something along the lines of what he (Aristotle) calls matter. For he does not criticize Plato for not identifying matter as one of the causes. Moreover, what Socrates says in the *Phd.* at 95 B–E regarding his sinews and bones indicates that he thinks of the participants as some kind of body. Finally, Aristotle has just made a (critical) reference to Plato's view on matter at *GC* II 1, 329a14–24.

some times but not at other times, he clearly does not mean that humans, for instance, can appear here or there or just anywhere, but rather that they do so in an orderly fashion. In this context it may be helpful to recall the famous Aristotelian slogan that it takes a human being to beget a human being.[28] Nor has Plato anything to say on the causal principle that makes something transition from the state of not participating in a Form to the state of participating in it and *vice versa*. Though there are admittedly Forms such as Health and Knowledge, it is not the case that ill or ignorant people get healthy and knowledgeable just by participating in the Forms of Health and Knowledge. Rather, the physician is often the cause of the return of the ailing patient to health, just as the teacher is often the cause of the pupil's learning (*GC* II 9, 335b20–24).

Plato's version of the Forms as causes in the *Phaedo* does not account for generation, certainly not with the clarity and specificity that Aristotle demands. One might grant that the *Phaedo* passage is not the place to explore the ramifications of introducing such a principle, or to investigate the details concerning the range of its efficacy. At the same time, however, it is fair of Aristotle to say that Plato does not identify the need for an efficient cause as a philosophical issue worthy of sustained analysis. Nor should one complain that Aristotle ought to do better than make the *Phaedo* the *locus* of a full scale attack on Plato's failure to observe the need for an efficient cause. The *Phaedo* does after all point to generation and corruption as the direct motivation for introducing the Forms. Equally importantly, it is not at all clear that Plato's overall conception of Forms and participants creates sufficient conceptual space for an efficient cause that could survive Aristotle's criticism in *GC* II 9.[29] Possibly, this very

[28] See Henry 2019: 83–84, who takes Aristotle's objection to be that, on Plato's view, the generation of a human being should continue ad infinitum and never come to an end. But this generation does come to an end when the process has reached an adult human being. This fits with Henry's view that in the context of generation, which is the concern of *GC*, Aristotle's focus is on the form in its role as a final cause of the substance's *becoming* and not in its role as the account of the essence of each substance. However, reading Aristotle's objection as does Henry is an overreach, and fails to take account of Aristotle's point that generation takes place sometimes and sometimes it does not (ποτὲ μὲν ποτὲ δ' οὔ).

[29] It would be grossly unfair of Aristotle to level so general an attack (cf. 335b8 "no one speaks [of the efficient cause]" (λέγει δ' οὐδείς), if he thought that the author of the Forms can deal with his criticism elsewhere. True, in *Phil.* 27 A–C, Plato distinguishes a fourth genus in addition to the limit, the unlimited, and the mixed (see Gill 2019: 71–90). He calls it a cause (αἰτία) and says that it is by its agency that everything that comes to be comes to be (*Phil.* 26 E1–4). This is too large an issue to discuss here, though still it may be said that the *Philebus* account does not provide enough information to explain how (for example) a substance begets a like substance, or how a sick patient becomes heathy, in a way that could survive Aristotle's criticism. The same may be said of the *Timaeus,* where the job of efficient cause is arguably handed to the Demiurge. (For a contrasting

shortcoming stands in the way of a lucid account of the Platonic Form as a final but also formal cause, and deserves the criticism that Aristotle delivers in *Meta.* I.[30] It is fair to say that generation requires a moving cause which Plato dreamt of but failed to provide (cf. *GC* II 9, 335b8).

At the same time, the possibility must not be discounted that Aristotle considers Plato's introduction of Forms as causes to be a step in the right direction, even though he does not explicitly say so. The preamble to the attack on the *Phaedo* may be saying so implicitly. Aristotle first identifies matter and the "for the sake of which (οὗ ἕνεκεν), the shape and form" as causes (335b6–7). Only after having done so does Aristotle say that "a third cause must also be added, of which everyone dreamt but no one put into words" (335b7–8, picking up 335a30–1), then moving on to the attack against the *Phaedo* (335b9–16, 18–24). This may suggest that, however unsatisfactorily, the cause "for the sake of which (οὗ ἕνεκεν), the shape and form" is already there in Plato, indeed in the *Phaedo* (in much the same way as 335b16–17 may suggest that the cause as matter is already there in the accounts of generation and corruption provided by Aristotle's "materialist" predecessors).

10.7 The Materialists

The main objective of this part of *GC* II 9 (335b24–336a12), as of the one discussed in the previous section, is to examine the adequacy of previous accounts of generation, with a special focus on whether they managed to identify the all-important cause, namely the efficient (or moving) cause. Attention is now directed to the materialists who singled out matter as the cause that alone explains change and generation adequately. They receive initial praise for being sensitive to the fact that matter "alters and transforms" (ἀλλοιοῦν καὶ μετασχηματίζον), a point that makes their account, Aristotle says, "more akin to physics" (φυσικώτερον) than that of Plato. By praising the materialists and implicitly (while later explicitly) acknowledging the role of matter in the production of motion, Aristotle highlights yet another shortcoming of Plato's account. By focusing on matter, the account of the materialists is close to the facts, something that cannot quite be said about the account of Plato "and his followers" (τῶν οὕτω

view, see Henry 2019: 84: "*GC* II 9 is not meant as a general criticism of Plato's settled views on generation across all his works. It is a more localized attack on the *Phaedo* because the *Phaedo* is an example of the sort of theory that relies exclusively on forms as causes of natural generation.")

30 See, for example, *Meta.* I 7, 998a34–b6, especially 998a34–35: "no one has given a distinct account (σαφῶς...ἀποδέδωκε) of the essence (τὸ [...] τί ἦν εἶναι καὶ τὴν οὐσίαν)."

λεγόντων, 335a25). Matter is the locus of alteration, and neglecting it amounts to failure to pay due attention to change, thus undermining Plato's account of the particular changes which generation and corruption are.

The verdict of the examination of Plato's account of generation is that it was nowhere close to identifying the efficient cause. Since this cause is considered central in the production of change, and matter is the locus of change, Aristotle's praise may appear to be a sign that the efficient cause was within the materialists' reach. However, already in the first lines of the discussion of the materialists, Aristotle says enough to make the reader skeptical about such a hasty conclusion (335b29). Surely, the materialists have done better than Plato in so far as they have focused on the seat of change in accounting for the change that generation is. However, this fact alone does not bring them any closer to identifying the efficient cause as such. And it is worth recalling that Aristotle opens his critical discussion of Plato *and* the materialists by lumping them together: they all dreamt of the efficient cause but failed to deliver it (335b8).

Right after the initial praise, Aristotle goes on to say that the materialists too are wrong (335b29). First, they are wrong in believing that matter alone can be an agent of change. According to Aristotle, they are unappreciative of the fact that matter's nature is to *be* acted upon and to *be* moved, not to move and act upon other things (335b29–31). Blinded perhaps by the change manifested in matter, they think that matter is also the agent that causes it. Still, though matter is an essential cause of change, that is because matter has the capacity to suffer or undergo change, not to put that capacity to work by itself. In fact it is impossible for matter to undergo change unless some agent acts on it. Hence, the first lesson to take away from Aristotle's discussion of the materialists is that the causal role undergoing or suffering change should not be conflated with that of the efficient (or moving) cause of change.

Aristotle adds that these thinkers are wrong *also* because they "leave out" (παραλείπουσι) "the more important cause" (τὴν κυριωτέραν αἰτίαν); they do so *because* (γάρ) they have "cut out" (ἐξαιροῦσι) "the essence and the form" (τὸ τί ἦν εἶναι καὶ τὴν μορφήν) (335a34–35).[31] Seen in context, this second charge seems to be that the materialists have attributed

[31] It may seem peculiar that Aristotle chides the materialist for omitting the cause that he just chided Plato for introducing (thus Williams 1982: 184). Still, Aristotle's criticism of Plato is not against the introduction of Forms as such. Rather, by Aristotle's lights, Plato has failed to make good use of the Forms in his theory. More directly, Aristotle thinks that Plato has failed to adequately explore and bring out the full range of their causal efficacy.

moving powers to matter that matter cannot have. They have done so because they have ignored the causal role of the form, which supplies origin of motion.

The authoritative cause that the materialists are said to have ignored is described as "the essence and the form." Neither of these terms refers to the efficient cause, which might incline the reader to hold that Aristotle's complaint is not that the materialists have failed to provide the (or maybe *an*) efficient cause, but only that they have failed to see the role that the form and the essence play as causes of generation. After all, in Aristotle's account of generation, the essence and the form are final causes and those (as we have seen) are not movers. Furthermore, this observation apparently coheres with Aristotle's next point, namely that the materialists (or at least some of them) have attributed moving powers to bodies, such as heat and cold, which he characterizes as "too tool-like" (λίαν ὀργανικάς) (336a1–3). In their view, Aristotle says, it is in the nature of heat to dissociate and of cold to bring together, and of each of the other powers from which the materialists propose to generate everything either to act or be acted upon (336a3–4).[32] In all of this, scholars have thought we have reason to hold that Aristotle implies that the materialists have come close to providing an efficient cause.[33] But it is unclear that any such view receives support from

[32] *Meta.* I 3, 984b3–11 makes probably the same point, though in greater detail, mentioning elements as the bearers of the contraries in question. Here too, at 336a6–7, *GC* II 9 refers to fire as the bearer of heat, while *GC* II 9, 336a11–12 makes fire an agent of change. Subject to the action of the cold, one may assume, elements come together, and that brings about generation, whereas subject to the action of the hot, they come apart, and that causes corruption.

[33] See, for instance, Henry 2019: 89. Henry does well to point out that the objections to the materialists advanced in *GC* II 9 have points of contact with the criticism leveled against them in *Meta.* I 3, 984l6–27. However, Henry says that *GC* II 9, 335b29–33, which according to his reading criticizes only one group of materialists, namely those who claim that matter *alone* is the agent of change, and for this reason lack an efficient cause, contains "essentially the same argument deployed at *Meta.* 984a16–27" (Henry 2019: 89). This is not quite right. *GC* II 9 is not concerned with distinguishing groups of materialists. Rather than paying attention to the historical development of the materialist view, as does the *Metaphysics* passage, *GC* II 9 addresses the materialist view quite generally as a theoretical system employing a specific theoretical approach to address the question of generation. By contrast, and by paying attention to the historical development within different lines of thought wrestling with the problem of generation, the *Metaphysics* passage deals first with thinkers of a materialist inclination who have come to realize that one principle alone cannot by itself produce motion. (Even Parmenides, who is not a materialist, realizes that one principle is not enough and introduces (in some sense) two (984b4–5)). Later thinkers increase the number of principles to *two* by introducing several material elements pointing also to the hot and the cold – thus arriving at the full-blown materialist view discussed in *GC* II 9. Notice that Aristotle says that they have increased the principles to *two* by introducing *several* elements and qualities (*Meta.* I 3, 984b5–8). Subsequent thinkers realize that not even by thus increasing the number of elements and qualities can they *generate* (γεννῆσαι) the "*nature of the things that are*" (τὴν τῶν ὄντων φύσιν) (984b9). As a result, they have turned to the *next kind* of cause (984b8–11). So the materialist approach that *GC* II

GC II 9. Fire, Aristotle says, both acts and is acted upon (336a6–7), which would commit the materialists to saying that fire is both active and passive, and, we might add, invite the following question: Which part of fire acts and which part is acted upon, and why? Aristotle's point is not simply that the materialists have only missed the formal and final cause, though they did have, or were somehow close to or touched upon, the efficient cause. Rather, his criticism is that having done away with "the cause as form" (336a2–3), the materialists have done away with the efficient cause altogether. Even if matter *could* be said "to impart motion and to act" (κινεῖν καὶ ποιεῖν) all by itself – and even this, in Aristotle's view, would be a mistake (335b29–31) – the fact remains that the motions and actions in question would be aimless, as there would be nothing directing them with an eye to fulfilling a purpose. Aristotle's point is that what produces motions like *that* could never be "the more commanding cause" of the generation of *substances*.

Aristotle does characterize the moving powers that some materialists attribute to matter as tool-like (336a1–2, 7–12); but he does not think that this fact brings them any closer to identifying the efficient cause. What makes something a tool is the fact that it may be used to complete specific tasks aimed at achieving specific ends. Tools play at most a *contributory* causal role. Moreover, in order for the tools to fulfill even this subsidiary causal role, they are in need of the presence of a final cause that governs and sets in motion an efficient cause, which alone puts them to proper use and makes them do the work as this has to be done in order to achieve the specific ends in question. In important respects even the use of the term "tool" to signify the moving power of matter as these materialists apparently did is a misnomer. To see this, we need only imagine the perfectly plausible scenario of our discarding some item as trash only to be told later that we have just disposed of a rare and expensive tool. We had no idea what the thing was *for* and so it appeared to us to be just trash. Indeed, Aristotle does not call fire, despite its power to heat, a tool. He only says that the causal power which the materialists have attributed to the elements is too tool-like (λίαν ὀργανικάς) (336a2) in its character. As a matter of fact, what he says is that however active the materialists make fire to be, they fail to see that fire moves things in a way that is inferior to the way

9 considers as a whole – with its various developments that the *Metaphysics* passage reviews historically – does not provide the cause that *GC* II 9 is primarily/programmatically concerned with, namely the efficient cause that generates *substances*.

that tools do (ὅτι χεῖρον ἢ τὰ ὄργανα) (336a12).[34] By attributing generative motion to fire, the materialists end up crediting tools with a primary causal role in the productive process. Toward the conclusion of his criticism of the materialists, Aristotle says that ascribing the generation of substances to the workings of fire is like ascribing the production of objects to the saw itself (336a7–11). But the saw, alone by itself, cannot move. What he thereby says, inadvertently perhaps, but still tellingly, is that in the context of the generation of substances, the power of, for instance fire to heat is actually no power to *generate substances* at all.[35]

We may want to use the tool metaphor in connection with the term "efficient cause" and say that the motions of the simple bodies and their simple powers amount to an *instrumental* efficient cause.[36] But we must not fail to bear in mind that this does *not* imply that in *GC* II 9 Aristotle attributes to any of the materialists he criticizes a notion of efficient cause, in any theoretically interesting sense, of the phenomena whose principles he is discussing in *GC* II 9: namely, the generation of substances. In Aristotle's view, the "commanding cause" of the generation of substances is no tool. And doing things randomly by itself does not make anything a tool, let alone a moving cause. Properly for something to be a moving *cause* in *GC* II 9 it must do the work that aims at achieving some specific end, and such is only the cause that supplies the form, which also specifies the end to be achieved.

[34] The idea of matter endowed with disorderly and directionless motions brings to mind Plato's notion of receptacle in the *Timaeus* (49 A1–4, 52 A8, D2–4). It *also* brings to mind the bones-and-sinews passage in the *Phaedo* (98 C2–D7), and the distinction between cause (αἰτία) and contributory cause (συναίτιος) advanced in the *Statesman* (281 E1–10).

[35] Nor is it a power that can generate elements. Aristotle holds that the elements are generable and corruptible and that their generation, as well as their corruption, are orderly, strictly specifiable process. The elements are composites of matter and contrarieties and the generation of any one of them depends on the corruption of another. The tool-like movement some materialists are said in *GC* II 9, to be ascribing to the elements provides zero conceptual space for such generation or, for that matter, corruption.

[36] See Henry 2019: 94.

11

On Generation and Corruption *II 10*

Spyridon Rangos

11.1 Introduction

The initial sentence of *GC* (I 1, 314a1–6) announces that the principal aim of the treatise is to distinguish the causes and the *logoi* of generation and perishing (τάς τε αἰτίας διαιρετέον καὶ τοὺς λόγους αὐτῶν) in a way that will be general enough to apply to all cases of things (ὁμοίως κατὰ πάντων) that come into being and pass away by nature. The treatise will also (ἔτι) deal with growth (αὔξησις) and alteration (ἀλλοίωσις) in an attempt to establish what each of them is, and whether generation is one and the same with alteration or not (as their different names suggest). From Aristotle's formulation and what follows in the main part of the treatise, it becomes evident that the subjects of quantitative and qualitative change have been included in this investigation because some earlier thinkers (e.g., Empedocles) tended to blur the distinction between generation and mere change.

At this early stage, it is not clear what Aristotle means by *logoi* and how they differ from "causes". Since αἰτίαι and λόγοι are syntactically connected by means of the strong conjunctive τε καί, the (otherwise attractive) possibility of καί being epexegetic is precluded. Most commentators think that *logoi* here signify "definitions" (i.e., of generation and perishing) and translate the passage accordingly.[1] But *prima facie* this is not the most natural way of understanding the phrase. To distinguish (διαιρετέον) the causes certainly means to tell their different kinds apart. Correspondingly, to distinguish the definitions would have to mean to tell apart the

* I would like to acknowledge my gratitude to the editors and the anonymous readers of the volume for their precious help in the improvement of this essay. I am also grateful to Argyro Lithari and Stasinos Stavrianeas who commented upon an earlier draft.

[1] For example, Joachim in Barnes 1984: I: 512; Williams 1982: 1; Brunschwig 2004: 31, 35; Burnyeat 2004: 7n1.

definition of generation from that of perishing, a rather trivial distinction of opposites.

Philoponus (*In GC* 7.22–8.19) testifies to an ancient controversy about the scope of Aristotle's distinction between αἰτίαι and λόγοι in the introductory sentence of *GC*. We do not need to follow either the view of the unnamed interpreters (τινές) he mentions (*In GC* 7.23–8.2), who detect an implicit distinction between *universal* and *proximate* causes, or his own suggestion that the distinction is between the *four* causes, on the one hand, and the *specific efficient* cause that is operative in each case, on the other (8.2–19). The important thing is that in late antiquity no one seems to have thought that the λόγοι mentioned at the beginning of *GC* should be understood as "definitions" pure and simple. And it is certainly significant that ancient commentators felt the need to distinguish between "causes" and *logoi* as if they were related but distinct concepts.

Marwan Rashed seems to be closer to Aristotle's intention when he translates λόγοι as "*raisons*," discerns "*l'importance de la finalité*" in the initial sentence of the treatise, and compares it with *PA* I 1, 639b16–19, where a similar distinction between αἰτίαι and λόγοι is implicitly introduced.[2] In that context, Aristotle, eager to rank the efficient and final causes of natural generation, writes:

> It seems that the one which we call "for the sake of which" is primary; for this one is a λόγος, and the λόγος is a beginning (ἀρχή) alike in the case of things constituted according to craft and those constituted by nature. (*PA* I 1, 639b14–16)

Aristotle's reason for the priority of the final over the efficient cause is that the former is a *logos* and as such a beginning. The context and the examples used (639b16–19) indicate that by *logos* Aristotle means an account of the desired *end*. The physician and the builder will have to arrive, either through reasoning (διανοίᾳ) or sense-perception (αἰσθήσει), at a definition (ὁρισάμενος) of what health or a house (respectively) is before they can cure a patient or build a house. For it is only by virtue of such an account that they can provide causal explanations about why they do things the way they do them, and why they must do them so (ἀποδιδόασι τοὺς λόγους καὶ τὰς αἰτίας οὗ ποιοῦσιν ἑκάστου, καὶ διότι ποιητέον οὕτως). Aristotle also says (639b19–21) that the final cause and the beautiful (τὸ καλόν) are more present (widely and/or intensely) in the works of nature than in the products of craft.

[2] Rashed 2005: 1, 85.

To come back to the initial sentence of *GC*, I suggest that Aristotle means the intended division of the various causes of generation and perishing to be supplemented by *teleological accounts* of generation and perishing, that is, not just definitions of what they are but accounts that will also include some explicit reference to the end(s) which those natural processes accomplish. If that suggestion is correct, then the *logoi* mentioned at the outset of *GC* point to accounts of generation and perishing that explain why it is better for some things to come into being and perish than not to come into being at all in the first place (nor perish, for that matter).[3]

Now, for the entire first book of *GC* and the largest part of the second, Aristotle explores generation and perishing primarily by means of hylomorphism.[4] To be sure, in *GC* I 3, 318a1–8, Aristotle says that the cause in the sense of the beginning of movement has been treated ἐν τοῖς περὶ κινήσεως λόγοις (a clear backward reference to *Phys.* VIII) where it has been established that one thing is for all time unmoved and another always moved. But since, he continues, the unmoved thing forms the subject of another and more primary philosophical discipline (*sc.* than physics), that is, first philosophy, it is the cause of generation and perishing which moves by being itself continuously moved that will be treated later – a forward reference to the efficient cause dealt with in *GC* II 10.

It is only at the outset of *GC* II 9 that a new beginning is announced (335a24–26). Here Aristotle speaks of neither causes nor accounts but of principles (ἀρχαί). It is nevertheless clear that the principles he has in mind are identical with the causes mentioned in the introductory sentence of the treatise. This new beginning specifies that generation and perishing occur only in the region around the center of the universe which, as we learn a few lines later (335a33), is the domain of possible (as opposed to necessary) existence. As announced at the outset of the entire treatise, the new beginning will inquire about principles/causes in a manner general enough to apply to all cases of generation alike (περὶ πάσης γενέσεως ὁμοίως). Aristotle says that the number and kinds of principles applicable to generable and perishable things are equal to the number and kinds of principles to be found in the domain of eternal and primary things (335a29). This can only mean that in the domain of generation and perishing the familiar scheme of the *four* causal principles is at work.

[3] The relation of ὁ λόγος τῆς ἑκάστου οὐσίας to τὸ οὗ ἕνεκα is also explicitly present in *GC* II 9, 335b6–7.

[4] Solmsen 1960: 379.

The same fourfold scheme of causes is implied by means of a backward reference to *Phys.* II 3 and 7 at the very end of the chapter (336a13). However, no explicit mention of the final cause is to be found in this or the following two chapters.

The aim of *GC* II 9 is to establish that in natural generation, no less than in skillful human production (335b27–33), there has to be a *third* cause (τρίτην, 335a30–31, 335b8), beside matter and form, that is, a moving cause (ὃ ἂν ᾖ κινητικόν, 335b28–29), which all previous philosophers dreamed of but no one states (335b7–8). But in the context of *GC* II 9 Aristotle is interested to show the inadequacy of earlier theories which tried to account for generation and perishing by means of either matter alone or matter and form.[5] The recapitulation of 336a14 makes it plain that matter and form have been sufficiently discussed. The specification of the moving (or efficient) cause of generation and perishing is the focus of the first part of *GC* II 10 (336a15–b26). The second part (336b26–337a17) supplements the three causes explored so far by the *fourth* cause, the final one, which was clearly included in the fourfold scheme implied at the beginning and the end of *GC* II 9 and, if the above suggestion about the scope of *logoi* is correct, was also signaled out for later treatment at the outset of the treatise. The last part of *GC* II 10 (337a17–33) is an appendix that establishes the uninterrupted continuity of time and movement (and, by implication, of generation) with reference to the uniqueness of the first rotating body, that is, the outermost sphere of the universe.[6] We may add that the eternal movement of this first rotating body derives its continuity from an unquenchable desire for the immobility of the first unmoved mover. Aristotle does not explicitly deal with this first principle of movement in the context of *GC* because he thinks that its proper treatment belongs to first philosophy rather than physics (*GC* I 3, 318a4–6). But he takes its existence for granted. Besides the indisputable reference to it in *GC* I 3, 318a4, there is also the hypothetical assumption of an unmoved mover in *GC* I 6, 323a12–14, 31–33, and the admission of a single first principle (ὑπὸ μίαν ἀρχήν) of movement said to be one (ἕν) and the same (τὸ αὐτό) and also *unmoved* (ἀκίνητον) in *GC* II 10, 337a19–22.[7] Moreover, the teleological account of generation and

[5] See Panos Dimas in this volume.

[6] Since the appendix is not discussed in this chapter, see Verdenius-Waszink 1966: 75–79, where an analysis of the argument, and especially of the convoluted and long first sentence (337a17–22), is to be found.

[7] To these indications we may add (i) the question about the identity of "the first mover" that Aristotle addresses to Empedocles in *GC* II 6, 334a7–8 and (ii) the backward references to *Phys.* VIII in *GC* II

perishing, which is the subject of the second part of *GC* II 10, makes reference to substance (οὐσία) in a context that could be taken to imply the existence not only of eternal substances in everlasting rotations but also of a primary substance exempt from all kinds of change, including circular locomotion, the change that affects substance the least (according to *Phys.* VIII 7, 261a20–23).

With the analysis of the moving and final causes of generation and perishing in *GC* II 10, the exploration of the causes/principles of generation and perishing announced at the outset of *GC* I 1, and reiterated in *GC* II 9, is fulfilled (337a15–17). The last chapter of the treatise (*GC* II 11) is a supplement on the modality of future events instigated by the proof, in the last part of *GC* II 10, of the necessary continuity of celestial movement.[8]

The present chapter is divided into three sections. The arguments for the sun's annual course as the moving cause of generation and perishing, as well as the consequences, both stated and unstated, of this claim are the topic of Section 11.2. An analysis of the teleological argument and the metaphysical assumptions it presupposes is the subject of Section 11.3. The concluding section illustrates how Aristotle's fourfold scheme of causes is meant to work.

11.2 The Moving/Efficient Cause of Generation and Perishing (336a15–b26)

The last sentence of *GC* II 9 indicates that what has been accomplished so far is a determination of the material and formal causes of generation and perishing. We now expect treatment of the other two causes, the efficient and the final. And this is what we get.

Aristotle assumes as proven (δέδεικται) the eternity of the circular movements of the heavenly bodies in order to derive from it the continuity of generation (336a15–18). Although he speaks of locomotion, and more literally of "moving by way of locomotion" (336a15), he clearly means the *circular* locomotion of the heavens. He also assumes as well stated in an earlier discussion – that is, in *Phys.* VIII 7 – the priority of locomotion over all other changes (336a18–20), that is, changes in the category of quality, quantity and even substance (τὸ πρώτην τῶν μεταβολῶν τὴν φορὰν ἀλλὰ μὴ τὴν γένεσιν εἰπεῖν). But he provides a short argument in favor of the

10 (336a15, 336a18–19, 337a18, 337a25), which imply knowledge of the whole content of this book (cf. Rashed 2005: 178).

[8] I refer the reader to Sean Kelsey in this volume.

causal dependence of generation on locomotion (336a20–23), and a more extended argument for the specification of the celestial locomotion that is the moving cause of generation and perishing (336a23–b9).

11.2.1 The Causal Role of Celestial Circular Motion

Aristotle's short argument (336a20–23) states that "it is far more reasonable" (πολὺ εὐλογώτερον) to consider a moving thing to be the cause of another thing's generation than to regard the generation of a thing (lit. "what-is-not") as the cause of the being for something that is. The reasonableness of this assertion is based on the fact (οὖν) that the moving thing is already something, whereas the entity to be generated does not yet exist. If the non-existence of the thing to be generated forms the minor premise of Aristotle's syllogism here, his tacit major premise seems to be that what does not exist does not cause anything at all. However, since it is not the non-existing thing as such but rather the process of generation that should be compared with the locomotion of the moving thing, and since the process of generation is obviously something, one might think that the argument is fallacious. Moreover, it is not difficult to see – even before the modern discovery of the internal combustion engine, which Williams brings to bear against Aristotle's claim[9] – that a process of generation *may* be the cause of another thing's locomotion: the generation of water out of airy mist in a cloud may cause the downward movement of sand or dust particles previously floating idly in the atmosphere.

A more charitable interpretation would take Aristotle's words not as building up to a conclusive argument about the causal priority of locomotion over generation but as an indication that this may indeed be the case. On such an interpretation, a moving thing is a qualified substance, that is, a substance with a property (locomotion), whereas a thing in the process of generation is neither a substance yet nor, for that matter, a qualified substance. And since, given Aristotle's ontology, it is substances and their properties that are causes, it would follow that it is far more reasonable to think of a moving substance as cause of a process of generation than the other way round.

Another interpretative option would benefit from a look at *Phys.* VIII 7, to which Aristotle refers at 336a19. There Aristotle distinguishes three notions of priority (260b17–19); subsequently in the chapter, he argues that locomotion is prior to generation according to all three of them.

[9] Williams 1982: 186.

Aristotle claims (261a20–23) that, of all changes, locomotion is the one which affects substance *the least* since it affects only a substance's place but none of its other qualitative or quantitative traits. As such locomotion is the only kind of change that is admissible in the domain of eternal being (260b29–30, cf. 26–29), and the things that it characterizes the most are self-movers (261a23–26), that is, the ultimate origin (ἀρχή) of all kinds of change.

When we come back to the beginning of *GC* II 10 with these distinctions in mind, it would seem that "existing" (τὸ ὄν) and "not existing" (τὸ μὴ ὄν) at 336a21–22 are metaphysically loaded terms denoting eternal and generable existence respectively. If so, then Aristotle means to claim that the generation and perishing that occur in the region around the center of the universe should be causally related to the only kind of change that obtains in the domain of eternal being, and that is circular locomotion. To reverse this causal sequence would entail that the sublunary processes of generation and perishing are the causes of the eternal revolution of the heavenly bodies, that is, that possible being is the cause of necessary being, an unacceptable assumption by Aristotle's lights.

11.2.2 The Sun's Approach and Retreat as Moving Cause of Generation and Perishing

If the processes of generation and perishing are to be causally traced to the domain of eternal being in the heavens, on the available astronomical knowledge of Aristotle's time (and if we exclude the revolutions of the five known planets which did not seem to have any visible effect on earthly things), three options should be considered: (i) the daily rotation of the outermost sphere of the fixed stars which brings about the alternation of day and night; (ii) the monthly cycle of the moon; and (iii) the annual course of the sun along the ecliptic which brings about the succession of the seasons.

Aristotle does not consider option (ii) at all. In *GA* IV 10, 777b16–35, however, he acknowledges the moon's causal contribution to generation and perishing and explains it by recourse to the fact that the moon receives its light from the sun and is, as he puts it, "another, as it were, smaller sun." It follows that the moon is only mediately related to generation and perishing.

On the empirical evidence of the growth of vegetation in spring and its decline in fall (336b15–19) – a fact with tremendous economic and religious significance for all basically agrarian communities such as the

Greek *poleis* – Aristotle assumes that the sun's annual course along the ecliptic is responsible for *all* kinds of generation and perishing in the region around the center of the universe, and not just for the generation and perishing of annual plants and animals. But it is perhaps significant for Aristotle's method here that the empirical evidence is mentioned *after*, and *as a supplement to or confirmation of*, what is an *a priori* comparison of options (i) and (iii) in 336a23–b9.

From the fact that generation and perishing are contraries Aristotle infers the necessary *contrariety* or *duality*, in either direction or unevenness, of the locomotion which causes them to occur (336a26–30). For, as he says twice (336a30–31, 336b9), the causes of contraries should themselves be contraries. The circular movement of the first heaven (ἡ πρώτη φορά) does not meet both requirements (continuity and contrariety or duality). For there is nothing uneven or dual about it. But the annual path of the sun in the "inclined circle," that is, along the ecliptic, does meet the requirements and is, therefore, considered responsible for generation and perishing (336a31–32).

Given the absolute uniformity of the movement of the outermost sphere of the universe, if that movement were an efficient cause at all, it would have to be the efficient cause of *either* generation *or* perishing but *not* of both at once. These options are silently ruled out by Aristotle, and it is not hard to see why. On the one hand, if the outermost rotation were the efficient cause of generation alone, then things would only come to be and never pass away, an assumption which (even without the testimony of the senses to the contrary), given the everlastingness and spatial limitedness of Aristotle's universe, would entail the simultaneous presence of an infinite number of extended things in a limited space – a logical absurdity (unless the generated things are progressively smaller in size *ad infinitum*). On the other hand, if the outermost rotation were the moving cause of perishing alone, then things would have passed away long ago (cf. *GC* I 3, 318a16–19): in fact, given the everlastingness of Aristotle's universe, they would have already passed away at any temporal point in the past. In either case, neither generation nor perishing would still be around, and the desire to account for them would lack the very thing whose presence it meant to account for in the first place.

From the above considerations it follows that either (i) the movement of the outermost sphere does not play an efficient role on generation and perishing or (ii) it is the moving cause of the *continuity* of the sun's annual course (336b3). On Philoponus' testimony (*In GC* 291.18–20), Alexander of Aphrodisias thought that Aristotle considered the revolution of the sphere of the fixed stars to be the *ultimate* cause of the continuity of generation and perishing. Philoponus' own view (*In GC* 291.21–28) is that the continuity of generation and perishing is due to the continuity of the sun's movement along the ecliptic. Both views have their modern supporters because Aristotle seems to waver between the two alternatives.[10] At 336a33 he says that locomotion "along the inclined circle" possesses both continuity and a duality of movements, whereas a few lines later (336b2–3) he explicitly assigns the continuity of generation and perishing to "the locomotion of the whole heaven" (ἡ τοῦ ὅλου φορά). What he probably means is that the sun's annual course is the *proximate* general cause of generation and perishing because it possesses both a duality of movements and continuity. But whereas the duality of movements is the *specific* property of the sun's annual course, the continuity of this revolution, like the continuity of all planetary movements, causally depends on, and is ultimately derived from, the continuity of the rotation of the outermost sphere of the universe. This is in fact what Aristotle explicitly argues for in the last part of *GC* II 10 (337a18–33). Therefore, I tend to side with Alexander and his modern supporters in favor of option (ii).

Since we know from *DC* I 4 that for Aristotle there is no contrary to circular motion, contrariety "in direction" (τῇ φορᾷ) at 336a30 must mean movement in a *straight* line that goes back and forth. But since this kind of movement in Aristotle's eyes cannot be considered continuous (since it involves rest), the option is silently ruled out. By contrast, contrariety "in unevenness" (τῇ ἀνωμαλίᾳ) might be a *circular* movement that is uneven either in speed or in another sense (cf. *Phys.* V 4, 228b21–28). I do not think that Aristotle intends, as Joachim suggests, variation in speed.[11] It seems more reasonable to think (with Williams and Rashed) that by "contrariety in unevenness" Aristotle intends the apparent irregularity of the sun's annual course.[12]

Considered in itself, the sun's annual course is uniform and as such proceeding always at the same speed. I think that the qualification "in the same state" in the formulation of the principle "same cause *in the same state* gives rise to same results" (336a27–28) is meant to explain in what sense the sun's annual course is regular and in what sense uneven. If the sun's annual course along the ecliptic is considered *in itself* it is in fact regular. For it is not reversed or made contrary to itself at any temporal point. As

10 Mugler 1966 and Migliori 1976 follow Alexander (*apud* Rashed 2005: 77n4); Williams 1982: 188–190 follows Philoponus, but his arguments are not convincing.

11 Joachim 1922: 257–258. 12 Williams 1982: 187, Rashed 2005: 173.

Philoponus (*In GC* 291.1–5) rightly suggests, the unevenness of the sun's movement that Aristotle has in mind is an unevenness *relative to us*, not an absolute unevenness. Its origin lies in the fact that the plane of the sun's annual path is inclined to the plane of the earth's equator (which is one and the same with the plane of the celestial equator) by an angle of, as we know, 23.5°. As a result of such inclination, the sun's daily course varies considerably throughout the year. In the northern hemisphere, the sun rises, reaches culmination, and sets by following a progressively more northern course in winter and spring, and a progressively more southern course in summer and fall, while the opposite holds for the southern hemisphere. Aristotle speaks of this changeable route as the sun's approach and retreat – perhaps as a concession to a popular view. However, it is not the linear *distance* of the sun from the earth (whose extremes, unknown to Aristotle, were later called "apogee" and "perigee" and are, in any case, almost irrelevant to seasonal variation) but the varying *angles* by which the sun rays reach any place on the surface of the earth that cause the succession of the seasons. And this is the sense in which the sun's annual course is uneven.

Since Aristotle claims (*GC* I 3, 318a23–25; *GC* II 10, 336b24) that the generation of one thing presupposes the destruction of another, one might wonder in what sense there is a period of generation (winter and spring) and a corresponding period of perishing (summer and fall). The perplexity did not escape the ancient commentators. Philoponus (*In GC* 289.27–290.5) provides a solution in accordance with Aristotle's own doctrine in *GC* I 3. He says that when Aristotle speaks of an annual period of generation, he means a season in which *superior* substances are born (τὴν τῆς κρείττονος οὐσίας γένεσιν) from the perishing of individuals of inferior kinds; and when he speaks of a period of perishing, he means the reverse.

Since the sun's so-called approach and retreat has been established as the moving cause of generation and perishing, it follows that generation and perishing according to nature happen in equal times (336b9–10). For the temporal period of the approach, which (we are meant to think) extends from the winter solstice to that of summer, is (roughly) equal to the temporal period of the retreat, which extends from the summer solstice to that of winter. The division of the solar year in two equal halves explains why the process from birth to maturation and acme equals in size the reverse process from acme to decline and extinction. Aristotle brings the evidence of the senses to bear on what has been so far a purely logical conclusion (336b15–19). When he says that "we see that as the sun approaches there is generation and as it retreats there is perishing, and each phase happens in equal time" he obviously means to draw our attention to the annual cycle of vegetation. But he also acknowledges exceptions of which the cause is to be found in the unevenness of matter (336b20–24).[13] Since the generation of one thing presupposes the perishing of another and *vice versa*, as an earlier chapter of the treatise has shown (*GC* I 3, 318a23–25), Aristotle thinks that particular processes of perishing are indeed speeded up because the matter of the perishable thing is already too degenerate to wait for the appropriate time for the thing's extinction. This speeding up of the process of perishing results in a corresponding speeding up of the correlative process of another thing's generation. It follows that generation and perishing appear to be uneven or irregular. But to Aristotle's mind, this evident unevenness or irregularity of generation and perishing should be seen against a background of regularity and order if it is to be properly understood. That is why in the immediately preceding section (336b10–15) Aristotle maintains that all living things possess naturally determined lifespans or temporal life limits (what he calls "the times and lives") which are all measured by a corresponding περίοδος. His basic concern in this section is to highlight the natural order pertaining to the domain of generation and perishing, and to account for it by virtue of the sun's so-called approach and retreat, that is, the solar year. Aristotle's emphasis on natural order paves the way for a transition from the efficient cause of generation and perishing discussed in the first part of *GC* II 10 to the final cause that is going to be the subject of the second part of the chapter. Before moving to that part, it is worth pausing to think about the relation of the sun's annual course to the determination of the lifespans of living things and, more generally, about the relation of the efficient cause of generation and perishing to the formal and material causes discussed earlier in the treatise.

11.2.3 The Sun's Approach and Retreat as the Universal, and Substantial Form as the Specific, Moving Cause of Generation and Perishing

Aristotle has established that the efficient cause of generation and perishing is the sun's so-called approach and retreat. He is proud to have discovered

[13] I think, in agreement with Rashed, that διὰ τὴν πρὸς ἄλληλα σύγκρασιν in 336b20–21 is sound and, *pace* Joachim, should not be obelized. The word σύγκρασις is attested prior to Aristotle (e.g., in Plato's *Phil.* 64 C) and the meaning it provides in the present context is defensible. But I do not think that the addition of <τὸ> is needed after διὰ in 336b24, as Joachim thinks and Rashed adopts (without angle brackets) in his text. The sentence ὥστε συμβαίνει διὰ τὴν τούτων γένεσιν ἄλλοις γίνεσθαι φθοράν says roughly the same as *GC* I 3, 318a23–25, διὰ τὸ τὴν τοῦδε φθορὰν ἄλλου εἶναι γένεσιν καὶ τὴν τοῦδε γένεσιν ἄλλου εἶναι φθοράν, but with a different syntax: in *GC* II 10, 336b24, γίνεσθαι is governed by impersonal συμβαίνει and διὰ is linked directly (i.e., not with an implied infinitive) with γένεσιν.

the third cause, which all previous philosophers were seeking in vain (*GC* II 9, 335b7–8). The inquiry for the efficient cause has been motivated by the inadequacy of earlier theories meant to account for generation and perishing with recourse to either the material cause alone (e.g., Empedocles) or a combination of the material and the formal causes (e.g., Plato) (*GC* II 9, 335b9–336a12). Against the former group of thinkers, Aristotle complains that they omit "the most important cause" (τὴν κυριωτάτην αἰτίαν), that is, the essence (τὸ τί ἦν εἶναι) in the sense of form (μορφήν). As he put it in a previous chapter (*GC* II 6, 333b7–14), the cause that explains why, either always or for the most part, human being comes to be from human being and wheat from wheat is neither fire or earth nor Friendship or Strife but ἡ οὐσία ἡ ἑκάστου. Against the latter group of thinkers, with a glance at the theory of Forms expounded in the *Phaedo*, Aristotle raises the following question (335b18–20): since both the Forms and the things which partake of them (τὰ μεθεκτικά) always exist, why is it that the Forms do not always generate? The sun's so-called approach and retreat is meant to provide the solution to this puzzle. It does not substitute the significance of "the most important cause" but supplements it. But if the sun's annual movement is the *unique* efficient cause of generation and perishing, all things should follow the cycle of birth, growth, maturation, decay, and extinction of annual plants. And this is obviously not the case. Aristotle knows that there are plants and animals that live longer or less than a solar year (336b14–15). His discovery of the efficient cause of generation and perishing in the sun's annual movement is the discovery, I submit, of the *universal* efficient cause of generation and perishing or, if you wish, of the most proximate efficient cause *that applies equally to all generated things*. We must think that the generative power of the sun's rays jointly works with the compound of matter and form that is the fertilized egg (and, perhaps, prior to that, provides the atmospheric conditions that are necessary or favorable for fertilization) and urges the potential implicit in that compound to its actualization and development. We must think that there is no essentially *destructive* power in the sun's rays: it is only the substantial *decrease* of the generative power of the sun's rays that accounts for the process of perishing. If the sun's approach and retreat is the universal moving cause of generation and perishing, in an important (though not completely accurate) sense, the most *specific* moving cause of generation and perishing is the substantial form itself. For that alone knows what to make of the movement that has been imparted to it by the sun. As a result of this impulse to actualization caused by the sun, the form, this "most important cause," undertakes the task to develop its implicit structure to maturity by providing the proximate matter with the shape that is appropriate to the form's own perfection. In what looks like a combination of the substantial form imparted to the child by the parent and the sun as efficient causes of generation Aristotle explicitly writes in *Phys.* II 2, 194b13: ἄνθρωπον γὰρ ἄνθρωπος γεννᾷ καὶ ἥλιος (cf. *Meta.* XII 5, 1071a13–17).

If the above interpretation is correct, it follows that the lifespans of the various plants and animals are not determined by the sun's movement but by the substantial forms themselves. What the sun does by means of its so-called approach and retreat is to instigate, maintain, and complete a process of generation, maturation, decay, and extinction that is in line with the *inherent* "number" or "period" that each form *already* possesses. By taking advantage of the ambiguity of περίοδος (lit. "going round," "way round"), which may mean both "rotation"/"revolution," that is, a circular *motion*, and "period," that is, a limited length of *time*, Aristotle in 336b12–15 brings the *lifespans* of all plants and animals, that is, their "periods" as already determined by the corresponding form, in relation to the annual *rotation* ("period") of the sun – and perhaps also to the other two basic *revolutions* ("periods") observed in the sky: the diurnal movement of the sphere of the fixed stars, and the monthly cycle of the moon. That it is those *three* basic revolutions, rather than the solar year alone, that Aristotle may have in mind in *GC* II 10, 336b13–15 is suggested by a passage in *GA* (IV 10, 777b16–35) referred to earlier, where the alternation of day and night, the lunar month with its subdivisions, and the solar year, as well as the time lengths that amount to multiples of them, are considered to be the naturally determined "periods."

At 336b11 Aristotle speaks of "the times and lives of each (ἑκάστων)" without specifying whether by "each" he means species or individuals. Philoponus (*In GC* 294.21–27) believes, rightly to my mind, that Aristotle intends species. But Aristotle's statement tallies well with either of the two major interpretations of οὐσία in *Meta.* VII – with both those who think (e.g., Frede–Patzig) that Aristotle accepts forms of individuals and those who think that he accepts only forms of indivisible species (e.g., Ross and the so-called Londinenses). In any case, Aristotle thinks that the approach and retreat of the sun, either once or for a number of times (πολλάκις), brings to fruition the potential lifespan that is *already present* in an individual living being as a result of it possessing either an individual form of its own or the common form of the species to which it belongs.

11.3 The Final Cause of Generation and Perishing (336b26–337a17)

Aristotle moves smoothly from the efficient to the final cause of generation and perishing by means of a single word: εὐλόγως (336b27). Having established that, due to the continuity of the sun's annual movement, generation and perishing "will never give out," Aristotle says that "this happens with good reason," and he provides a teleological argument meant to establish the claim (336b27–34).

11.3.1 Aristotle's Teleological Argument

The argument says that since to be is better than not to be, and Nature always desires what is better, God, having realized that things further removed from the beginning cannot be, completed the whole by means of the remaining manner, that is, by making generation continuous.

Literally understood, the argument presupposes two entities, Nature and God, of which the first always desires the best and the second brings about its achievement. Moreover, it seems to presuppose a temporal beginning of the universe, that is, a time, say t_1, in which some things exist but others do not, and a later time, say t_2, in which, on top of the things that already exist from the beginning, some other things come into being by way of continuous generation and perishing. But since we know that (i) Aristotle considered the universe to be eternal, in the sense that at any time t_1 it is formally identical with itself at any other time t_2 (hence, that it is formally identical with itself at *all* times) and (ii) God does not play any creative role in Aristotle's cosmology, it follows that the language of the argument is figurative.[14] I suggest that the argument should be reconstructed somewhat along the following lines:

(i) In all things (ἐν ἅπασι) nature desires what is better (βέλτιον).
(ii) To be is better (βέλτιον) than not to be.
(iii) To be in the fullest or absolute sense is impossible (ἀδύνατον) for all things (ἅπασιν) since some of them (i.e., sublunary things) are further removed from the first principle of reality (διὰ τὸ πόρρω τῆς ἀρχῆς ἀφίστασθαι).
(iv) To be by way of everlasting coming-to-be and generation is the best approximation (ἐγγύτατα) to being in the way of Substance (τῆς οὐσίας), that is, to being in the fullest or absolute sense.

[14] Cf. Henry 2019: 210.

∴ Nature provided the things that were further removed from the first principle of reality (i.e., sublunary things) with being by way of everlasting coming-to-be and generation as the best approximation to being in the fullest or absolute sense.

In the above reconstruction, I have made several identifications. First, I have taken the expression ὁ θεός in 336b32 to be a tag for τὴν φύσιν in 336b28. Second, I consider τοῦτο in 336b30 to refer to τὸ εἶναι which, when taken over by τοῦτο, means not just as *any* way of being whatsoever but being in the fullest or absolute sense. Third, I have taken τῆς οὐσίας in 336b33–34 to have such a strong connotation as to be rendered "Substance" (with a capital S), that is, "eternal and immutable substance." This is, I think, required from the context. Aristotle contrasts and compares everlasting coming-to-be and generation – two terms which appear to be synonymous here[15] – with ἡ οὐσία in a way that this last term must mean not just "substance" or "being" but "eternal and immutable substance" or "eternal and immutable being," if the contrast and the comparison are to make sense. The opposition is not between one kind of eternity (say, everlastingness) and another kind of eternity (say, timelessness) – for this opposition is not Aristotelian and, in any case, it is not present in the *GC* – but rather between one way of being (being in the sense of coming-into-being, reaching the acme, and passing-away, namely, being by way of reproduction and merely generic identity over time) and another way of being (being by way of remaining the self-same entity forever, i.e., being by way of changelessness and numerical identity over time). Since this other way of being is not the way of being that characterizes all substances but a way of being that characterizes only those higher substances in which no mere capacities or potentialities, but only actual activities or active actualities are present, it stands to reason to understand οὐσία here as a collective term for such always active substances, that is, for immutable and eternal substances. Finally, I have taken ἀρχή in 336b31 to refer not to a temporal *beginning* but to the very *first principle* of being. In other words, I have understood ἀρχή in a metaphysically loaded sense as ontological rather than temporal, and I have subsequently identified this ontological principle with the hierarchically primary eternal and immutable substance of Aristotle's *Phys.* VIII and *Meta.* XII, that is, with the first unmoved

[15] To regard καὶ τὴν γένεσιν as providing the subject of τὸ γίγνεσθαι, as Joachim (1922: 265) thinks, is possible but not necessary, and it complicates things. I have opted to take καί as epexegetic.

mover. Correspondingly, the distance (πόρρω) of which Aristotle speaks in 336b30 should not be understood as temporal but as ontological, and perhaps also as spatial in the sense that the *ontological difference* of generable from immutable substance is reflected in the *spatial remoteness* of the center from the circumference of Aristotle's spherical universe.

In the above reconstruction, I have interpreted Aristotle's argument in a manner that refers both to the way of being of generable and perishable things and to the continuity, hence the everlastingness, of the processes of generation and perishing.[16] The two ideas are simultaneously present in Aristotle's mind. He seems to think, at one and the same time, (a) that the best approximation to ungenerated and imperishable substance is a thing's generation and subsequent perishing (for to be in any kind of way is better than not to be at all) and (b) that the best approximation to the best way of being, namely the eternal and immutable way of being, is a continuous and everlasting process of generation.

In several instances where the principle "Nature does nothing in vain" appears in the Aristotelian *corpus* (present here as premise (i) in the form "Nature desires what is better") it is supplemented with the qualification "but always the best from the available possibilities" (ἐκ τῶν ἐνδεχομένων *vel sim.*). Though not explicitly stated, this qualification is also present in this argument. It is present most conspicuously in premise (iii), which explicitly provides a restriction to the available possibilities (336b30); it is also present in the expression τῷ λειπομένῳ τρόπῳ at 336b31; finally, it may be suspected in premise (iv), which states that being by way of everlasting coming-to-be and generation is closest (ἐγγύτατα) to, but (we have to assume) not identical with, eternal and immutable substance (τῆς οὐσίας).

That "being is said in many ways" is a familiar Aristotelian doctrine and is recalled at line 336b29. In the present context, the polysemy of "being" is not meant primarily to refer to quality, quantity and the other kinds of predication distinguished in the *Categories*. It refers primarily, I submit, to the ways of being a substance which it is the aim of the *Metaphysics* to elucidate. I cannot enter here into the heated debate about the interpretation of the *Metaphysics* as a whole, but a few things might be said.

Aristotle thinks that if there is a first philosophy, over and above physics, this philosophy must be a science of being *qua* being. He, moreover, thinks that the question about being *qua* being inevitably leads to the question of substance; for substance is the focal reference of "being" in the

[16] In other reconstructions of the argument (e.g., Quarantotto 2015), the stress is on continuity.

sense that all other categories are predicates and attributes of substance. It follows that the question about being *qua* being, that is, the most general and universally applicable question, is ultimately about substance.

Aristotle thinks that sensible substances are compounds of matter and form and as such are perishable. But neither matter as such nor form as such are perishable. The generability and destructibility of sensible substances are to be explained with reference to the polarity of capacity and activity. The primary locus of capacity is matter, and the primary locus of activity is form. If there is to be a kind of substance that is intrinsically independent of, and altogether separate from, matter, this kind of substance must be pure *activity* without any non-actualized capacity; for capacity implies what is possible to be or not to be, and what is possible to be or not to be may, at some time, not be. Now, since any activity not based on a corresponding capacity necessarily is what it is, the primary kind of substance must be a *necessary* (rather than contingent) and *simple* (rather than compound) substance. Aristotle thinks that the conditions that should be fulfilled for such a primary kind of substance to exist are indeed fulfilled in the case of divine *intellection* (νόησις or νοῦς). Intellection is a kind of thinking different not only from the process of learning but also from discursive reasoning (διάνοια). It is, moreover, utterly free from pain and frustration, pleasant, and completely self-fulfilling. As such it represents, or rather *is*, the most excellent form of *life*. With the completion of the analysis of substance, Aristotle has guided us to a higher level of being, the necessary and simple substance whose eternal life consists in intellection.

It is not clear how far the details of the *Metaphysics* doctrine about substance are presupposed in *GC* II 10. What seems rather clear is that Aristotle's teleological argument implies grades or levels of being, for one, and the so-called "principle of plenitude," for another. I shall take them in turn, though they are intimately related.

That there are levels or gradations of being in Aristotle's *Metaphysics* is a plausible idea that has become more attractive thanks to the defense it has received from Owens.[17] I think, with Diana Quarantotto,[18] that the teleological argument of *GC* II 10 supports this idea. Τοῦτο in 336b30 refers to τὸ εἶναι in the previous line, and what follows says that being is impossible to obtain in all things. But then Aristotle says that the "remaining manner" in which the whole was in fact completed consists in continuous generation. Now, we may *either* think that continuous

[17] Owens 2005. [18] Quarantotto 2015.

generation is *not* a way of being at all *or* that it *is* a way of being that is inferior to the best way of being. From what Aristotle says about substance in the *Categories* and the *Metaphysics* it is evident that he opts for the second alternative; for he clearly believes that there *is* such a way of being as sensible and generable substance, which actually is the focal reference of sensible being quite generally, and which it is the end of generation to bring about. Moreover, by stating that eternal coming-to-be or generation is closest to Substance, Aristotle clearly regards the everlastingness of the process of coming-to-be as the best approximation to the eternal stableness of immutable and eternal substance. But a continuous process of things being generated cannot be similar to a substance unless the things that undergo that process are also, in some relevant respect(s), similar. I infer, again, that Aristotle understands the polysemy of "being" in the present context (336b29) not only or primarily in the sense of the various categories but rather in the sense of grades of *substantial* being, such as the duality of generable/sublunary and unchangeable/superlunary substance. Notice that the search for unchangeable substances, announced in several places of the *Metaphysics* (III 2; VI 1; VII 2; XII 1; XIII 1) and dealt with in *Meta.* XII 6–10, is instigated not only by the general desire to know the full scope of reality but also by the specific desire to account fully for sensible, generable and perishable substance.[19] For without the postulation of unchangeable substances the being of generable substances remains only partially understood.

Aristotle uses, in a limited way, the so-called "principle of plenitude" and the related "principle of continuity" to explain why being in the sublunary sphere is such as it empirically appears to be.[20] The phrase συνεπλήρωσε τὸ ὅλον (336b31–32) indicates that god (or nature) has a desire to *complete* the whole universe with as many ways of being (and, we may add, kinds of being, see below) as possible. And the phrase οὕτω γὰρ ἂν μάλιστα συνείροιτο τὸ εἶναι (336b32–33) indicates that the everlasting continuity of the process of generation is in fact the best way for reality to string together, that is, for the different ways of being to relate to, rather than remain disjoint from, one another. I say that Aristotle uses the principles of plenitude and continuity *in a limited way* because the focus of his interest here is the great divide between generable and ungenerable substance. But within the domain of generable substance, which is treated here as a unitary field, there are inanimate things, such as the four simple

[19] Frede 2000: 27. [20] Lovejoy 1936: 52–56.

bodies, as well as living things, such as plants and animals; and Aristotle knows that there is no clear-cut demarcation line between them:

> Nature proceeds little by little from things lifeless to animal life in such a way that it is impossible to determine the exact line of demarcation, nor on which side thereof an intermediate form should lie. Thus, next after lifeless things comes the plant, and of plants one will differ from another as to its amount of apparent vitality; and, in a word, the whole genus of plants, whilst it is devoid of life as compared with an animal, is endowed with life as compared with other corporeal entities. Indeed, as we have just remarked, there is observed in plants a continuous scale of ascent towards the animal. So, in the sea, there are certain objects concerning which one would be at a loss to determine whether they be animal or vegetable. For instance, certain of these objects are fairly rooted, and in several cases perish if detached: thus the pinna is rooted to a particular spot, and the razor-shell cannot survive withdrawal from its burrow. Indeed, broadly speaking, the entire genus of testaceans have a resemblance to vegetables, if they be contrasted with such animals as are capable of progression. (*HA* VIII 1, 588b4–12; trans. by D' Arcy Thompson as reprinted in Barnes 1984: I, 922)

If we bring the above passage to bear on the teleological argument of *GC* II 10, then Aristotle's fully fledged doctrine might be the following. The region around the center of the universe is far removed from the celestial realm, where imperishable substances dwell, and its matter is such that it cannot retain the stamp of the form imparted to it for good. It wearies out. The substitute for the inability of sublunary matter to retain the form, any form, is everlasting generation, that is, the process whereby an individual substance leaves behind it another individual identical to it in form but numerically different (cf. *GC* II 11, 338b12–19). This is a general rule that applies to all kinds of things in the region around the center of the universe. But even within it there is a great deal of variation ranging from rational animals on the top of the hierarchy to inanimate simple bodies at the bottom. Since (for reasons having to do with the available matter) it is not possible for all things to be endowed with the highest kind of being, some things have inferior kinds of being such that no possibility of being, even the lowest, remain unrealized. In this way the universe is full of all kinds of beings. The natural kinds contained therein form a continuous whole with no clear boundaries between the lower and the higher among them. What brings them together as a class, that is, the class of generable and perishable substance, is the fact that they are compounds of matter and form and as such are perishable. But they have the natural ability to reproduce themselves and, through it, the extinction of the individual does not entail the destruction of the species. Their generic everlastingness is the best substitute for the impossibility of their being individually immortal.

The idea goes back to Plato's *Symposium* (207 C–208 B) where Diotima teaches Socrates that erotic love is for the sake of procreation, and procreation for the sake of immortality. Since it is impossible for an essentially mortal individual, she claims, to retain its identity and persist for ever, as it would wish, the only available way for it to gain an immortality of sorts is procreation. Thus, procreation is a substitute for the real but usually hidden object of erotic desire, which is personal immortality, and erotic passion itself is nothing but a natural urge for the *Ersatz* satisfaction of this unattainable goal.

This idea is also present in *DA* II 4, 415a26–b7 where Aristotle uses explicitly Platonic language, and says that the final goal of reproduction is participation in eternity and divinity as far as this is possible (ἵνα τοῦ ἀεὶ καὶ τοῦ θείου μετέχωσιν ᾗ δύνανται).[21] The phrase recalls not only Diotima's teaching but also perhaps the famous ὁμοίωσις θεῷ κατὰ τὸ δυνατόν of *Theaet.* 176 A. Aristotle does not restrict the natural desire for eternal and divine life to the urge for procreation but states that, quite universally, "whatever is done in accordance with nature is done for the sake of *that*" (ἐκείνου ἕνεκα πράττει ὅσα πράττει κατὰ φύσιν). But then, as if realizing the essential vanity of this desire, Aristotle distinguishes two kinds of final cause, the end itself and the beneficiary. For mortal things the end of immortal, godlike existence is unattainable. Nevertheless, mortal things benefit from their unattainable desire since by means of it they reach an attainable goal: the immortality of the species.

Aristotle's teleological argument states the reason for the perpetual existence of generation and perishing in the region around the center of the universe, which is the plenitude and continuity of Being. Had there not been generable and perishable life and the natural cycle of the simple bodies, Being would not have exhausted all its possibilities.

11.3.2 The Cycle of Generation and Perishing as an Imitation of Celestial Circular Motion

After the exposition of the final cause of generation and perishing by means of the teleological argument, Aristotle draws an inference about

[21] Henry 2019: 208–217 believes that the continuity-of-generation argument of *GC* II 10 should not be seen in relation to the reproduction argument of *DA* II 4. Henry's immediate concern is the interpretation of the sexual-differentiation argument in *GA* II 1, 731b20–732a11, which to his mind has more things in common with *GC* II 10 than with *DA* II 4. He may be right in that. Still, I think that it is different aspects of a *single* idea that Aristotle expounds in all three passages, especially since sexual differentiation is explicitly said in *GA* II 1, 731b22–24 to have its teleological origin from above (ἄνωθεν), i.e., in the divine realm.

the reciprocal transformations that occur in the region around the center of the universe (337a1–7). He claims that in all those cases where *A* changes into *B*, *B* into *C*, *C* into *D*, and so on, until the final product in the series turns out to be *A* again – as it happens, for instance, with the sequence water-air-fire-water of simple bodies – the process of transformation imitates (μιμεῖται) circular movement since it ends up with, hence comes back to, the same kind of thing from which it got started.

As a parallel passage from *Meteor.* I 9, 346b35–347a8, indicates, Aristotle thinks that the way up and the way down of elemental transformation is a continuous process of generation and perishing which resembles rectilinear locomotion in opposite directions. As such it cannot be regular; for it involves terminal points where the process is reversed. But since the beginning and the end of the process coincide, as in the famous Heraclitean maxim about the circle (*DK* 22 B 103), this process, though strictly speaking straight, can also be seen as circular and, hence, as an imitation of the sun's annual rotation which is in fact its efficient cause. Contrary to the negative overtones which the concept of imitation seems to bear when ascribed to the Pythagoreans, in our passage the notion of imitation is undoubtedly positive, and it aims to bridge the gap between the lower realm of generation and perishing and the upper realm where selfsame substances subsist in endless rotations.[22]

A similar doctrine expounded with the same intention of bridging the gap between the imperishable things (τὰ ἄφθαρτα) in the sky and the changeable things (τὰ ἐν μεταβολῇ ὄντα) on earth is present in *Meta.* IX 8, 1050b28–30. But Aristotle's point there is different: by saying that changeable things, such as earth and fire, imitate (μιμεῖται) imperishable things what he has in mind is not the circularity of elemental transformation but the uninterruptedness of elemental activity (καὶ γὰρ ταῦτα ἀεὶ ἐνεργεῖ) which is a result of the fact that simple bodies possess motion by their own nature and in themselves (καθ' αὑτὰ γὰρ καὶ ἐν αὑτοῖς ἔχει τὴν κίνησιν). It is clear, in any case, that in both *GC* II 10 and *Meta.* IX 8 the concept of imitation is used to highlight relevant respects in which the lower realm of

[22] In *Meta.* I 6, 987b11–14, Aristotle reports that the Pythagoreans said that things are by way of imitation of numbers (οἱ μὲν γὰρ Πυθαγόρειοι μιμήσει τὰ ὄντα φασὶν εἶναι τῶν ἀριθμῶν). In his view, Plato has only changed the name of the relation from *mimesis* to *methexis* (Πλάτων δὲ μεθέξει, τοὔνομα μεταβαλών) and has claimed that things exist by way of participation in Forms. But neither the Pythagoreans nor Plato cared to explain what "imitation" or "participation" is. For Aristotle, the notion of *methexis* is an empty concept and a poetic metaphor (*Meta.* I 9, 991a20–22 = *Meta.* XIII 5, 1079b24–26), and such a judgement can reasonably be extended to Pythagorean *mimesis*.

temporal and unstable being longs for and foreshadows the perfect orderliness of eternal being. When employed as a metaphysical concept meant to illustrate the subordination of lower to higher being as well as the connection between the two realms, imitation is not a mental activity, as in the more literal sense of *mimesis* in the *Poetics*. The simple bodies imitate the circular movements of the celestial realm by a certain kind of necessity that is inscribed in their natures.

Having said that, we should not reach the premature conclusion that because no mental activity is involved in *mimesis* when used in physical or metaphysical contexts to denote the dependence of the sublunary region to the world above, Aristotle should think of it as a mere metaphor. In *Phys.* II 8, 199b26–28, Aristotle says that it is absurd (ἄτοπον) for people to deny that things happen for the sake of something (τὸ μὴ οἴεσθαι ἕνεκά του γίγνεσθαι) if they do not see the moving cause to have deliberated (ἐὰν μὴ ἴδωσι τὸ κινοῦν βουλευσάμενον); for even craft/art, he hastens to add, does not deliberate (καίτοι καὶ ἡ τέχνη οὐ βουλεύεται). What he means, I submit, is that deliberation about the appropriate means for the construction of a certain product, and the mental hesitation about the proper course of action that is implicit in deliberation, are signs of a less-than-perfect craftsman/artist. The accomplished possessors of craft/art act with a kind of automatism that flows unobtrusively from their having the relevant technical *hexis* in their souls, as when an excellent dancer is dancing without a single thought crossing her mind.[23] In stark opposition to modernist approaches, Aristotle would think, I believe, that our inability or unwillingness to see the operation of teleology in nature is a sure sign of an anthropocentric or anthropomorphic, and thus limited, conception of finality. Such an all-too-human conception of teleology tends to forget that there might be a higher and more easily flowing kind of finality than the mental setting of goals, and the occasionally copious deliberation about the proper means for their attainment. When used in metaphysical or physical contexts, I infer, "imitation" is not a mere metaphor: it tries to grasp from an objective perspective the utterly non-mental, but no less potent for that, urge of all natural things to resemble imperishable substances by bringing their limited being in line with the highest kind of being, as far as this is, in each case, possible.

[23] Sedley 2007: 177–181 has argued that for Aristotle it is not the craftsman *as a person* but the form of the relevant craft in his/her soul that does not deliberate. I tend to side with Broadie's (Broadie 2007: 93–95, 98–99) "perfectionist" interpretation.

11.4 Concluding Remarks

It is now time to address the topic touched upon in this chapter from a broader perspective.

In the *DC*, Aristotle goes out of his way to prove that there must be a simple body (the so-called "aether," *DC* I 3, 270b16–25, *Meteor.* I 3, 339b22–28), over and above the four simple bodies attested on earth, tellingly called by Aristotle "first body,"[24] if the eternal substances of the heavens are to possess appropriate, that is, incorruptible, matter (*DC* I 2–3). To Aristotle's mind, the postulation of such a body, besides the evident four of our experience, is required from the observable transformation of earth, water, air, and fire into one another. In *GC* II 9 Aristotle says that the (as it were) material cause in the domain of generation and perishing is what is capable of existing and not existing. What he must mean is that the four simple bodies, which in their multiple combinations provide the material for all sublunary substances (*GC* II 8), are in themselves of such a structure as to render things based on them essentially unstable and to a certain extent also contingent. The celestial realm, by contrast, is essentially characterized by necessity, and as such it has to possess a material cause that is itself ungenerated and incorruptible. The structure of the first body allows only one kind of change: change of place by means of a moving process (regular and circular locomotion) that brings the moving body, time and again, to the very same spot. Increase and decrease, generation and perishing, and qualitative alteration are not to be found in the heavens.

In *Phys.* VIII and *Meta.* XII the multiple and endless rotations of the celestial bodies are all stirred and coordinated by a first moving cause. To prevent an infinite regress, Aristotle assumes that the first moving cause must be itself unmoved. Thus, he is led to postulate a first unmoved mover which imparts movement by its own admirable perfection. The kind of simple and intellective life that this first mover leads fuels in other things a desire to become as much like it as possible. Since, when conditions are favorable, desire always prompts movement to the desirable thing, the eternal substances that are less perfect than the first unmoved mover try to quench their desire for the kind of life that the primary mover leads by means of endless rotations. They thus imitate the pure activity of intellection on the lower level of circular locomotion.

[24] Falcon 2005: 113–121.

Unlike the assumed order of things in Plato's *Timaeus*, there is no place in Aristotle's cosmology for a Creator-God and/or a Cosmic Soul. Aristotle's Nature is a collective term used to indicate the sum of particular natures in the sublunary realm. Similarly, Aristotle's God, when brought forward to account for events in the sphere of generation and perishing, is another name for Nature as a collective term. Now, if each particular nature tries to accomplish the end inscribed in its form, and for the most part succeeds in doing so, two questions arise: (i) What is the precise relation between the final and formal causes that stir the development of each particular substance? Is the distinction between them only verbal or does it amount to a real difference? (ii) How does the coordination and apparent harmony of natures in the sublunary sphere occur? If each nature aims at the accomplishment of its own limited end, how does global harmony arise?

To tackle the second question first: since Aristotle does not accept creationism or divine providence, he cannot appeal to any direct kind of divine involvement in the lower world. Nor does he seem to accept a pre-established harmony across natures as Leibniz later did. Aristotle's solution to the puzzle of cosmic harmony is to be found in *Meta.* XII 10, 1075a11–25, where he claims that the universe possesses goodness and the best (τὸ ἀγαθὸν καὶ τὸ ἄριστον) as an army does: both in the sense of an immanent order, and in the sense that there is a general, separate from the army itself, who is the cause of this order. What Aristotle means, I suggest, is *not* that the general imposes the order he has in mind to the army as an efficient cause but that his presence stirs in each and every individual soldier a desire to do their best. The order comes about as a result of all individual wishes to excel as soldiers. But without the presence of the general the desires of the soldiers would not be fueled. The general stands for the final cause of the army's order, and the individual wishes to excel for its efficient causes. The fact that the general transcends the army, since he is not a part of it but its principal, sheds light on the way natural order depends on a cause that is beyond nature. The significance of the analogy lies in the separate existence of an entity (κεχωρισμένον τι) which is the ultimate cause, and thus provides the ultimate explanation, of an order which is immanent in the parts that produce it. It is in this way, Aristotle claims, that the various kinds of living things (e.g., fish, birds and plants), and indeed everything in the universe, are arranged to one end (πρὸς γὰρ ἓν ἅπαντα συντέτακται).

Aristotle, of course, acknowledges that the lower one gets in the realm of nature the more room one finds for chance and disorderly outcomes (another indication that he espouses gradations of substantial being). His second analogy in the same passage of the *Metaphysics* is that of a household where the free persons do, for the most part, what is best for the whole, but the slaves and the brutes behave in a (more or less) disorderly way, with little concern for the common good (μικρὸν τὸ εἰς τὸ κοινόν). The higher one gets in the hierarchy of generable and perishable things while the less one finds contingency and chance to be the rule; and the transcendence of this entire realm of possible being opens up a vista of necessary excellence that culminates in the activity of divine intellection.

In Aristotle's eyes, without the postulation of a first unmoved mover, cosmic order would remain unaccountable. Later in the same chapter, while addressing his predecessors and the inadequacy of their theories, Aristotle specifies that none of them explains why some things are perishable and some imperishable (1075b13) and why generation will always happen and what its cause is (1075b16–17). Aristotle believes that his postulation of a primary unmoved mover solves all these problems. For, as he writes (1070b24–27), "if there is to be nothing else besides sensible things, there will be no first principle, no order, no generation, no celestial movements, but each principle will always hang on another principle, as in the accounts of all the poets who speak of gods and all the physicists."

When we return to *GC* II 10 with these considerations in mind, we see that, though reserved for explicit treatment in first philosophy explorations rather than physical treatises (*GC* I 3, 318a5–6), the first unmoved mover is present in the background of a chapter which deals with the continuity of generation and perishing and accounts for it from the perspective of its efficient and final causes. The sun's annual revolution in the ecliptic, the efficient cause of climatic variation along the seasons, and of generation and perishing as a result, has the first unmoved mover as its own final cause. To resume the army analogy, we may say that the sun's sphere does its best to resemble the immobility of the first unmoved mover and as a result of this effortless endeavor – for there is no toil and pain in the superlunary realm – it moves regularly and uninterruptedly, thus causing continuous and uninterrupted generation and perishing in the region around the center of the universe. On their part, sublunary things receive the movement, which the sun's approach provides, and use it to stir the potential of their substantial forms to development, maturation, and perfection. Substantial forms are inherited by means of procreation according to the principle ἄνθρωπος ἄνθρωπον γεννᾷ (cf. *GC* II 5, 335b7). While the sun retreats, the diminution of the sun's effect on any particular terrestrial place causes a corresponding decrease in the development of the very same substances or their decline and final extinction. In annual plants and animals, the entire circle of generation and perishing is completed in one year. In other natural kinds, it takes a longer or shorter period. In all cases, the extent of a life cycle is determined not by the sun but by the substantial form itself. It is also the substantial form that determines the kinds of alterations that the simple bodies present in the food undergo in order to provide the necessary proximate matter for the proper development of an organism. By saying that the substantial form determines the lifespan of each individual, I mean that it determines the number of solar years that the individual has to live through in order to exhaust the life potential of its nature. But since each sublunary substance is a compound of form and corruptible proximate matter, it occasionally happens that the proximate matter, owing to a faulty constitution, degenerates earlier than expected and thus brings death to the individual before the completion of its full life cycle. It follows that, quite apart from a violent death, which is always an open possibility in the domain of contingency, there is a great deal of variation even in the cases of a natural passing away.

Let us now move on to the first question, which is concerned with the relation of the final to the formal cause. In *GC* II 9, 335b6–7, Aristotle writes that in generable things "the cause by way of that-for-the sake-of-which is the shape and the form (ὡς δὲ τὸ οὗ ἕνεκα ἡ μορφὴ καὶ τὸ εἶδος), and this is the definition of the essence of each thing (ὁ λόγος ὁ τῆς ἑκάστου οὐσίας)." To say that the form is the final cause is to identify the end of each thing with its own perfect development according to the formula inscribed in its substantial form. But then, one might ask, what is the relation of this self-contained form as goal (an achievable end) with nature's desire for the best and, therefore, for being in the fullest or absolute sense (an impossible end), if nature is nothing but the collective name for each substantial form? In other words, how should we understand that one and the same substantial form desires at one and the same time both its own perfection and something impossible that transcends it?

I suggest that Aristotle's answer would take the following course. The perfection of each individual substance entails procreation, and procreation is an *Ersatz* immortality by proxy that takes the place of the truly desirable personal immortality whenever the latter is impossible. It follows that in each and every substance there are, so to speak, two final causes at work. One sets an achievable end: individual perfection according to the limitations set by the substantial form. The other sets an impossible end: unlimited individual perfection, that is, perfection beyond the limits set by the substantial form, which ultimately amounts to a desire for the absolutely best kind of being, that is, the intellective activity of the first unmoved mover. It is this last final cause that explains how "the god," without getting involved with things external to itself at all, still less with the state of affairs in the sublunary region, and without being an efficient cause at all, "completed the whole" (*GC* II 10, 336b31–32). God's presence, by being what it is, that is, unqualified perfection of being, stirs a desire for the absolutely best kind of being to all those things which come either directly or very indirectly and dimly, by means of a large number of intermediate stages, to an acknowledgement of its presence. If even inanimate natural things such as the four simple bodies imitate, by their continuous cycles of mutual transformations, the eternal revolutions of the heavens, it is no wonder that living things do the same by means of their more developed mechanisms of procreation. I said earlier that in 336b32 ὁ θεός is a tag for each and every individual nature. If all individual natures share the same desire for immortality and for the best kind of being, it is no wonder that, in the infinity of time of Aristotle's universe, they must have *always already* devised the best possible ways to achieve their goal, that is, the actual mechanisms of their procreation and collective survival. But if so, then ὁ θεός, who is said to have "completed the whole in the remaining way, i.e., by making generation perpetual" (*GC* II 10, 336b31–32), is not only another name for the collectivity of Nature but also a rather loaded term that points to something well beyond the sum of individual natures. It is from desire for the very best kind of being that all successful individual natures have, from time immemorial, achieved their limited best. For as Aristotle writes, in a different context (*NE* VII 13, 1153b32): πάντα γὰρ φύσει ἔχει τι θεῖον.

12

On Generation and Corruption *II 11*

Sean Kelsey

12.1 Introduction

My object in this essay is to try to provide some basic exposition of this difficult chapter. This has seemed to me worth doing because though the upshot of the chapter is by and large clear, the argumentative details are often very hard even to make out – at least if I may judge from the time and trouble it has taken me to make up my own mind, with many changes back and forth along the way. I hope that, even if the particular proposals I make prove incorrect, still they will be useful as a point of reference for readers whose minds are clearer and keener and more sure-footed than mine is.[1]

The question Aristotle poses in this chapter is whether there is anything that comes to be which does so necessarily, and his answer, briefly put, is that there would be, if (and only if) there were anything whose coming to be was everlasting, which there would be, if (and only if) there were anything whose coming to be was cyclical, which (in point of fact) there is (e.g., solstices, and anything else whose coming to be is due to the rotation of the heavens) (cf. *GC* II 11, 338a14–b5). Aristotle does not say why this question merits discussion at this juncture.[2] One plausible

* Thanks to Sarah Broadie, Arnold Brooks, Andrea Falcon, Jessica Gelber, James Lee, Marko Malink, Fred Miller, Daniel Nolan, Spyros Rangos, and Bryan Reece for comments on earlier drafts of this paper (and/or another incorporating parts of this one). Thanks also to Nick Denyer for his patience and generosity replying to some queries from out of the blue.

1 There is not much literature on this chapter, though see Joachim 1922: 270–277, Mansion 1946: 75–89, Verdenius-Waszink 1966: 79–83, Barnes 1976: 290–292, Williams 1982: 197–210, Rashed 2005: 178–181, Buchheim 2010: 151–155, 556-78, and Delcomminette 2018: 309–322. The chapter is also discussed in the Ἀπορίαι καὶ λύσεις attributed to Alexander (71.3–72.8 and 87.22-89.24 Bruns, translation in Sharples 1992) and in Philoponus' commentary (In *GC* 302.11–303.20, translation in Kupreeva 2005), on both of which see Sharples 1979.

2 Compare Joachim: "With the treatise on the causes Aristotle has completed the task which he originally proposed to himself (cf. *35a24–37a33). The present chapter, therefore, is to be regarded as an appendix" (Joachim 1912: 270).

suggestion is that the question arises from a "fact" which, though laid down and (indeed) partially explained earlier, is not fully explained until *GC* II 10:[3] namely, that generation is "connected" (συνεχής) and does not "give out" or "fail" (διαλείπειν) – if I may put it this way, that the world never "rests."[4] The question might arise from this fact, because (in Aristotle's view) the fact that the world never rests implies that it cannot rest, which might seem to imply that at least some things *do* come to be of necessity. This latter implication would merit discussion, inasmuch as (again in Aristotle's view) many things that come to be plainly do *not* come to be of necessity (*GC* II 11, 337b3).

The chapter as a whole naturally divides into three sections: first an introduction, announcing the question to be considered; then the main body of the chapter, given to answering that question; and, finally, a brief concluding section, given to a difficulty raised by that answer. I take these sections in turn.

12.2 The Question (*GC* II 11, 337a34-b13)

The chapter begins by announcing the question to be considered in what follows: briefly put, *is there anything that will be of necessity?* (337b1–3)

I begin with some comments about what question this is. First, Aristotle is not here asking about all beings, but just about those that come to be (τὰ γιγνόμενα). This is clear because later, in the course of clarifying the question, he takes it for granted that some beings are incapable of not being (337b11–12, cf. *GC* II 9, 335a33–34); it is plain that any such being will be in the future and that of necessity. His question, then, is not whether *anything* will be of necessity, but whether anything that comes to be will – as he sometimes puts it, whether any such thing *comes* to be (γενέσθαι) of necessity (337b8–13). Second, the question is not limited to the coming to be of substances; it also encompasses coming to be other things: for example, coming to be of some quality or in some place. This is clear from the example of solstices (337b12–13), whose coming to be is a matter of the sun's coming to be in the place of its summer or winter

[3] So Falcon in this volume.

[4] See *GC* II 11, 337a34–b1 and II 10, 336a23–b4, the latter recalling *GC* I 3, 317b34–318a25. (Generation would not be "connected" if the world ever came to a standstill; it would "give out" or "fail" if the world ever came to a standstill and never got going again. When I speak of the world "resting," I mean its having come to a standstill (whether temporary or permanent). Aristotle argues (in effect) that these scenarios are much the same, when he argues, in *Phys.* VIII 1, from the fact that there is now motion, that there always has been and always will be.)

"turning" (τροπή). Third, when Aristotle asks whether anything comes to be "of necessity," he is not asking whether there is anything whose coming to be is even conditionally necessary, as (say) the coming to be of foundations is necessary, on the condition that there will in fact be a house.[5] Instead, he is asking whether the coming to be of anything is necessary (period), without qualification, *sans phrase* or *tout court* (ἁπλῶς) (337b10–11). Taking these points together, the question posed in this chapter is whether anything comes to be anything of necessity (period).

The question is prompted by the observation that "in things moved continuously in respect of generation or of alteration or of change in general, the next one is and comes to be, this after that, without fail" (337a34–b1). This observation is about sequences of motion or change; I take it that these sequences are "continuous" just in case they are not punctuated by intervals of rest, and that they do not "fail" just in case there always *is* a "next." Aristotle does not say why this observation gives rise to his question; in the lines that follow, what he goes on to explain is, not why we must ask whether *anything* will be of necessity, but rather why we need not ask whether *everything* will (337b3–13). Perhaps his idea is that the fact that a sequence of motion or change never does fail implies that it cannot fail, which might seem to imply that some things do come to be of necessity – despite the fact, as he says, that a great many things plainly do *not* come to be of necessity (337b3).[6]

12.3 The Answer (*GC* II 11, 334b14–338b5)

The bulk of the chapter is given to settling the question just described in the affirmative: there *are* things that come to be which do so of necessity. Aristotle begins his discussion abruptly; rather than first outline and explain his overall strategy, he simply dives right in. I have thought it less tendentious to follow his lead in this. The discussion naturally divides into four sub-sections. I take them in turn.

12.3.1 Conversion of Conditional Necessities (337b14–25)

Aristotle begins his treatment by raising, without notice or explanation, a new question. The question he had just raised was whether anything

[5] The example is taken from 337b14–15; for the contrast, see 337b26–27.

[6] The connection between "always" and "necessary," averted to below at 337b35, is argued for in *Cael.* I 12; there is an excellent discussion of this argument in Denyer 2000: 143–161. For a different view of the form of the argument employed there see Malink-Rosen 2012: 179–261.

comes to be of necessity (period). The question he raises in this section is about certain merely conditional necessities: namely, ones in which "it is necessary that the earlier come to be if the later will be" (337b14).[7] In particular, the question is whether these necessities "convert" (ἀντιστρέφω, 337b24): that is, "whether it *follows*" (ἆρ' οὖν), if it is necessary, for example, that foundations come to be if there will be a house, "that it is also necessary that a house come to be if foundations have" (337b14–16).

Aristotle's answer to this question comes in two parts. The first part is negative: ordinarily, necessities of this sort do *not* convert (337b16). The second part enters an exception: they do not convert, *unless* it is also necessary, "period" (ἁπλῶς), that the later will be – in which case they *do* convert (337b16–18). Aristotle does not argue for the first point, at least not here in *GC* II 11 (though see perhaps *Apo.* II 12, 95a30–b1). But he does argue for the second point – the exception – in the lines that immediately follow (337b18–23). Indeed, this point appears to be *the* result he is particularly after in this section. For it is this point, and this point alone, that he repeats in the concluding reprise: "wherever, therefore, it is necessary for the later to be, it converts, that is (καί), always (ἀεί), the earlier having come to be, it is necessary that the later come to be" (337b23–25).[8]

In the rest of this section I discuss, first, how Aristotle argues for this result, and second, why the point is important – how it contributes to settling the question posed at the chapter's outset.

The argument is contained in the following passage, which I have divided into sections for ease of reference.

[7] Is the scenario envisaged one in which, necessarily, if the later will be, the earlier will have come to be? Or one in which, if the later will be, then, necessarily, the earlier will have come to be? Put another way, does "it is necessary" (ἀνάγκη) govern the entire conditional or is it part of the apodosis? Though grammatically it appears to be part of the apodosis (ἀνάγκη + infinitive, γενέσθαι), I have taken it that logically it governs the entire conditional (so too, e.g., Sharples 1979: 32, Williams 1982: 199, Delcomminette 2018: 313) – and ditto throughout. My reason is that this chimes better with the examples which follow: "e.g., if a house, foundations, and if this, clay" (337b14–15). For necessary in these examples, once there *is* a house, are the very things that were necessary all along: not foundations, or clay, but rather the "connections," between clay and foundations and foundations and houses. That this is Aristotle's view of these examples is shown by his later saying that what is necessary – unlike foundations and clay – is "everlasting" (ἀίδιον) (337b29–338a3). (Note that this appears to imply that Aristotle *rejects* 'the necessity of the past' (*pace* Williams 1982: 199). But note too that this is compatible with his accepting it that "what is done cannot be undone": that is, with his accepting it that, necessarily, if something was, it never will be that it never was. In thinking about these matters I have been helped by Denyer 2000 and van Rijen 1989.)

[8] Two brief comments. First, I take καί ("that is") as epexegetical; there would be a point to this, as this is the first occurrence in this chapter of the language of "conversion." Second, I take ἀεί ("always") as equivalent to "in every case"; the point is that it *follows* that, necessarily, the later comes to be if the earlier has.

(a) For the earlier stood to the later thus, that, if that will be, it is necessary that the former [have come to be] beforehand. (b) If therefore it is necessary that the later come to be, it is also necessary that the earlier [come to be], and (c) if the earlier, it is therefore necessary that also the later – (d) not because of that, but because it was posited that the later will be of necessity. (e) Wherever, therefore, it is necessary for the later to be, it converts – that is, always, the earlier having come to be, it is necessary that the later come to be. (*GC* II 11, 337b18–25)

In this passage Aristotle begins with a reminder (a): we are supposing that, necessarily, if the later will be, the earlier [will have come to be] beforehand. He then performs, in close succession, two inferences ((b), (c)), followed up with a comment on the second inference (d). Finally, he concludes with a restatement of the point he is particularly after ((e), "the concluding reprise"). My principal focus will be on the inferences performed in (b) and (c). I take them in turn.

Regarding the first inference, I take it that the conclusion is the apodosis of (b) (as opposed to the entire conditional). Read this way, the argument would go about as follows:

Premise: (1) Necessarily, if the later will be, the earlier comes to be.
Premise: (2) Necessarily, the later comes to be.
Conclusion: (3) Necessarily, the earlier comes to be.

Here premise (1) is the point we were reminded of in (a). Premise (2) is the protasis of (b), while the conclusion (3) is its apodosis. The alternative is to suppose that the conclusion is the entire conditional: "if (2) then (3)." But the objection to this is that it leaves (2) – the protasis of (b) – unasserted, and (therefore) unavailable for use in the subsequent inference performed in (c). This is problematic, because Aristotle will comment in (d) that the subsequent inference goes through "because it was posited (ὑπέκειτο) that it [i.e., the later] will be of necessity" – that is to say, because it was posited that (2). Suppose then we ask, "*when* was it posited that (2)?" The only answer I can see is "in the course of the inference performed in (b)." But in that case (b) is better described, not as deducing the conditional, *if* (2) *then* (3), but rather as (conditionally) deducing (3) from (2).[9]

This brings us to the second inference, the one performed in (c). The text here is very compressed: "and if the earlier, also the later therefore is necessary." How are we to fill out the protasis of this conditional? How are we to fill out its apodosis? What is the conclusion – the entire conditional,

[9] Taken this way, the inference appears to go through, being (roughly) of the form: (i) necessarily, if p then q, (ii) necessarily, p, therefore (iii) necessarily, q.

or just its apodosis? In thinking about these questions I will be guided by the result Aristotle is particularly after in this section: namely that, in certain cases, certain conditional necessities convert (337b14–25). Recall that the necessities at issue are ones in which (as we are reminded in (a)) (1) necessarily, if the later will be, the earlier comes to be. The particular result Aristotle is after is that (1) does not convert *unless* it is also the case that (2) necessarily, the later comes to be – in that case (1) *does* convert (337b16–18). Put another way, the result Aristotle is after is that inferences of the following sort go through:

Premise: (1) Necessarily, if the later will be, the earlier comes to be.
Premise: (2) Necessarily, the later will be.
Conclusion: (1*) Necessarily, if the earlier comes to be, the later will be.

Suppose then we ask: "where, in the passage we are discussing, is (1*) deduced from (1) and (2)?"[10] We may proceed by a process of elimination: not in (e) (which is just a reprise), nor in (a) (which is just a reminder), nor in (b) (this is evident), nor in (d) (which is just a comment on the inference performed in (c)). So, if Aristotle does deduce (1*) from (1) and (2) in the passage we are discussing, he does so in the bit I have labeled (c).

Such a reading requires us to take (c) (minus the "therefore" – τοίνυν) as a statement of (1*), which (putting back the "therefore") is therein deduced from (1) and (2) (which were also the premises of the inference performed in (b)). This then is how I propose to take it.[11] This requires (I think) three things. First, it requires that the protasis of (c) be filled out with a form of γίγνεσθαι (as opposed to: a form of γίγνεσθαι plus ἀνάγκη).[12] Second, it requires that ἀνάγκη in (c) logically govern the entire conditional (as opposed to its being part of the apodosis).[13] Third, it requires that the inferential particle (τοίνυν) mark that the conclusion of the inference herein performed is the entire conditional (as opposed to just its apodosis). This last strikes me as difficult, though I note that Philoponus remarks, "it [i.e., τοίνυν] must be read transposed

[10] This question is fairly put, as appears confirmed by the "for" (γάρ) in (a) and the "therefore" (ἄρα) in (e).

[11] Taken this way, the inference appears to go through, for the reason Aristotle gives in (d): roughly put, if it is necessary that *q*, it follows that, necessarily, if *p* then *q*.

[12] So, e.g., Joachim (as in the Revised Oxford Translation, Vol. 1: 553), Williams 1982: 57, Rashed 2005: 81, Delcomminette 2018: 312. Perhaps in support of this is that the alternative would make the protasis of (c) reiterate the apodosis of (b), in which case we might have expected "and if this" (καὶ εἰ τοῦτο) rather than "and if the earlier" (καὶ εἰ τὸ πρότερον).

[13] This would be in keeping with our handling of ἀνάγκη in other conditionals in this run of text (see note 7 above). (Exception: ἀνάγκη occurs twice in (b): εἰ τοίνυν ἀνάγκη γενέσθαι τὸ ὕστερον, καὶ τὸ πρότερον ἀνάγκη. Here we took the second ἀνάγκη as part of the apodosis.)

(ὑπερβατῶς)," that is to say, as though (c) read, "'and if the earlier, therefore, the later is also necessary' (καὶ εἰ τὸ πρότερον τοίνυν, καὶ τὸ ὕστερον ἀνάγκη)" (In *GC* 307.25–26)[14] – perhaps that would make this easier.

In any case it remains to consider why the local result Aristotle is after in this section might be important. That result (again) is that certain conditional necessities convert (namely, those in which, necessarily, if the later will be then the earlier will have come to be) – not as a rule, but in cases where it is also necessary (period) that the later will be. How is this result supposed to bear on the original question (namely, whether anything comes to be of necessity (period))? Though at this juncture we can only speculate, still it may be useful to have on the table a concrete proposal, to be borne out (or not) by what follows. My proposal, then, is that the local result is supposed to bear on the original question by indicating *a* line of argument by which that question may be settled in the affirmative. Distinctive of this line of argument is that it proceeds, from facts present or past, to the necessity of other, subsequent, future facts.[15] In particular, such an argument would proceed from facts of a kind that are necessary on the hypothesis of other, subsequent facts: as, for example, summer solstices might be thought necessary on the hypothesis of subsequent winter solstices. To be sure, as Aristotle himself virtually insists, such an argument will not succeed unless it is necessary – necessary (period) – that there *be* such subsequent facts (e.g., subsequent winter solstices). It is also true that this would appear to make such an argument worthless, by making it presuppose the conclusion it is supposed to establish. Here I would suggest that (for the moment) we be patient. Perhaps there are other ways of making such an argument go through. To anticipate, suppose it turned out, not only that earlier summer solstices are necessary on the hypothesis of subsequent winter ones, but also that earlier winter solstices are necessary on the hypothesis of subsequent summer ones – in Aristotle's language, suppose that summer solstices and winter solstices come to be from one another "cyclically." In that case we might argue – as we will see, Aristotle does argue – first, that summer solstices and winter solstices are *always* in process of coming to be, and second, that they are (therefore) *necessarily* in process of coming to be. The upshot would be an argument, from facts present or past (e.g., a summer solstice), to the

[14] Trans. after Sharples 1979: 38n13 and Kupreeva 2005: 102.

[15] Compare *Phys.* VIII, where Aristotle argues, from the fact that there *is* motion now, that there always will be motion (and always has been).

necessity of other, subsequent, future facts (e.g., a winter solstice). Such an argument would proceed, not from the *premise* that it is necessary (period) that there will be a winter solstice, but rather from the "observations" that there have been solstices and that solstices come to be from one another cyclically. (The argument fails, of course, but not because it premises its conclusion; it fails because it does not follow, from the fact that there have been solstices, and that solstices come to be cyclically, that solstices are always in process of coming to be.)

12.3.2 Necessary and Everlasting (337b25–338a3)

So far we have the claim that in certain cases certain conditional necessities convert. That is, in cases where, necessarily, if the later will be, the earlier will have come to be, it follows (conversely) that, necessarily, if the earlier has come to be, the later will be – not as a rule, but if it is also necessary – necessary (period) – that the later will be. Aristotle's thesis in this next section is that this condition is satisfied if and only if the later is coming to be always (338a2–3).

Aristotle arrives at this result by process of elimination. First, he considers cases where "they go to infinity downward":

> Now, if the series of things coming to be goes to infinity downward (εἰς ἄπειρον εἶσιν ἐπὶ τὸ κάτω), it will not be necessary (period) that *this* one of the later ones come to be, but on a hypothesis; for there will always be another in front (ἔμπροσθεν) which necessarily [comes to be], because of which it is necessary that *that* one come to be. The result is that if there is no beginning of the infinite series of things coming to be, neither will there be any first because of which it will be, necessarily, that it comes to be. (*GC* II 11, 337b25–29)

Envisaged in this passage, as I understand it, are cases where (say) an A must come to be if there will be a *B*, and a *B* if a *C*, and a *C* if a *D*, and so on without end.[16] (Note that, in such cases, if it *were* necessary (say) that an *A* come to be, this would be conditional on the necessity of there later being a *B*. That is, at least here, scenarios in which an *A* must come to be if

[16] I formulate these conditionals in general terms, because that seems to me to be the natural way of taking the example Aristotle uses to illustrate his points (namely, as concerning the coming to be and being of *a* house, as opposed to some particular house). So, for example, when Aristotle says about these cases that "it will not be necessary (period) that this one (τόδε) of the later ones come to be" (337b26–27), I take the demonstrative τόδε to point to a particular one of the things further "down" (ἐπὶ τὸ κάτω), e.g., a *B*, or a *C*, or a *D*, or an *E*, and so on. I interpret all the pronouns in this section in this way.

there will be a *B* are scenarios in which it is the case not only that (i) necessarily, if there will be a *B*, an *A* (will have) come to be, but also that (ii) if it is necessary that *A* (will have) come to be, it is also necessary that there will be a *B*.)[17] Aristotle says about such cases that none of the later ones will come to be of necessity (337b25–27). His reason is that, in such cases, each later one will also be earlier than something else, something "because of which it is necessary that it come to be" – in which case "there will be no first because of which it will be necessary that it come to be" (337b27–29). To illustrate, suppose the question were whether a *C* will come to be of necessity. *Ex hypothesi* it is only necessary that a *C* come to be if it is *also* necessary that there will be a *D*. Thus, the question of whether it is necessary that a *C* come to be will turn on the question of whether it is necessary that a *D* come to be, which will turn on the question of whether it is necessary that an *E* come to be – and so on without end. If I may put it this way, if there is no "first," there is nothing in the series the necessity of whose coming to be is not "inherited" from its successor, in which case there will be no necessity to "inherit" in the first place.

Next Aristotle considers cases "where there is a limit":

> Not but that even where there is a limit (ἐν τοῖς πέρας ἔχουσι) will it be true to say that it is necessary (period) that it come to be: for example, a house, whenever foundations come to be. For whenever they come to be, unless it is always necessary that this is coming to be, it will follow that what admits of not always being always is. (*GC* II 11, 337b29–33)

Envisaged in this passage are cases where (say) an *A* must come to be if there will be a *B*, and perhaps so on, except that there *is* a last item in the series.[18] He says that even here it will not be true to say, once an *A* comes to be, that it is necessary that the last come to be – not unless it is always necessary that the last be *coming* to be (γίνεσθαι) (337b32). For example, suppose that foundations must come to be if there will be a house and that it stops there. Even here it will not be true to say, once foundations come to be, that it is necessary that a house come to be – not unless it is always necessary that a house be coming to be. His reason is that otherwise "it will follow that what admits of not always being" – namely, the genesis of a house – "always is" (337b33). I suppose he is thinking that if it is not *always* necessary that a house be coming to be, then the genesis of houses

[17] In symbols, scenarios in which not only (i) $(Fb \rightarrow FPa)$ but also (ii) $\Box FPa \rightarrow \Box Fb$.

[18] Aristotle calls this last a "principle" (ἀρχή) and a "first" (πρῶτον) (337b28–29), i.e., a first "in the direction down" (ἐπὶ τὸ κάτω) (337b25).

must "admit of not always being." But why should it follow that the genesis of houses "always is"? Perhaps his idea is that, first, if it is necessary that a house (will have) come to be (γενέσθαι), it is necessary that a house (will have been) in process of coming to be (γίνεσθαι), and second, that if it is ever necessary that a house is in process of coming to be, it is always the case that a house is coming to be.[19]

Left to consider are cases where it *is* always necessary that (e.g.) a house be in process of coming to be:

> Rather, it must always be in the course of coming to be, if its coming to be is of necessity (ἀλλὰ δεῖ τῇ γενέσει ἀεὶ εἶναι, εἰ ἐξ ἀνάγκης ἐστὶν αὐτοῦ ἡ γένεσις); for "of necessity" and "always" are together, because what is of necessity is not such as to not be. The result is that if it is of necessity, it is everlasting, and if everlasting, of necessity; and if, then, its coming to be is of necessity, its coming to be is everlasting, and if everlasting, of necessity. (*GC* II 11, 337b33–338a3)

In such a case it obviously will be true to say, once foundations come to be, that it is necessary that a house be in process of coming to be (γίνεσθαι); for *ex hypothesi* that is true to say always. That said, however, it is certainly not obvious that it will also be necessary that a house (will have) come to be (γενέσθαι) – that is, will have *finished* coming to be, and so eventually *be*. Nor is it clear that Aristotle says that this too will be necessary. For what he says is, rather, that if the coming to be of a house *is* [ever] necessary, then a house "must always be in process of coming to be" (337b33–35). And he continues in this vein, speaking not of when it is necessary that a house *come* to be (γενέσθαι), but rather of when "its coming to be is of necessity" (ἐξ ἀνάγκης ἐστὶν αὐτοῦ ἡ γένεσις), which (he says) it will be just in case its coming to be is "everlasting" (ἀίδιον) (338a2–3).[20]

To conclude. The previous section had focused our attention on certain conditional necessities, in which something must come to be if something else will be. Aristotle had said that these necessities convert just in case it is also necessary that the later come to be (γενέσθαι). The upshot of the present section is that it is necessary that the later be in process of coming to be just in case (i) there is a "last" and (ii) its coming to be is everlasting.

[19] The second point stems from the doctrine that what is necessarily the case is the case always.

[20] Just by itself, the phrase "its coming to be" (αὐτοῦ ἡ γένεσις) is ambiguous: it could designate either (i) a process underway (i.e., its *coming* to be (γίγνεσθαι)) or (ii) a process completed (i.e., its *having* come to be (γενέσθαι)). I have plumped for (i), because I do not see how Aristotle could be thinking that a process completed could be everlasting.

(Whether this is enough to guarantee that such a last will ever *finish* coming to be remains to be seen.)

12.3.3 Everlasting and Cyclical (338a3–17)

In this section Aristotle proceeds to do two things. First, he argues that the coming to be of something is necessary just in case it is cyclical (lit. "comes round in a circle" and "returns") (338a3–11). He then applies to this the result of the section discussed above in Section 12.2.1 (337b14–25), claiming that in such cases, first, "it will be necessary that it convert, e.g., if *this* of necessity, therefore the earlier too," and second, "it is also necessary that the later come to be" (338a11–14). New in this section is really just the first point, that the coming to be of something is necessary just in case it is cyclical. The argument for this point is as we would expect: the coming to be of something is necessary just in case it is cyclical, because (i) it is necessary just in case it is everlasting and (ii) it is everlasting just in case it is cyclical. Having argued for (i) in the previous section, Aristotle here concentrates on (ii). The discussion is conducted without particular examples; to make things concrete, I will use the example of winter and summer solstices.[21]

The argument begins with a division into cases. The coming to be of winter solstices, for example, must either "have a limit" or not (338a5–6): that is, I take it, it must either have a beginning and/or end, or it must have neither a beginning nor an end. Disposing of the first case is straightforward enough. If the coming to be of winter solstices has a beginning or end, then it either once was, or now is, or eventually will be, that no winter solstice is in process of coming to be. But in that case the coming to be of winter solstices will not be everlasting (338a10) – that is, it will not always be the case that a winter solstice is in process of coming to be.

Suppose then that the coming to be of winter solstices does not have a limit; in that case, Aristotle says, it must either be "straight" (εἰς εὐθύ) or "cyclical" (κύκλῳ) (338a6). What alternatives are these? I take it that if the coming to be of winter solstices were cyclical, this would *not* require the coming to be of *each* winter solstice to begin from and end in *that* winter solstice. For that would require (say) that last winter solstice recurs,

[21] I continue to speak in terms of generalities, taking the new point in this section to be that what is necessary just in case it is cyclical is the coming to be of (e.g.) *a* winter solstice (see footnote 16 above).

repeatedly, again and again.[22] Rather, I take it, the coming to be of winter solstices will be cyclical just in case it begins from and ends in *a* winter solstice: begins from one, ends in the next. If that is right, then the coming to be of winter solstices would be straight just in case their coming to be was not like that: that is, did *not* begin from and end in a winter solstice.

Aristotle claims that if the coming to be of something is straight, it is not everlasting. His reason is that in that case "there would be no principle at all (neither down, taken as toward what will be, nor up, taking as toward what has come to be), but there must be a principle" (338a7–9). The idea that there would be no principle is the idea that there would be no *temporal* beginning or end. But no temporal beginning or end of *what*? Presumably, of the whole series of conditional necessities that the coming to be of a winter solstice belongs to. To illustrate, suppose that there must be a fall equinox if there will be a winter solstice, and a winter solstice if there will be a spring equinox; the case we are to consider is in one in which this continues *ad infinitum*, in both directions, without "cycling back" (ἀνακυκλεῖν) or "returning back" (ἀνακάμπτειν): for example, a case in which there must be a *B* if later there will be an *A*, and an *A* if a fall equinox, and a fall equinox if a winter solstice, and a winter solstice if a spring equinox, and a spring equinox if an *X*, and an *X* if a *Y*, and so on in both directions without end. Now, in such a case, Aristotle claims, the coming to be of winter solstices would not be everlasting – rather, he says, "it is necessary that there be a principle" (338a9–10). Why? Earlier he argued that, in cases where "it goes to infinity down," it is not necessary that any of the later ones come to be, because *ex hypothesi* the necessity of that is conditional on it being necessary that something else come to be later (337b27–29). Perhaps he is now thinking, in a similar vein, that in cases where there is no principle, it is not the case that (e.g.) a winter solstice is coming to be, because *ex hypothesi* that is conditional on the fact that something later (e.g., a spring equinox) is coming to be (the coming to be of a winter solstice being part and parcel of the coming to be of a spring equinox). Put another way, perhaps he is thinking that, if there is no principle, then winter solstices will not *always* be coming to be because they will not *ever* be coming to be, there being no "first" thanks to which, because *it* is coming to be, a winter solstice is coming to be too.

It remains to consider cases where the coming to be of winter solstices is cyclical. So, for example, suppose that (1) a summer solstice must come to be if there will be a winter one and that (2) a winter solstice must come to

[22] Though see 338b14–15, discussed below in Section 12.4.

be if there will be a summer one, and so on, in both directions, without end. In such a case, Aristotle claims, "it will be necessary that it convert," with the result that the coming to be of both winter solstices and summer solstices will be everlasting and continuous (338a11–14). "It converts" *ex hypothesi*: for *ex hypothesi* we are supposing both that (1) and that (2). Moreover, again *ex hypothesi*, if a winter solstice ever *is* coming to be, that will be because a (later) summer solstice is then coming to be too, and likewise vice versa. The result will be that if there is now, or once was, a solstice coming to be, there always was and always will be a solstice coming to be, in which case the coming to be of solstices, being everlasting, will also be necessary (period). Thus Aristotle concludes that, among things that come to be, "what is of necessity (period) is found in cyclical movement and genesis, i.e., if [their coming to be is] cyclical, it is necessary that each be coming to be and have come to be, and if necessary, their coming to be is cyclical" (338a14–18).

12.3.4 Application (338a17–b5)

So far Aristotle has argued that the coming to be of something is necessary just in case it is cyclical, on the grounds that in that case it is everlasting, and if everlasting then necessary. His first point in this section is that this result is a happy one: "Since the cyclical movement, namely, that of the heavens, turned out to be everlasting for other reasons too, it indeed makes sense that any movements that are of this [. . .] are coming to be and will be of necessity" (338a17–b1). His next point is that, given all this, this is true not only of the movements of the heavens, but also of any movements caused by them: "For if what is being moved cyclically is always moving something, it is necessary that the movement of these things be cyclical too – e.g., since the upper movement [is cyclical], the sun cyclically [moves cyclically] *this* way, and since [it moves cyclically] that way, for that reason the seasons come to be cyclically and return, and, seeing as these come to be [cyclically] that way, again those caused by them [come to be cyclically]" (338b1–5). This result is also a happy one, for it is just this, apparently, that we in fact find: the cyclical movement of the heavens causes the sun's movement along the ecliptic, which in turn causes changes of season, which in turn causes other changes, all of which proceed cyclically.

Stepping back, we may represent the reasoning of the main body of the chapter (337b14–338b5, discussed in Section 12.3.1–12.3.4) about as follows. Suppose that (1) a summer solstice has just come to be, and further that (2) the coming to be of summer and winter solstices is cyclical. But (3) if the coming to be of solstices is cyclical, it is everlasting, and (4) if everlasting, necessary. But (5) if the coming to be of solstices is necessary, we may infer, from the fact that there has been a summer solstice, that there will be a (subsequent) winter solstice. From these points it then follows that (6) it is necessary that there be another solstice. In this reasoning, the inference-rule relied upon in (5) is argued for at 337b14–25; premise (4) is argued for in the next section, 337b25–338a3; premise (3) is argued for in the section before that, 338a3–17; and "facts" of the sort posited in premises (1) and (2) are acknowledged in the present section, 338a17–b5. What makes Aristotle's reasoning in this chapter maddening is that it is only in retrospect (if then!) that we can see the *point* of his beginning as he does, viz. by establishing the inference-rule relied upon in (5). The point of establishing that rule is to open up a line of argument which moves from facts present or past (a la (1), (2)) to the necessity of some other, subsequent, future fact (a la (6)).

12.4 The Difficulty (*GC* II 11, 338b6–19)

In this concluding section Aristotle raises and answers a difficulty presented by his conclusions thus far. Earlier he argued that if the coming to be of something is necessary, it must be cyclical, because only thus will it be everlasting. But what then of (e.g.) men and animals? For by contrast with the coming to be (e.g.) of air and waters, their coming to be, though everlasting, is apparently *not* cyclical.

Aristotle answers this difficulty by distinguishing between two ways in which the coming to be of something can be cyclical: by beginning and ending in the same *individual* being, and by beginning from and ending in the same *kind* of being. Having made this distinction, he claims that a coming to be will be cyclical in the first way just in case its "subject" (lit. ἡ οὐσία ἡ κινουμένη) is indestructible. Note that this coming to be will not be the coming to be of substance; it is this that permits us to say, or rather requires us to say, that the very thing with which the coming to be began has returned to its original situation: for example, that the very thing with which the coming to be of a solstice began, for example, the sun, in a certain place, for example, the place of its winter turning, has returned: that is, we have, again, the very same *being*, the sun, in the very same *place*, the place of its winter turning. To be sure, I did suggest earlier that the coming to be of solstices could hardly begin and end in one and the same solstice; but that was because I was then speaking as though solstices were substances (in which case what I suggested would have been true). In fact, however, the being of a solstice is not the being of a substance, *sans phrase* or *tout court*; rather it is the being of a particular substance, namely, the sun, not *sans phrase* or *tout court*, but *in a place*. Similarly, neither is the coming to be of a solstice the coming to be of a substance, *sans phrase* or *tout court*; it is rather a substance coming to be *in a place*. Consequently, if *that* substance is indestructible, and likewise (if I may say so) *that* place, then if the coming to be of solstices is cyclical, one and the same solstice *will* recur – that is, it will again be the case that *that very substance* is in *that very place*.[23]

So much for why the coming to be of some kinds of being begins and ends in the very same being (not just in kind, but also in number); what of those other kinds of being, whose coming to be begins and ends just in beings of the same *kind* – is the genesis of particular ones of them necessary, *sans phrase* or *tout court*? Aristotle remarks that though it is necessary (e.g.) that *I* come to be if my child will, this necessity does not convert: it is not necessary that my child will have come to be if I have. The reason, I take it, is that the continuation of the human race does not in fact depend on the birth of any of my children, nor indeed on the birth of any of us; that is, though it is indeed necessary, in Aristotle's view, that the race never fail,[24] it is *not* necessary that any *one* of us come to be, *not even on the hypothesis that* the race never fails. Put another way, though it is necessary that I come to be if certain things will, for example, if my children will, the coming to be of my children is not necessary (period); moreover, though that the race will not fail *is* necessary (period), it is not necessary that *I* come to be if the race will not fail. The first necessity does not convert, because its consequent is not necessary; the second does not convert, because it is not (in fact) necessary (not even conditionally). The result is that for beings like us, whose coming to be begins and ends just in beings of the same *kind*, there is not a one whose coming to be is necessary (period).

[23] Compare perhaps *GC* II 10, 337a25–32.

[24] See, e.g., *GA* II 1, 731b24–732a1.

PART II

Translation

CHAPTER 13

Aristotle
On Generation and Corruption Book II

This translation is a joint work. Each author translated her or his chapter. Andrea Falcon and Sean Kelsey revised the entire translation with the goal of making it as uniform as possible.

1

[328b26] So then, with regard to mixture, contact, and acting and being acted upon, we have said how they belong to the things that change naturally. And further, with regard to simple generation and perishing, we have said how they occur, what they are of, and by what cause they occur. And similarly also with regard to alteration, we have said what being altered is and how it differs from them. But it remains to consider the so-called elements of bodies. The reason is that, for all the naturally constituted substances, their generation and perishing are not without the perceptible bodies. And of these some people say that the underlying matter is one, positing, for example, air or fire or something intermediate between these, which is both a body and separable, while other people say that it is more than one in number, some speaking of fire and earth, while others speak of these and of air as a third, and others also speak of water as a fourth of these, as does Empedocles. From these things, when they are combined and separated or altered, they say that the generation and perishing of things result.

[329a5] So let us agree that it is correct to say that the primary ones are principles and elements – the things from which, by their changing either by combination and separation or by some other kind of change, generation and perishing result. However, those who posit a single matter beyond the aforementioned things, and this as bodily and separable, are in error. For, without perceptible contrariety, it is impossible for this body to exist.[1] For it is necessary for this *apeiron*, which some say is the principle, to be

[1] Reading αἰσθητῆς at 329a11 (with HJ[1]V). Rashed prints αἰσθητόν (with EMW and Ḥunayn). The textual issue is discussed in Essay 1.

light or heavy or cold or hot. And the account in the *Timaeus* has no determinacy. For he has not said clearly, of the "all-receiver," whether it is separate from the elements. And nor even does he make any use of it, despite having said that it is something underlying the so-called elements, prior to them, as gold is related to golden products. (And further, this too is not well said, when said in this way. Instead, the things of which there is alteration are like this; but with regard to things of which there is generation and perishing, it is impossible to refer to them as that from which they have come. And yet he says that truest by far is to say that each of them is "gold.") But he carries out the analysis of the elements, which are solids, as far as plane figures. But it is impossible for the "nurse" and the primary matter to be the plane figures.

[329a24] But we say that there is a certain matter of the perceptible bodies, yet this is not separable, but is always with contrariety, from which the so-called elements come to be. And about them we have made more precise determinations elsewhere. However, since it is also in this way that the primary bodies are from the matter, it is necessary for us to make determinations about these too, regarding as a principle and as primary the matter which is inseparable but which underlies the contraries; for neither is the hot the matter for the cold, nor is this the matter for the hot, but the underlying thing is the matter for both. So that, first of all, the potential perceptible body is a principle; second, the contrarieties [are principles] – I mean, for example, heat and cold; and only thirdly [are] fire, water and such things [principles], for these change into one another, and it is not as Empedocles and others say (for there would be no alteration), while the contrarieties do not change. But nonetheless, even so, it should be said which and how many are principles of body; for the other philosophers posit them and make use of them, and say nothing as to why these or this many.

2

[329b6] Since, then, we are searching for the principles of perceptible body, and this is tangible, and tangible is what is perceived by the sense of touch, it is clear that not all contrarieties produce forms and principles of body, but only those that relate to touch. Indeed, bodies differ according to some contrariety or other, and in particular according to tangible contrariety. It is for this reason that neither whiteness or blackness, nor sweetness or bitterness, and similarly none of the other perceptible contrarieties produce any element. Yet sight is prior to touch, so that its subject should also be prior; however, the object of sight is not a quality of tangible body *qua* tangible, but with respect to some aspect that is different and (if so it be) prior by nature.

[329b16] It is, then, of tangible bodies themselves that we must determine which are the primary differentiae and contrarieties. Here are the pairs of contrarieties that relate to touch: hot-cold, dry-moist, heavy-light, hard-soft, viscous-brittle, rough-smooth, coarse-fine. Of these, heavy and light are neither active nor passive; for things are not called heavy or light because they act on something else or are acted upon by something else, whereas the elements must act on or be acted upon by one another, since they mix and change into one another.

[329b24] But things are called hot and cold because they have the capacity to act, moist and dry because they have the capacity to be acted upon. For hot is what associates things of the same kind (for dissociation, which is what they say fire does, is associating things of the same kind, since the result is to expel what is foreign), while cold is what brings together and associates things of the same kind and things of different kinds alike. Moist is what, though easily delimited, is not determined by limits of its own; dry is what is easily determined by its own limit, though difficult to delimit.

[329b33] The fine and coarse and viscous and brittle and hard and soft and the other differentiae are derived from these. For since it is of the moist to take the shape of its container, on account of its having no limit but being easily delimited and following what touches it, and the fine takes the shape of its container (for its parts are fine and what possesses small parts takes the shape of its container, because whole touches whole and the fine is like that in the highest degree), it is clear that the fine derives from the moist, the coarse from the dry. Again, the viscous derives from the moist (for the viscous is a moist thing that has been affected somehow, as happens for instance with oil), the brittle from the dry; for brittle is what is completely dry, to the point that it has also solidified due to the lack of moisture. Furthermore, the soft derives from the moist; for soft is what yields into itself without changing its position, which is exactly what the moist does. That is why it is not the moist that derives from the soft but rather the soft from the moist. The hard derives from the dry; for hard is what has solidified and what has solidified is dry.

[330a12] Dry and moist are said in many ways. For both the moist and the damp are opposed to the dry, and again both the dry and the solidified are opposed to the moist. All these things, however, derive from the dry and the moist that are said to be primary. For the dry is opposed to the damp and whereas what has foreign moisture on its surface is damp (what has foreign moisture deep inside is sodden), what has been deprived of this is dry; it is clear, then, that the damp will derive from the moist, and

that the dry opposed to it will be derived from the primary dry. Again, the moist and the solidified are derived in the same manner. For what has its own moisture deep inside is moist (what has foreign moisture deep inside being sodden) and what has been deprived of this is solidified. Hence these too derive from the dry and the moist.

[330a24] It is evident, then, that all the other differentiae are reduced to the first four, whereas the latter are not reduced to fewer differences. For neither is the hot what is moist or what is dry, nor is the moist what is hot or what is cold, nor are the cold and the dry [produced] by either one another or by the hot and the moist. As a result it is necessary that there be these four.

3

[330a30] Since the elements are four, and the pairings of four are six, and the contraries are not by nature to be joined with each other (for it is impossible for the same thing to be hot and cold, and again dry and moist), it is clear that the pairings of the elements will be four: hot and dry and hot and moist, and again cold and moist and cold and dry.

[330b2] And [these pairings] correspond according to argument to the apparent simple bodies, fire, air, water, and earth. For fire is hot and dry, air is hot and moist (for air is a sort of vapor), water is cold and moist, and earth is cold and dry, so that the distribution of differentiae to the primary bodies is reasonable and the number of them is in accordance with argument. For everyone who makes the simple bodies elements makes them either one or two or three or four.

[330b9] So, all those who say there is only one, and then generate the others by means of condensation and rarefaction, end up making the principles two: the rare and the dense or the hot and the cold. For these are fabricators, while the one underlies in the manner of matter.

[330b13] But those who make two straightaway, just as Parmenides makes fire and earth, make the intermediates – air and water – mixtures of these. And, similarly, the ones who maintain three, just as Plato does in the divisions; for he makes the middle a mixture. And they mean almost the same thing, those who make two and those who make three – except that while some cut the middle in two, others make it only one. But some, for example Empedocles, maintain four straightaway. Yet he also draws them together into two, for he sets all the other things against fire.

[330b21] Yet fire and air and each of the things mentioned are not simple, but mixed. But the simple things, though like these, are not

however the same as them. For instance, if one of them resembles fire, it is fiery, not fire, and what resembles air is airy; and similarly also in the case of the others. But fire is excess of heat, just as ice is excess of coldness; for congealing and boiling are excesses, the one of coldness and the other of heat. If then ice is congealing of the moist and cold, then also fire will be boiling of the dry and hot. And that is why nothing comes to be from either ice or fire.

[330b30] Since there are four simple bodies, two each are of each of the regions:[2] fire and air of what is borne toward the boundary, earth and water of what is borne toward the center. And earth and fire are both extremes and most unmixed, while water and air are middles and quite mixed. And each of two is contrary to each: fire is contrary to water and air to earth, for these are put together from contrary affections. Nevertheless, strictly, since they are four, each is of one affection: earth of dry rather than cold, water of cold rather than moist, air of moist rather than hot, and fire of hot rather than dry.

4

[331a7] Since it was determined before that generation for the simple bodies is out of one another, and, at the same time, too, it is evident according to perception that they come to be (for otherwise there would be no alteration, for alteration is in respect of the qualities of tangible objects), one must say what is the manner of their change into one another, and whether every one of them is capable of coming to be out of every one, or whether some are capable of this, others incapable.

[331a12] It is clear, then, that all simple bodies by nature change into one another. For generation is into contraries and out of contraries, and all elements have contrariety toward one another because of their differentiae being contraries; for in the case of some both differentiae are contraries, for example, in the case of fire and water (for the former is dry and hot, the latter wet and cold), whereas in the case of others only one is, for example, in the case of air and water (for the former is wet and hot, the latter wet and cold). Therefore, considered generally, it is clear *that* every one of

[2] Reading τόπων (FHL and Philoponus) rather than πρώτων (FHJ[1]VLMW and the Arabic translation).

them is of a nature to come to be out of every one, though as to *how* [they come to be out of one another], considered individually, it is not difficult to see even now.

[331a22] For though all will be out of all, there will be a difference in being slower and faster and being easier and harder. For as many as have counterparts relative to one another, the change of these will be quick, but as many as do not have them, it will be slow, because it is easier for one to change than for many. For example, out of fire there will be air when one quality changes (for if the former was hot and dry, the latter hot and wet, the result if dry should be overcome by wet will be air); again, out of air there will be water if hot should be overcome by cold (for the former is hot and wet, the latter cold and wet, so that when hot changes there will be water). Also, in the same manner, out of water earth and out of earth fire; for both have counterparts relative to both. For water is wet and cold, and earth cold and dry, so that when wet is overcome there will be earth. And again, since fire is dry and hot, and earth cold and dry, if cold is destroyed, fire will be out of earth. Therefore, it is clear both that generation for the simple bodies will be in a circle and that this manner of change is easiest because the counterparts are present in consecutive simple bodies.

[331b4] However, though it is possible for water to come to be out of fire and earth to come to be out of air, and again, for air and fire to come to be out of water and earth, it is more difficult because the change is of more things. For it is necessary, if fire will be out of water, that both cold and wet be destroyed, and again, if air will be out of earth, that both cold and dry be destroyed. And similarly, too, if water and earth will be out of fire and air, it is necessary that both things change. So this generation takes longer; however if one affection of each pair is destroyed, the change though easier is not into one another, but rather earth and air will be out of fire and water, and fire and water out of air and earth. For whenever the cold of water and the dry of fire is destroyed, air will be (for the hot of the latter and the wet of the former remain), but whenever the hot of the fire and the wet of the water is destroyed, earth will be, because there will remain the dry of the former and the cold of the latter. And, similarly, fire and water will be out of air and earth, for whenever the hot of air and the dry of earth is destroyed, water will be (for the cold of the one and the wet of the other remain), but whenever the wet of air and the cold of earth is destroyed, fire will be, because the hot of the former and the dry of the latter will remain, which very things are fire. But the generation of fire is confirmed also by perception: for flame is fire most of all, but this is burning smoke, and smoke is out of air and earth.

[331b26] However, in respect of the consecutive elements, it is not possible, with the destruction of one contrary in each of the elements, that there occur a change into any of the bodies, because either the same or contrary qualities remain in both. But neither of these options admits of a body's coming to be. For instance, if the dry of fire were destroyed, the wet of air would be destroyed, for in both the heat remains; but if the heat out of each were destroyed, the contraries dry and wet would remain. Likewise also in the other cases: for in all consecutive elements there exists one same and one contrary quality. Hence, it is clear at the same time that whereas those changing out of one into one come into being when one has been destroyed, those changing out of two into one come into being when more qualities have been destroyed. So we have established that all come to be out of every one, and in what manner the transition into one another occurs.

5

[332a3] Nevertheless, let us also consider them in the following way. For if the matter of natural bodies is, as some suppose, water and air and the like, there must be one or two or more of them.

[332a6] Certainly they cannot be one; for example, they cannot all be air or water or fire or earth, if indeed change is into contraries. For if they were all air, then if air remains, there will be alteration, but there will not be generation. At the same time, it does not even seem to be thus, such that water is simultaneously also air or anything else. So there will be a contrariety and difference, one part of which [it] will have, as fire has it. Nevertheless, fire will not be hot air: for that sort of thing is alteration and is not observed; at the same time if air again results from fire, there will be the hot changing into its contrary. So, this will belong to air, and air will be something cold. Hence, it is impossible for fire to be hot air: for the same thing will be simultaneously hot and cold. So, both will be something else that is the same, and some other common matter. The same account holds for all of them; there is not one from which they all are. there is not one of them from which everything comes to be.

[332a20] Nor is there anything else beyond these, such as an intermediate between air and water or air and fire, which is coarser than air and fire, but finer than the others; for air and fire will be this with a contrariety. But one of the contraries is a privation; hence, it is not possible for this ever to be isolated, as some say regarding the *apeiron* and the surrounder. Thus,

in the same way, it is any one of these or it is nothing. If then, there is nothing perceptible prior to these, [it] would be all these.

[332a27] Accordingly, it is necessary either that they always remain and do not change into one another, or that they change into one another, and either all of them, or some but not others, as Plato wrote in the *Timaeus*. That all of them must change into one another has been proven previously; and it has been stated previously that they do not come to be from one another with equal speed, since those that have a counterpart come to be from one another more quickly, and those that do not have a counterpart come to be from one another more slowly. If, then, the contrariety according to which they change is one, they must be two: for the matter is the intermediate which is imperceptible and inseparable. Since they are observed to be more, there must be at least two [contrarieties according to which they change]. [But the contrarieties] being two, they cannot be three, but four, as they are observed to be: for the pairings are this many, for though there are six combinations, two cannot come to be because they are contrary to one another. These matters have been discussed previously.

[332b5] Since they change into one another, it is evident from the following that it is impossible for one of them either at the extremity or in the intermediary to be the principle. On the one hand, the principle will not be at the extremities, because all of them then will be fire or earth, and this theory amounts to saying that all of them are from fire or earth. On the other hand, it is not an intermediate, as some suppose air changes into water, and water changes into air and earth, but the extremities do not further change into one another: for change must stop and must not proceed straight out to infinity in both directions. For in that case, infinite contrarieties will belong to one of them. Let *E* stand for earth, *W* for water, *A* for air, and *F* for fire. Now, if *A* changes into *F* and *W*, a contrariety will belong to *AF*. Let these be white and black. If *A* changes again to *W*, there will be another; for *W* and *F* are not the same. So, let this be dry and moist; let *D* stand for dry and *M* stand for moist. Accordingly, if white remains, then moist and white will belong to water, but if not, then water will be black; for the change is into contraries, so it is necessary for water to be either black or white. Let it be the first. Now, similarly, *D*, dryness, will belong to *F*. So, there will also be a change from *F*, fire, into water, for a contrary belongs to it: for fire was first black, then dry, and water moist, then white. It is clear, then, that the change into all of them will be out of one another, and in these cases at least, that the remaining contraries will belong to E, earth, and the two counterparts will be black and moist: for these have not yet been coupled.

[332b30] That it is not possible to proceed to infinity, which we were about to discuss before we went through this, is clear from the following. For if fire, which is *F*, again changes into something else and does not revert, for example, into *X*, then some contrariety other than those mentioned will belong to fire, *F*, and *X*: for it is assumed that none of *E W A F* are the same as *X*. So, let *F* be *K*, and let *X* be *J*. *K*, then, will belong to *E W A F* because they change into one another. This, however, has not yet been proven; but it is evident, that if *X* again changes to something else, another contrariety will belong to *X* and fire, *F*. Similarly, with continual addition a contrariety will belong to the previous bodies, so that if infinite, infinite contrarieties will also belong to one thing. But if this is the case, then there will be neither definition nor generation: for it is necessary, if they come to be from one another, to proceed through so many contrarieties, and still more, so that there will never be change into certain things, for example, if the intermediates were infinite (but this is necessary, if indeed the elements are infinite); furthermore, fire will not come to be from air, if the contrarieties are infinite. And all of them turn out to be one: for all of them above *F* must belong to those below, and these to those above, so that all of them will be one.

6

[333a16] One may well marvel at those who claim that the elements of bodies are more than one so that they do not change into one another, as Empedocles does, and ask how it is admissible for them to assert that the elements are comparable. Yet Empedocles says, "*For these are all equal.*"

[333a20] If what is meant is that they are comparable in quantity, all the things which are comparable must possess some identical something by which they are measured; for example, if out of one cup of water come ten cups of air, both would have to be an identical something, since they are measured by the same thing. If, however, they are not comparable in quantity (in the sense that from so much of one comes so much of the other), but in power (for instance, a cup of water and ten cups of air have the same cooling power), even then they are comparable in quantity, not *qua* quantity but *qua* having such a power. It would be possible to compare their powers, not by a quantitative measure, but by an analogy: for example, "as *x* is hot so *y* is white." But "as" in quality signifies likeness and in quantity equality. Indeed, it is manifestly absurd that the bodies, though incapable of changing into one another, are comparable, not by analogy, but by a measure of their powers; namely, so much fire and many

times that much air being equally or similarly hot. For the same thing, which is more, will, because it is of the same kind, have such a proportion.

[333a35] Moreover, growth is impossible according to Empedocles, except by addition. For fire grows by fire, "*earth grows its own bodily frame, and aithêr, aithêr,*" but these are additions. Things which grow do not, it is believed, grow in this way.

[333b3] It is much more difficult to give an account of coming to be which occurs by nature. For things which come to be naturally all come to be in a particular way always or for the most part; those that depart from the always or for the most part come to be spontaneously or by chance. What, then, is the cause of the human being always or for the most part coming to be from the human being, or wheat, rather than an olive, from wheat? And won't it be bone if things are put together in a particular way? For nothing comes to be when things come to be as chance has it, according to what he says, but only when they do so by a certain formula. What, then, is the cause of this? For it is surely neither fire, at any rate, nor earth. But neither is it Love or Strife, for the former is a cause only of aggregation, the latter of segregation. In fact, the cause is the substance of each thing, not merely "*mixing and separation of things mixed,*" as he says. Chance "*is the name given to these happenings,*" not proportion; for things can be mixed by chance. The cause, then, of things which exist naturally is their being such and such, and this is the nature of each thing, about which he says nothing. So there is nothing "*About Nature*" in what he says. Moreover, it is this that is the well-being and the good of each thing; but he commends only mixture. And yet it is not Strife but Love that separates the elements which are by nature prior to god, although they are also gods.

[333b22] Again, his account of motion is too shallow. It is not enough to say that Love and Strife are the reason why things move, unless to be by Love means to move with this type of movement and to be by Strife to move with this other one. What was needed was to give definitions, or assumptions, or demonstrations, be they precise, relaxed, or in some other fashion. Again, since the bodies are seen to be moved both from constraint (namely contrary to nature) and in accordance with nature – for instance, fire moves upward without constraint, though by constraint downward – and that which is in accordance with nature is contrary to that which is by constraint, and there is such a thing as by constraint, so there is also such a thing as being moved in accordance with nature. Is it this movement, then, which Love initiates, or not? No, for, on the contrary, earth's natural movement is downward and resembles separation, and Strife rather than Love is the cause of movement in accordance

with nature; so, in general, Love would be contrary to, rather than in accordance with, nature. But unless Love or Strife moves them, there is no movement or rest of the bodies themselves at all. But this is absurd. Moreover, that they do move is seen. For, though Strife separated, *aithêr* was borne upward not by Strife. But, he sometimes says it is as if by chance, "*for it was in this way that it chances to run, but often [it chanced to run] differently*," while at other times he says it is the nature of fire to be borne upward, but the *aither*, he says, "*sank down under the ground with long roots.*" At the same time, he also says that the universe is in the same state now under the rule of Strife as it was formerly under the rule of Love. What, then, is the first mover and cause of movement? It is surely not Love and Strife. But of what movement are these causes, if that other is the principle?

[334a10] An additional absurdity is that soul is constituted by the elements or is one of them. For how will the alterations of soul occur, such as being musical and once again unmusical, or remembering or forgetting? For clearly, if the soul is fire, it will come to possess only those properties that are appropriate to fire *qua* fire; if it is a mixture, only bodily properties. But none of these is bodily. But these matters are the work of another inquiry.

7

[334a15] But regarding[3] the elements out of which the bodies are composed, for those who hold that there is something common, or that the elements change into one another, if one of these is the case, then the other follows necessarily. Those who, by contrast, do not make coming to be [of the elements] from one another, not even as from each, except in the way that bricks come to be from a wall – [for them] it is baffling how, from those [elements] there will be bone and flesh and any of the others. But what was said involves difficulty even for those who generate [the elements] from one another: in what way does something else, over and above them, come to be from them? I mean, for example, water can come to be from fire and fire from it, for there is something common underlying; but also flesh comes to be from these, and marrow too. How then do these come to be?

[334a26] For the former, who speak like Empedocles, what way will it be [*sc.* in which flesh, bone and the like come to be from the elements]?

[3] Reading περί (with LMWVH, Philoponus, and Joachim) rather than ἐκ (with JVE^{1}F^1 and Rashed).

For it must be composition, just like a wall from bricks and stones. And this mixture, then, will be from elements that are preserved, but lying next to one another in small bits. In this way, then, flesh and each of the others will come to be. It follows, then, that fire and water do not come to be from any part of flesh whatsoever, in the way that, from wax, a sphere might come to be from this part, a pyramid from some other one, but either could also have come to be from either part. But this indeed occurs in this way: both from any part of the flesh whatsoever. For those who speak in the former way, however, this is not possible; rather, like stone and brick from a wall, each comes to be from a different place and part.

[334b2] How there will be something from both of them, for example, from hot and cold or fire and earth, similarly involves difficulty even for those who posit a single matter for these. For if flesh is from both and is neither of them, nor yet a composition of things preserved, what remains except that the matter is what is from them? For the destruction of the one produces either the other or the matter.

[334b7] Is it, then, since something can be more and less hot and cold, that whenever something should be the one without qualification in actuality, it will be the other in capacity, but whenever it is not completely hot or cold, but rather in the one case hot-cold, so to speak, and in the other cold-hot, because things being mixed destroy the excesses of one another, then it will be neither the matter nor either of the contraries in actuality without qualification, but in between, and according to its being in capacity more hot than cold or *vice versa* – according to this ratio – twice as hot in capacity as cold, or three times, or in some other such way?

[334b16] The other things [*sc.* flesh, bone, and the like], then, will be from the contraries or the elements, these having been mixed, and the elements will be from those [*sc.* flesh, bone, and the like], which are in capacity [the elements] in a way; not however in that way, namely as the matter [is in capacity the elements], but in the way stated. And what is coming to be [something] is in this way mixture, but in the former way matter.

[334b20] Since the contraries are also affected according to the earlier account (for what is in actuality hot is in capacity cold, and what is in actuality cold is in capacity hot, so that if they are not commensurate, they will change into one another, and similarly in the case of the other contraries), the elements also change primarily in this way, and from them flesh and bone and such things [come to be], the hot becoming cold, the cold hot, when they go toward the intermediate; for neither [hot nor cold] is there [in the intermediate], though the intermediate is wide and not

indivisible. And in the same way, dry and wet and such things produce flesh and bone and the others according to intermediacy.

8

[334b31] All mixed bodies – as many as exist in the region around the center – are composed of all simple bodies. Earth is present in all of them because each is found above all and mostly in its natural place. Water is present because the composite body must be bounded and water alone of the simple bodies is easily bounded. Another reason is that earth cannot remain together without what is wet, but this is what holds it together. If what is wet were completely extracted from earth, earth would fall apart. Thus, earth and water are present on account of these reasons. Air and fire are present because they are contraries to earth and water. Earth is contrary to air, whereas water is contrary to fire (in the way that it is possible for a substance to be contrary to another substance). Thus, since instances of coming to be are from contraries, and one or the other extreme of the contraries is present, the other must be present as well. As a result, every simple body will be present in every composite body.

[335a9] Evidence of this seems to be the nourishment of each of them. Everything is nourished by the things out of which it is composed, and everything is nourished by more than one thing. Even those that might be thought to be nourished by only one thing – plants by water – are nourished by more than one. Earth is in fact mixed with water. This is why farmers endeavor to irrigate by mixing water and earth.

[335a14] Since nourishment is akin to matter and what is nourished is the shape, namely the form, taken together with the matter, it is altogether reasonable that, although they all come to be from one another, it is fire alone of the simple bodies that is nourished, as the ancients also say. The reason is that fire alone and above all is akin to the form because it is naturally borne toward the boundary. Each of the simple bodies is naturally borne toward its own region, and the shape and the form of all things is found in their boundaries.

[335a23] Thus, it has been established that all bodies are composed of all the simple bodies.

9

[335a24] Since some things are generable and perishable, and generation takes place in the region around the center, we must say concerning all

generation alike how many principles there are and which ones they are. For it will be easier for us to study particular cases in this way, namely when we have first a grasp of issues concerning what is general.

[335a28] Now, the principles of generable and perishable things are equal in number to and the same in kind as those of the eternal and primary things. One of them is as matter, and another as form. However, it is necessary that there be in addition a third one. That is because the first two are not enough for the generation of things, just as they are not enough for the primary things.

[335a232] Cause by way of matter for the things that are generable is what is capable of existing and not existing. For some things exist of necessity, namely the eternal things, whereas others of necessity do not exist. Of these, the former are incapable of not existing, whereas the latter are incapable of existing, since it is impossible for anything to be otherwise than it is necessary for it to be. Some things, however, are capable both of existing and not existing, which is exactly what that which is generable and perishable is. For this kind of thing exists at one time but does not exist at another time. Hence, it is necessary that generation and perishing be found in what is capable of existing and not existing. That is why this [*sc.* what is capable of existing and not existing] is cause as matter in generable things, while cause as that-for-the-sake-of-which is the shape and the form. That is the account of the essence of each thing.

[335b6] However, we ought to add also the third cause, the one everyone dreams of but no one states. Some thought that the nature of the Forms is an adequate cause of generation, just as did Socrates in the *Phaedo*. After censuring everyone else for saying nothing on this issue, Socrates lays it down that of the things that are some are Forms and others are participants of the Forms, and that while each thing is said to be in accordance with the Form, it comes to be by participating in it and perishes by losing it. As a result, he thinks that if this is true, the Forms are by necessity causes of generation and perishing. But others think that [the cause of generation and perishing] is the matter itself, since it is from it that motion originates. But neither spoke well.

[335b17] If the Forms are causes, why do they not always generate things uninterruptedly rather than sometimes doing so while at other times not, since there exists always both Forms and participants? Moreover, in some cases, we see the cause to be different. For instance, it is the doctor who produces health and the person who is knowledgeable who produces knowledge, although health itself and knowledge exist, as well as their

participants. And the same is true of the other things done in accordance with a capacity.

[335b24] If someone were to say that it is the matter that generates by way of its motion, this person would be speaking more like a student of nature than those who speak in this way. For what brings about alteration and change in shape is more truly a cause of generation, and in all things, likewise in the domain of nature and in that of craft, we are used to calling the producer whatever imparts motion. Nevertheless, they too do not speak correctly: it belongs to matter to be acted upon and to be moved, whereas imparting motion and acting upon belong to another capacity. This is evident in the things that come to be both by art and by nature: water does not itself produce an animal out of itself, nor does wood produce a bed: rather, it is the art that does it in the latter case. Hence, they do not speak correctly for this reason, and also because they leave out a more important cause; they omit the essence and the form.

[336a1] Moreover, they also attribute capacities to bodies in virtue of which these generate in too instrumental a manner, eliminating the cause in accordance with the form. Since, as they say, it is the nature of the hot to dissociate and of the cold to associate, and of each of the other things to act or be acted upon, they say that it is from these [*sc.* the hot and the cold], and by means of these, that all the other things are generated and perish. But it is apparent that even fire itself is both moved and acted upon. Moreover, what they do is very much like someone who would attribute the cause of generation to the saw and each of the instruments. In fact, it is necessary that [the wood] be divided if someone saws and be smoothted if someones planes, and likewise with the other tools. As a result, no matter how much fire acts and imparts motion, they nevertheless fail to see how it does so and that it is less effective than the tools.

[336a13] We have previously spoken in general about the causes and have now made distinctions about matter and form.

10

[336a14] Moreover, since we have shown that moving by way of locomotion is eternal, it follows of necessity that, these things being so, generation also goes on continuously. For locomotion will perpetually produce generation by bringing near and bringing away the generating principle. At the same time, it is evident that the earlier statement, when we said that of changes the primary one is locomotion rather than

generation, was also correct. For it is far more reasonable that what-is be the cause of the generation of what-is-not than that what-is-not be the cause of the being of what-is. In fact, what is moving locally *is*, whereas what is coming to be *is not*. That is why locomotion is prior to generation.

[336a23] Since we have laid it down and proved that both generation and perishing are continuous for things, and since we claim that locomotion is the cause of coming-to-be, it is clear that, if locomotion is one, it is not possible for both generation and perishing to occur because they are contraries (for the same thing in the same state always produces by nature the same effect, so that there will always be either generation or perishing); so the motions must be more than one, and they must be contrary in either direction or unevenness; for the causes of contraries are contraries. That is why the cause of generation and perishing is not the first locomotion but that along the inclined circle; for both continuity and a duality of movements are present in this one. Indeed, if there is always going to be continuous generation and perishing, it is necessary, first, that something be always moving, so that these changes do not give out, and, second, that its movements be two, so that of the two changes it is neither the one nor the other alone that happens. Thus, the locomotion of the whole heaven is the cause of continuity, whereas the inclination is the cause of its approaching and retreating. Indeed, this results in it getting far away at one time and it coming close at another. And since the distance is unequal, the movement will be uneven; so, if by coming close and being near it generates, by retreating and getting far away this same principle destroys; and if by coming close many times it generates, by departing many times it destroys – for the causes of contraries are contraries. And both the perishing and the generation according to nature occur in equal times.

[336b10] That is why the times as well as the lives of each thing have a number and are determined thereby. For there is order in all things, and every time and life is measured by a period, though not all of them by the same period but some by a shorter and some by a longer period; for some the period, that is, the measure, is the year, for others greater, for others shorter. The testimonies of our senses, too, appear to agree with our arguments. For we see that as the sun approaches there is generation and as it retreats there is perishing, and each phase happens in equal time; in fact, the time of generation and the time of perishing according to nature are equal. But it often happens that things perish in a shorter time because of the reciprocal commixture of things; for since matter is uneven and not everywhere the same, it is also necessary for cases of generation to be uneven, that is, some to be faster and some slower, so that it comes about

that as a result of the generation of these things other things perish.[4] As we have stated, generation and perishing will always be continuous, and they will never give out due to the cause we mentioned.

[336b26] And this happens with good reason. For since in all things we claim that nature always desires what is better, and since being is better than not being (in how many ways we mean "being" has been stated elsewhere), and since this is impossible to obtain in all things because they are far removed from the origin, the god has completed the whole in the remaining manner, namely by making generation perpetual – for in this way being would be strung together the most, since everlasting coming-to-be and generation are closest to substance. The cause of that, as we have often stated, is circular locomotion; for this alone is continuous.

[337a2] That is why the other things that change into one another with respect to their properties and powers, for instance the simple bodies, imitate circular locomotion; for when air comes to be out of water, and fire out of air, and water back again out of fire, we say that generation has revolved in a circle because of its coming back to its beginning. Therefore, even locomotion in a straight line is continuous by imitating circular locomotion. At the same time, from these observations the solution to what perplexes some people is evident, namely why, although each body moves toward its proper place, over an infinite time bodies have not come apart. The cause of this is in fact their changing into one another. For if each body remained in its own place and were not changed by the adjacent body, they would have already come apart. But they change on account of the locomotion's being double. And since they change it is not possible for any of them to remain in any assigned place whatsoever. From what has been said it is clear, then, that there is such a thing as generation and perishing and owing to what cause, and what the generable and perishable thing is.

[337a18] Now, it is necessary that (i) there *be* something that imparts motion,[5] if there is to be motion, as we have earlier stated elsewhere; and if there is always to be motion, that (ii) there must be something *eternal*; and, if there is to be continuous motion, that (iii) this thing be *one and the same*, *unmoved*, *ungenerated*, and *unalterable*; and if the circular motions are more than one, that (iv) all of them, though many, somehow be under a

[4] Reading at 336b24 (with all MSS) διὰ τὴν τούτων instead of διὰ τὸ τὴν τούτων (adopted by Rashed).

[5] Reading at 337a18 εἶναί τι τὸ κινοῦν (with Joachim and Mugler) instead of εἶναί τι (Rashed).

single principle; for since time is continuous, it is necessary for motion to be continuous, if it is impossible for time to be without motion; time is, then, the number of some continuous motion, hence of *circular* motion, just as we established in the introductory discourses. But is motion continuous because what is in motion is continuous or because what it moves in is continuous – I mean, for instance, the place or the property? It is in fact evident that motion is continuous because what is in motion is continuous (for how else could the property be continuous except by the very thing to which it belongs being continuous? And if what is in motion is continuous also because what it moves in is continuous, this continuity belongs only to place, since it has some magnitude). Of the things moved only what is moved in a circle is continuous in such a way as to be continuous with itself. This, then, namely the body that is moved in a circle, is what makes motion continuous, and the motion is what makes time continuous.

11

[337a34] Since we see, in things moved continuously in respect of generation or of alteration or of change in general, the next one being and coming to be, this after that, without fail, we must consider whether there is anything that will be of necessity, or nothing, but rather everything admits of not coming to be. For that some do is clear, and straightaway "will be" and "is going to be" are different for this reason. For whereas it is bound to be true at some point, of what it is true to say will be, that it is, nothing prevents what it is now true to say is going to be from not coming to be; for someone going to walk might not walk. And in general, since some beings admit also of not being, it is clear that so it will be for things that come to be too – that is, it is not necessary that *this* will be. Are they then all like that? Or are they not all like that, but rather some come to be necessarily, period, and just as, in the case of being, some cannot not be, though others can, so it is regarding coming to be too? For example, is it necessary that solstices come to be and are not such as not to admit [of coming to be]?

[337b14] Well, if it is necessary that the earlier come to be if the later will be – for example, if a house, foundations, and if this, clay – is it then also necessary that a house come to be if foundations have? Or not yet, unless it is also necessary, period, that the latter come to be? But in that case it is also necessary, foundations having come to be, that a house come to be. For the earlier stood to the later thus, that, if that will be, it is

necessary that the former [have come to be] beforehand. If therefore it is necessary that the later come to be, it is also necessary that the earlier [come to be], and if the earlier, it is therefore necessary that also the later – not because of that, but because it was posited that the later will be of necessity. Wherever, therefore, it is necessary for the later to be, it converts – that is, always, the earlier having come to be, it is necessary that the later come to be.

[337b25] Now, if the series of things coming to be goes to infinity downward, it will not be necessary (period) that *this* one of the later ones come to be, but on a hypothesis; for there will always be another in front which necessarily [comes to be], because of which it is necessary that *that* one come to be. The result is that if there is no beginning of the infinite series of things coming to be, neither will there be any first because of which it will be, necessarily, that it comes to be.

[337b29] Not but that even where there is a limit will it be true to say that it is necessary (period) that it come to be: for example a house, whenever foundations come to be. For whenever they come to be, unless it is always necessary that this is coming to be, it will follow that what admits of not always being always is. Rather, it must always be in the course of coming to be, if its coming to be is of necessity; for "of necessity" and "always" are together, because what is of necessity is not such as to not be. The result is that if it is of necessity, it is everlasting, and if everlasting, of necessity; and if, then, its coming to be is of necessity, its coming to be is everlasting, and if everlasting, of necessity.

[338a3] If, therefore, something's coming to be is necessary, period, it is necessary to come round in a circle and return. For it is necessary that its coming to be either have a limit or not, and if not, that it be either straight or cyclical. But of these, if it really will be everlasting, it cannot be straight, because there is no manner of beginning, neither "down," taken as toward what will be, nor "up" toward what has come to be; but it is necessary that there be a beginning; nor, being limited, [can it] be everlasting. That is why it is necessary that it be cyclical.

[338a11] Therefore, it will be necessary that it convert: for example, if *this* of necessity, therefore the earlier too. But indeed if this, it is also necessary that the later come to be. And this, then, always, continuously – for saying it is through two or many makes no difference. Therefore what is of necessity, period, is found in cyclical movement and coming to be; and, if cyclical, it is necessary that each be coming to be and have come to be, and if necessary, their coming to be is cyclical.

[338a17] Since the cyclical movement – that is, the movement of the heavens – turned out to be everlasting for other reasons too, it indeed

makes sense that any movements that are of this, and any that are because of this, come to be and will be of necessity. For if what is being moved cyclically is always moving something, it is necessary that the movement of these things be cyclical too – for example, since the upper movement [is cyclical], the sun [moves] cyclically *this* way, and since [it moves cyclically] that way, for that reason the seasons come to be cyclically and return, and, seeing as these come to be [cyclically] that way, again those caused by them [come to be cyclically].

[338b6] Then indeed why ever is it that, while some manifestly do come to be in this way, as waters and air come to be cyclically (that is, if there will be a cloud, it is bound to rain, and if it will rain, there's also bound to be a cloud), human beings and animals do not return into themselves, so that the same comes to be again? (For it is not necessary that you come to be if your father did, though [it is that] him if you), but this coming to be seems to be straight [rather than cyclical].

[338b11] The beginning of the investigation is again the same: whether everything returns in the same way, or not, but while some in number, others only in form. Well, as for all those whose substance (the one being moved) is incorruptible, it is clear that they will be the same in number too (for the movement follows what is being moved), but all those [whose substance is] not, but rather [is] corruptible, it is necessary that they return in form but not in number. That is why water from air and air from water, the same in form, not in number; but if even these [are the same] in number, still, those whose substance comes to be are not, seeing as they are such as to admit of not being.

References

ANAGNOSTOPOULOS 2021 A. Anagnostopoulos, "Mixture, Generation and the First *Aporia* of Aristotle's *GC* 1.10." In *Phronesis* 66: 139–177.

ANAGNOSTOPOULOS 2009 G. Anagnostopoulos (ed.), *A Companion to Aristotle*. Blackwell, Oxford.

BARNES 1976 J. Barnes, "Aristotle, Menaechmus, and Circular Proof." In *Classical Quarterly* 26: 278–292.

BARNES 1984 J. Barnes, *The Complete Works of Aristotle*. The Revised Oxford Translation. 2 vols. Princeton University Press, Princeton.

BEERE 2006 J. Beere, "Potentiality and the Matter of Composite Substance." In *Phronesis* 51: 303–329.

BEERE 2016 J. Beere, "Receptacle as a Material Base for Bodies in Plato's *Timaeus*." In Buchheim *et al.*: 201–226.

BODNÁR 1997 I. Bodnár, "Movers and Elemental Motion in Aristotle." In *Oxford Studies in Ancient Philosophy* 15: 81–117.

BOGEN 1996 J. Bogen, "Fire in the Belly: Aristotelian Elements, Organisms, and Chemical Compounds." In Lewis-Bowen 1996: 183–216.

BOLTON 2009 R. Bolton, "Two Standards of Inquiry in Aristotle's *de Caelo*." In Bowen-Wildberg 2009: 51–82.

BOSTOCK 1995 D. Bostock, "Aristotle on the Transmutation of the Elements in *De Generatione et Corruptione* I 1–4." In *Oxford Studies in Ancient Philosophy* 13: 217–229 (reprinted in Bostock 2006: 19–29).

BOSTOCK 2006 D. Bostock, *Space, Time, Matter, and Form: Essays on Aristotle's Physics*. Oxford University Press, Oxford.

BOWEN-WILDBERG 2009 A. C. Bowen and Ch. Wildberg (eds.), *New Perspectives on Aristotle's De caelo*. Brill, Leiden.

BROADIE 1982 S. Waterlow [Broadie]. *Nature, Change, and Agency in Aristotle's Physics*. Oxford University Press, Oxford.

Broadie 2004 — S. Broadie, "*GC* I 4: Distinguishing Alteration." In de Haas-Mansfeld 2004: 123–150.

Broadie 2007 — S. Broadie, "Nature and Craft in Aristotelian Teleology." In S. Broadie, *Aristotle and Beyond: Essays on Metaphysics and Ethics,* Cambridge University Press, Cambridge: 85–100 (originally published in Devereux-Pellegrin 1990: 389–403).

Bruns 1892 — Alexander of Aphrodisias. Ἀπορίαι καὶ λύσεις. In *Commentaria in Aristotelem Graeca*. Supplementary volume. 2.2, edited by I. Bruns. Reimer, Berlin.

Brunschwig 2004 — J. Brunschwig, "*On Generation and Corruption* I 1: A False Start?" In de Haas-Mansfeld 2004: 25–63.

Buchheim 2010 — Th. Buchheim, *Aristoteles, Über Werden und Vergehen*. Übersetzt und erläutert von Th. Buchheim. Akademie Verlag, Berlin.

Buchheim *et al.* 2016 — Th. Buchheim, D. Meissner, and N. Wachsmann (eds.), ΣΩΜΑ: *Körperkonzepte und körperliche Existenz in der antiken Philosophie und Literatur.* Felix Meiner, Hamburg.

Burkert 1959 — W. Burkert, "STOICHEION: Eine semasiologische Studie." In *Philologus* 103: 167–197.

Burnyeat 2002 — M. F. Burnyeat, "*De Anima* II 5." In *Phronesis* 47: 28–90.

Burnyeat 2004 — M. F. Burnyeat, "Aristotle on the Foundations of Sublunary Physics." In de Haas-Mansfeld 2004: 7–24.

Castelli 2018 — L. M. Castelli, *Aristotle: Metaphysics Book Iota.* Oxford University Press, Oxford.

Charles 2000 — D. Charles, "*Metaphysics* Lambda 2." In Frede-Charles 2000: 81–110.

Charles 2018 — D. Charles, "*Physics* I 7." In Quarantotto 2018: 178–205.

Charlton 1970 — W. Charlton, *Aristotle: Physics Books I and II.* Clarendon Press, Oxford.

Charlton 1983 — W. Charlton, "Prime Matter: A Rejoinder." In *Phronesis* 28: 197–211.

Cherniss 1935 — H. Cherniss, *Aristotle's Criticism of Presocratic Philosophy*. The Johns Hopkins University Press, Baltimore.

Code 2008 — A. Code, "Aristotelian Colors as Causes." In Føllesdal-Woods 2008: 235–242.

Cohen 1996 — S. M. Cohen, *Aristotle on Nature and Incomplete Substance*. Cambridge University Press, Cambridge.

Cohen 2012 — S. M. Cohen, "Alteration and Persistence: Form and Matter in the *Physics* and *De Generatione et Corruptione*." In Shields 2012: 205–226.

CONNELL (forthcoming) S. Connell (ed.), *The Cambridge Critical Guide to Aristotle's Parts of Animals*. Cambridge University Press, Cambridge.

COOPER 2009 J. M. Cooper, "Two Notes on Aristotle on Mixture." In J. M. Cooper, *Knowledge, Nature, and the Good: Essays on Ancient Philosophy*. Princeton University Press, Princeton.

CROWLEY 2005 T. J. Crowley, "On the Use of *Stoicheion* in the Sense of 'Element'." In *Oxford Studies in Ancient Philosophy* 2005: 367–389.

CROWLEY 2008 T. J. Crowley, "Aristotle's 'So-called Elements'." In *Phronesis* 53: 223–242.

CROWLEY 2013 T. J. Crowley, "*De generatione et corruptione* 2.3: Does Aristotle Identify the Contraries as Elements?" In *Classical Quarterly* 63: 161–182.

CROWLEY 2019 T. J. Crowley, "On the 'Perceptible Bodies' at *GC* II 1." In *Archai* 27: 1–26.

CRUBELLIER 2000 M. Crubellier, "*Metaphysics* Lambda 4." In Frede-Charles 2000: 137–160.

DANCY 1978 R. M. Dancy, "On Some of Aristotle's Second Thoughts about Substances: Matter." In *Philosophical Review* 87: 373–413.

DE HAAS 1999 F. de Haas, "An Encounter with a Third Kind of Potentiality." In Thijssen-Braakhuis 1999: 21–46.

DE HAAS-MANSFELD 2004 F. de Haas and J. Mansfeld (eds.) *Aristotle's On Generation and Corruption I*. Symposium Aristotelicum. Clarendon Press, Oxford.

DELCOMMINETTE 2018 S. Delcomminette, *Aristote et la nécessité*. Librairie Philosophique J. Vrin, Paris.

DENYER 2000 N. Denyer, "Aristotle on Modality." In *Proceedings of the Aristotelian Society* 74: 143–161.

DEVEREUX-PELLEGRIN 1990 D. Devereux and P. Pellegrin (eds.), *Biologie, logique et métaphysique chez Aristote*. Éditions du CNRS, Paris 1990.

DIMAS *ET AL.* 2019 P. Dimas, R. E. Jones, G. R. Lear (eds.), *Plato's Philebus: A Philosophical Discussion*. Oxford University Press, Oxford.

EBREY 2015 D. Ebrey (ed.), *Theory and Practice in Aristotle's Natural Science*. Cambridge University Press, Cambridge.

EBREY 2020 D. Ebrey, "Aristotle on the matter for Birth, Life, and the Elements." In Taub 2020: 79–101.

FALCON 2005 A. Falcon, *Aristotle and the Science of Nature: Unity without Uniformity*. Cambridge University Press, Cambridge.

FALCON 2016 A. Falcon, "The Subject Matter of Aristotle's Physics." In Buchheim *et al.* 2016: 423–436.

FALCON-LEUNISSEN 2015 A. Falcon and M. Leunissen, "The Scientifics Role of *Eulogos* in Aristotle's *Cael II* 12." In Ebrey 2015: 217–240.

FINE 1995 K. Fine, "The Problem of Mixture." In *Pacific Philosophical Quarterly* 76: 266–369.

FØLLESDAL-WOODS 2008 D. Føllesdal and J. Woods (eds), *Festschrift for Julius Moravcsik*. College Publications, London.

FORSTER–FURLEY 1955 Aristotle, *On Sophistical Refutations, On Coming-to-Be and Passing-Away, On the Cosmos,* edited and translated by E. S. Forster and D. J. Furley. Harvard University Press, Cambridge.

FREDE 2000 M. Frede, "Introduction." In Frede-Charles 2000: 1–52.

FREDE 2004 D. Frede, "*On Generation and Corruption* I 10: On Mixture and Mixables." In *de Haas-Mansfeld* 2004: 289–314.

FREDE-CHARLES 2000 M. Frede and D. Charles (eds.), *Aristotle's Metaphysics Lambda.* Proceedings of the Symposium Aristotelicum. Clarendon Press, Oxford.

FREY 2007 Ch. Frey, "Organic Unity and the Matter of Man." In *Oxford Studies in Ancient Philosophy* 32: 167–204.

FREY 2015*a* Ch. Frey, "Two Conceptions of Soul in Aristotle." In Ebrey 2015: 137–160.

FREY 2015*b* Ch. Frey, "Capacities and the Eternal in *Metaphysics Θ.8* and *De caelo.*" In *Phronesis* 60: 88–126.

FREY (forthcoming) Ch. Frey, "Analogical Unity and Biological Kinds in *Aristotle's Parts of Animals.*" In Connell (forthcoming).

FURTH 1988 M. Furth, *Substance, Form, and Psyche: An Aristotelian Metaphysics.* Cambridge University Press, Cambridge.

GANNAGÉ 2005 E. Gannagé, *Alexander of Aphrodisias: On Aristotle's On Coming-to-Be and Perishing 2.2–5.* Bloomsbury, London.

GILL 1989 M. L. Gill, *Aristotle's Substance: The Paradox of Unity.* Princeton University Press, Princeton.

GILL 2019 M. L. Gill, "The Fourfold Division of Beings: *Philebus* 27 C–31B." In Dimas *et al.* 2019: 71–89.

GILL-LENNOX 1994 M. L. Gill and J. G. Lennox (eds.), *Self-Motion.* Princeton University Press, Princeton.

GRAHAM 2010 D. Graham (ed. and trans.), *The Texts of Early Greek Philosophy.* Part I. Cambridge University Press, Cambridge.

HARTE 2002 V. Harte, *Plato on Parts and Wholes: The Metaphysics of Structure*. Oxford University Press, Oxford.

HENRY 2014 D. Henry, "The Birds and the Bees: Aristotle on the Biological Concept of ἀνάλογον." In *Proceedings of the Boston Area Colloquium in Ancient Philosophy* 29: 145–169.

HENRY 2015 D. Henry, "Substantial Generation in *Physics* I 5–7." In Leunissen 2015: 144–161.

HENRY 2019 D. Henry, *Aristotle on Matter, Form, and Moving Causes: The Hylomorphic Theory of Substantial Generation*. Cambridge University Press, Cambridge.

IERODIAKONOU *ET. AL.* 2019 K. Ierodiakonou, P. Kalligas, V. Karasmanis (eds.), *Aristotle's Physics Alpha*. Symposium Aristotelicum. Oxford University Press, Oxford.

JAEGER 1948 W. W. Jaeger, *Aristotle: Fundamentals of the History of his Development*. Clarendon Press, Oxford (2nd edition).

JOACHIM 1922 Aristotle, *On Coming-to-be and Passing-away*, edited by H. H. Joachim Clarendon Press, Oxford.

JUDSON 1994 L. Judson, "Heavenly Motion and the Unmoved Mover." In Gill-Lennox 1994: 155–172.

KAHN 1960 Ch. Kahn, *Anaximander and the Origins of Greek Cosmology*. Columbia University Press, New York.

KALFAS 2011 V. Kalfas, *Aristotelis: Peri Geneseos kai Phthoras*, Translation, Introduction, Notes by V. Kalfas. Athens.

KARBOWSKI 2014 J. Karbowski, "Empirical *Eulogos* Argumentation in *GA* III 10." In *British Journal for the History of Philosophy* 22: 25–38.

KARBOWSKI 2016 J. Karbowski, "Justification 'by Argument' in Aristotle's Natural Science." In *Oxford Studies in Ancient Philosophy* 51: 119–160.

KELSEY 2003 S. Kelsey, "Aristotle's Definition of Nature." In *Oxford Studies in Ancient Philosophy* 25: 59–87.

KING 1956 H. R. King, "Aristotle without *Prima Materia*." In *Journal of the History of Ideas* 17: 370–389.

KRIZAN 2013 "Elemental Structure and the Transformation of the Elements *On Generation and Corruption* II 4." In *Oxford Studies in Ancient Philosophy* 45: 195–224.

KRIZAN 2018 M. Krizan, "Mixing and Homoeomers in *GC* II 7." In *Oxford Studies in Ancient Philosophy* 54: 187–226.

KUPREEVA 2005 I. Kupreeva (ed.), Philoponus, *On Aristotle's "On Coming-to-Be and Perishing 2.5–11."* Cornell University Press, Ithaca.

KURLAND 1958 S. Kurland (ed.), *Averroes* on Aristotle's *De generatione et Corruptione.* on Aristotle's*Middle Commentary and Epitome.* Medieval Academy of America, Cambridge.

LAKS–MOST 2016 A. Laks and G. W. Most, *Early Greek Philosophy. Volume 5: Western Greek Thinkers.* Harvard University Press, Cambridge.

LEUNISSEN 2014 M. Leunissen, "Commentary on Henry." In *Proceedings of the Boston Area Colloquium in Ancient Philosophy* 29: 170–181.

LEUNISSEN 2015 Leunissen (ed.), *Aristotle's Physics: A Critical Guide.* Cambridge University Press, Cambridge.

LEWIS 1996 E. Lewis (ed.), Alexander of Aphrodisias, *On Aristotle "Meteorology 4."* Duckworth, London.

LEWIS 2008 F. A. Lewis, "What's the Matter with Prime Matter?" In *Oxford Studies in Ancient Philosophy* 34: 123–146.

LEWIS-BOWEN 1996 F. A. Lewis and R. Bolton (eds.), *Form, Matter, and Mixture in Aristotle.* Blackwell Publishers, Malden.

LORENZ 2019 H. Lorenz, "*Physics* I 7: The Principles of Natural Things–Two of Three?" In Ierodiakonou *et. al.* 2019: 262–285.

LOVEJOY 1936 A. O. Lovejoy, *The Great Chain of Being: A Study of the History of an Idea.* Harvard University Press, Cambridge.

MALINK–ROSEN 2012 M. Malink and J. Rosen "A Method of Modal Proof in Aristotle." In *Oxford Studies in Ancient Philosophy* 42: 179–261.

MANSION 1946 S. Mansion, *Le jugement d'existence chez Aristote.* Éditions de l'Institut supérieur de philosophie, Louvain.

MCCABE 1994 M. M. McCabe, *Plato's Individuals.* Princeton University Press, Princeton.

MIGLIORI 1976 M. Migliori (ed.), Aristotele, *La generazione e la corruzione.* Loffredo editore, Napoli.

MUGLER 1966 Aristote, *De la génération et la corruption.* Texte établi par Ch. Mugler. Les Belles Lettres, Paris.

NATALI 2004 C. Natali, "*On Generation and Corruption* I 6." In de Haas-Mansfeld 2004: 195–217.

O'CONNOR 2015 S. O'Connor, "The Subjects of Natural Generations in Aristotle's *Physics* I 7." In *Apeiron* 48: 45–75.

OWENS 2007 J. Owens, *Aristotle's Gradations of Being in Metaphysics E-H.* St. Augustine's Press, South Bend.

QUARANTOTTO 2015 D. Quarantótto, "A Dynamic Ontology: On How Aristotle Arrived at the Conclusion that Eternal Change Accomplishes Ousia." in Leunissen 2015: 162–185.

QUARANTOTTO 2018 — D. Quarantotto (ed.), *Aristotle's Physics Book I: A Systematic Exploration*. Cambridge University Press, Cambridge.

RASHED 2005 — Aristote, *De la génération et la corruption*. Text établi et traduit par M. Rashed. Les Belles Lettres, Paris.

ROBINSON 1974 — H. M. Robinson, "Prime Matter in Aristotle." In *Phronesis* 19: 168–188.

SCALTSAS 2009 — T. Scaltsas, "Mixing the Elements." In Anagnostopoulos 2009: 242–259.

SCHARLE 2008 — M. Scharle, "Elemental Teleology in Aristotle's *Physics* II 8." In *Oxford Studies in Ancient Philosophy* 34: 147–183.

SCHARLE 2009 — M. Scharle, "A Synchronic Justification of Aristotle's Commitment to Prime Matter." In *Phronesis* 54: 326–345.

SCHARLE 2015*a* — M. Scharle, "Man from Man but not Bed from Bed: Nature, Art, and Chance in *Physics* II." In *Leunissen* 2015: 88–106.

SCHARLE 2015*b* — M. Scharle, "'And These Things Follow': Teleology, Necessity, and Explanation in Aristotle's *Meteorologica*." In Ebrey 2015: 79–99.

SEDLEY 2007 — D. Sedley, *Creationism and its Critics in Antiquity*. University of California Press, Berkeley and Los Angeles.

SEDLEY 2016 — D. Sedley "Empedoclean Superorganisms." In *Rhizomata* 4: 111–125.

SHARPLES 1979 — R. W. Sharples "'If what is earlier, then of necessity what is later?' Some Ancient Discussions of Aristotle, *De generatione et corruptione* II 11." In *Bulletin of the Institute of Classical Studies* 26: 27–44.

SHARPLES 1992 — Alexander of Aphrodisias, *Quaestiones*. Vol. 2, 2.16–3.15. Edited by R. W. Sharples. Cornell University Press, Ithaca.

SHARVY 1983 — R. Sharvy, "Aristotle on Mixtures." In *Journal of Philosophy* 80: 439–457.

SHIELDS 2012 — Ch. Shields, *The Oxford Handbook of Aristotle*, Oxford University Press, Oxford.

SOLMSEN 1958 — F. Solmsen, "Aristotle and Prime Matter: A Reply to Hugh R. King." In *Journal of the History of Ideas* 19: 243–252.

SOLMSEN 1960 — F. Solmsen, *Aristotle's System of the Physical World*. Cornell University Press, Ithaca.

TAUB 2020 — L. Taub (ed.), *The Cambridge Companion to Ancient Science*. Cambridge University Press, Cambridge.

THIJSSEN-BRAAKHUIS 1999 — J. M. M. H. Thijssen, and H. A. G. Braakhuis (eds.), The Commentary Tradition on Aristotle's *De generatione et corruptione:* Ancient, Medieval, and Early Modern. Brepols, Turnhout.

TRICOT 1951 — J. Tricot, Aristote, *De la génération et de la corruption.* J. Vrin, Paris (2nd edition).

VAN RIJEN 1989 — J. van Rijen, *Aspects of Aristotle's Logic of Modalities.* Kluwer Academic, Dordrecht and Boston.

VERDENIUS-WASZINK 1966 — W. J. Verdenius and J. H. Waszink, *Aristotle on Coming-to-Be and Passing-Away: Some Comments.* Brill, Leiden (2nd edition).

VITELLI 1897 — Philoponus, *In Aristotelis Libros de Generatione et Corruptione Commentaria.* In *Commentaria in Aristotelem Graeca*, vol. 14.2, edited by G. Vitelli. Reimer, Berlin.

WEDIN 2013 — M. Wedin, "Aristotle on the Impossibility of Anaximander's *apeiron: On Generation and Corruption*, 332a20–25." In *Phronesis* 58: 17–31.

WEHRLI 1969 — F. Wehrli, *Eudemus von Rhodos.* In *Die Schule des Aristoteles: Texte und Kommentar.* Heft 8. Schwabe, Basel (2nd revised edition).

WEISBERG-WOOD 2004 — M. Weisberg and R. Wood. "Interpreting Aristotle on Mixture: Problems about Elemental Composition from Philoponus to Cooper." In *Studies in History and Philosophy of Science* 35: 681–706.

WILDBERG 2004 — Ch. Wildberg, "*On Generation and Corruption* I 7: Aristotle on *poiein* and *paschein.*" In de Haas-Mansfeld 2004: 219–242.

WILLIAMS 1982 — *Aristotle's De generatione et corruptione.* Translated with Notes by C. J. F. Williams. Clarendon Press, Oxford.

WILSON 2014 — M. Wilson. *Structure and Method in Aristotle's Meteorologica.* Cambridge University Press, Cambridge.

Subject Index

Index of Passages

Works by Aristotle are listed as they appear in Bekker, not in alphabetical order. Plato's dialogues are listed in the order they appear in Stephanus.